THE LIQUID CRYSTALS BOOK SERIES

ADSORPTION PHENOMENA AND ANCHORING ENERGY IN NEMATIC LIQUID CRYSTALS

THE LIQUID CRYSTALS BOOK SERIES

Edited by
G.W. GRAY, J.W. GOODBY & A. FUKUDA

The Liquid Crystals book series publishes authoritative accounts of all aspects of the field, ranging from the basic fundamentals to the forefront of research; from the physics of liquid crystals to their chemical and biological properties; and, from their self-assembling structures to their applications in devices. The series will provide readers new to liquid crystals with a firm grounding in the subject, while experienced scientists and liquid crystallographers will find that the series is an indispensable resource.

PUBLISHED TITLES

Introduction to Liquid Crystals: Chemistry and Physics
By Peter J. Collings and Michael Hird

The Static and Dynamic Continuum Theory of Liquid Crystals:
A Mathematical Introduction
By Iain W. Stewart

Crystals that Flow: Classic Papers from the History of Liquid Crystals
Compiled with translation and commentary by Timothy J. Sluckin, David A. Dunmur, Horst Stegemeyer

Nematic and Cholesteric Liquid Crystals: Concepts and Physical Properties Illustrated by Experiments
By Patrick Oswald and Pawel Pieranski

Alignment Technologies and Applications of Liquid Crystal Devices
By Kohki Takatoh, Masaki Hasegawa, Mitsuhiro Koden, Nobuyuki Itoh, Ray Hasegawa, and Masanori Sakamoto

THE LIQUID CRYSTALS BOOK SERIES

ADSORPTION PHENOMENA AND ANCHORING ENERGY IN NEMATIC LIQUID CRYSTALS

Giovanni Barbero
Luiz Roberto Evangelista

Taylor & Francis
Taylor & Francis Group

Boca Raton London New York Singapore

A CRC title, part of the Taylor & Francis imprint, a member of the Taylor & Francis Group, the academic division of T&F Informa plc.

Published in 2006 by
CRC Press
Taylor & Francis Group
6000 Broken Sound Parkway NW, Suite 300
Boca Raton, FL 33487-2742

Printed in the United States of America on acid-free paper
10 9 8 7 6 5 4 3 2 1

International Standard Book Number-10: 0-8493-3584-1 (Hardcover)
International Standard Book Number-13: 978-0-8493-3584-6 (Hardcover)
Library of Congress Card Number 2005048508

Library of Congress Cataloging-in-Publication Data

Barbero, G. (Giovanni)
Adsorption phenomena and anchoring energy in nematic liquid crystals / G. Barbero and L.R. Evangelista.
p. cm. – (The liquid crystals book series)
includes bibliographical references and index.
ISBN 0-8493-3584-1
Liquid crystals. 2. Adsorption. 3. Liquid crystals – Surfaces. I. Evangelista, L. R. II. Title. III. Series.

QD923-B357 2005
530.4'29 – dc22 2005048508

T&F informa
Taylor & Francis Group
is the Academic Division of T&F Informa plc.

Visit the Taylor & Francis Web site at
http://www.taylorandfrancis.com

and the CRC Press Web site at
http://www.crcpress.com

Lettor, tu vedi ben com'io innalzo,
la mia matera, e però con più arte
non ti maravigliar s'io la rincalzo.
Dante (Purgatorio **IX**, 70-72).

This book is dedicated to the memory of
Marisa Cuniberto (1953-2003).

Contents

PREFACE

Liquid crystals have been known for more than one hundred years, but only in the second half of the last century have the bulk properties of these materials been extensively investigated. The efforts resulted in a good phenomenological description of the bulk properties of the nematic phase. For what concerns the surface properties of the nematic phase, the situation can be considered as less clear. Despite a great number of phenomenological results, a satisfactory understanding of the origin of the surface energy is missing. As in any other field of condensed matter, the surface and interfacial properties of nematic liquid crystals are rather complex. From the practical point of view it is important to know the alignment of a nematic liquid crystal sample when it is in contact with a solid substrate. The uniform alignment of liquid crystals in this case is crucial for display applications and other liquid crystalline devices. Therefore, the liquid crystal surface properties and, in particular, the characteristics features of the anchoring of the liquid crystals are very important for the performance of the liquid crystal devices since the strength of the anchoring affects the threshold characteristics of the sample.

However, the precise nature and the origin of the anchoring energy in nematic liquid crystals are still a subject of many fundamental and experimental studies and cannot be considered as a solved problem. In fact, this quantity is expected to be a property independent of the bulk. On the contrary, recent measurements show that, in many samples, the anchoring energy can be thickness dependent and can be also dependent on the bias voltage.

To explain the thickness dependence of the anchoring energy found in some nematic liquid crystal samples the phenomenon of the selective ion adsorption has been recently invoked. The influence of the selective ion adsorption on the anisotropic part of the anchoring energy strength has been discussed by several authors in the last years. According to this point of view, the adsorption phenomenon is responsible for an ionic separation inside the liquid. To this charge separation is connected an electric field distribution across the sample. The coupling of this field with the dielectric and flexoelectric properties of the liquid crystal gives rise to a dielectric energy density, localized near to the limiting surfaces, on mesoscopic thicknesses. This energy can be considered as a surface energy, which renormalizes the anisotropic part of the interfacial energy characterizing the interface nematic liquid crystal substrate. It has been established that this contribution to the anchoring energy can explain its thickness dependence as well as its bias voltage dependence.

The purpose of this book is to present a theoretical treatment for the adsorption phenomena and their effects on the anchoring energy in nematic liquid crystals. The approach is developed in successive steps, each one incorporating a further degree of generalization with respect to the previous one, in order to achieve a final and unified point of view.

The first part of the book is dedicated to the discussion of some phenomenological descriptions for the origin of the anchoring energy in nematics. There are good textbooks dealing with fundamental properties of nematics. For this reason, in Chapter 1, we present some of the essential concepts to describe the nematic phase in terms of a molecular approach (Maier-Saupe theory) as well as in terms of a macroscopic (continuum) theory. The presentation is limited to the ideas that will be useful for the remaining part of this book. Of particular importance in this regard is the effect of an electric field on the elastic energy density of a nematic sample. In Chapter 2, we introduce the concept of anchoring energy, discussing also possible forms for a phenomenological treatment of the problem. In Chapter 3, the bulk and surface elastic effects of quadrupolar interactions in liquid crystalline media are discussed. The idea is to apply the conceptual approach presented in the two preceding chapters to a problem of elastic theory in which intervene bulk and surface effects in a delicate manner. In Chapter 4, the temperature dependence of the surface free energy is analyzed in details, with recent theoretical predictions compared with experimental results in the low frequency range.

The second part of the book is dedicated to the adsorption phenomena. In Chapter 5, the general characteristics of the phenomenon are presented. The fundamental concepts of coverage and adsorption energy are discussed. An example dealing with the adsorption of magnetic grains in magnetic fluids is worked out in details. A statistical interpretation of the two-level approximation for the adsorption phenomenon is presented, by emphasizing the non-local character of the adsorption energy in this framework. The connection of the surface adsorption with anchoring transitions is investigated. In Chapter 6, the selective ion adsorption phenomenon is investigated in the framework of a Poisson-Boltzmann approach. In particular, the problem of the ionic adsorption in nematic samples is addressed in an introductory manner. The thickness dependence of the anchoring energy is investigated by means of an approximated model, working in the exponential approximation for the electric field profile inside the sample. The second part of the chapter deals with the destabilization effect of the surface electric field of ionic origin on the molecular orientation of a nematic sample. Of particular relevance is the analysis dedicated to the phenomenon of spontaneous Fréedericksz transition. The problems are again focused in the framework of the exponential field approximation, whose limits of applicability and validity are extensively discussed in Chapter 7. There, the effect of external electric fields on the nematic anchoring energy is also investigated. We establish the fundamental formulae that permit the calculation of the anchoring energy of dielectric origin, once the electric field distributions are known. A complete model for the determination of the electric field of ionic origin in nematic liquid crystals is presented in Chapter 8, with an extensive analysis of several cases of practical importance. Possible extensions of the model for adsorption of two species of charges and different surfaces are presented. In Chapter 9, a Fermi-like description for the adsorption phenomena is presented, and its predictions are compared with the classical approach discussed in the preceding chapters. A common feature of Chapters 5 and 6 is that they are dedicated to the static aspects of the adsorption phenomena. To complete the analysis of the surface adsorption, Chapter 10 is entirely devoted to the dynamical aspects of the ion adsorption. The drift-diffusion phenomenon in the context of liquid crystalline materials is analyzed. The presentation is focused on the problem of molecular reorientation dynamics in a nematic cell, when the adsorption phenomenon of neutral (dyes) or charged (ions) particles takes place. Finally, in Chapter 11 the role of the diffuse layer of the ionic charge on the impedance spectroscopy of a cell of a liquid is analyzed in two steps: firstly, by assuming that the ions have the same mobility, and that the electrodes are perfectly blocking and, then, generalizing the analysis to the case in which the mobility of the positive ions is not the same as the negative ones. Further, the adsorption phenomenon is explicitly taken into account and it is shown that it is responsible for an increasing of the real part of the impedance of the cell.

The authors are in debt to many colleagues and students, who participated in the successive developments of the subject during recent years. It is our pleasure to thank the collaboration of N. V. Madhusudana (Bangalore), A. K. Zvezdin, Z. Gabbasova and G. Sayko (Moscow), P. Galatola, A. Bourdon, and A. Bee (Paris), S. Faetti (Pisa), L. Komitov and P. Jägemalm (Gothenburg), A. G. Petrov and M. P. Petrov (Sofia), W. A. Popa-Nita and A. L. Alexe-Ionescu (Bucharest), G. Cipparrone, P. Pagliusi, G. Strangi, N. Scaramuzza, and C. Versace (Calabria), A. M. Figueiredo Neto (São Paulo), D. Olivero (Torino), R. S. Mendes, S. Fontanini, L. C. Malacarne, H. A. Pereira and F. Batalioto (Maringá). We are grateful to C. Oldano, A. Strigazzi, M. Becchi, and S. Ponti (Torino), I. Lelidis (Amiens), E. A. Oliveira (São Paulo), M. Simões (Londrina) and A. J. Palangana (Maringá), for the extremely encouraging discussions in the last years. We are also grateful to numerous students who read and commented on various preliminary versions of the manuscript. Finally, we are grateful to our sponsors: INFM (Sezione Torino), Capes, CNPq and Fundação Araucária (Brazilian agencies) and Fondazione CRT (Italy).

1

NEMATIC LIQUID CRYSTALS: FUNDAMENTALS AND BULK PROPERTIES

In this chapter, we discuss the main characteristics of the nematic liquid crystalline phase. First, the scalar and tensor order parameters are introduced. The molecular self-consistent approach of Maier and Saupe to describe the nematic-isotropic phase transition is presented, followed by an extension of the model to the case of dimerised systems, in which internal degrees of freedom have to be taken into account. The possibility to treat the phase as a continuum from the elastic point of view is analyzed. The bulk and surface elastic energy densities are also presented in the simplest approximation. The coupling of the anisotropic properties of the nematic phase with external fields is discussed in some details. Two illustrative examples are developed to fix the concept of uniformly and non-uniformly aligned samples. The order transition (Fréedericksz) is investigated for a sample in the shape of a slab, in the hypothesis of strong anchoring.

1.1 Nematic order parameters

Nematic liquid crystals (NLC) are organic materials formed by elongated molecules presenting long-range orientational order in the absence of translational order [1]–[11]. The molecules of NLC are highly anisotropic and are commonly modelled as rigid rods or ellipsoids of revolution, characterized by a long molecular axis $\mathbf{m}$. Nematics are the simplest of the liquid crystalline phases, and the intermolecular interaction energy responsible for the phase tends to align the long molecular axis parallel to a common direction $\mathbf{n}$. This *director* field $\mathbf{n}$ coincides with the optical axis of a uniaxial crystal. The physical properties of a nematic sample depend on the spatial distribution of the director $\mathbf{n}$, which is the average molecular orientation of the molecules forming the nematic phase.

In the absence of external electric field, the polarization of the nematic phase is zero, which means that the configurations represented by $\mathbf{n}$ and $-\mathbf{n}$

are completely equivalent, i.e.,

$$\langle \mathbf{n} \cdot \mathbf{m} \rangle = 0, \tag{1.1}$$

where $\langle ... \rangle$ indicates statistical average. Furthermore, to characterize the molecular alignment of the phase, it is useful to introduce a microscopic order parameter which measures the dispersion of **m** around the direction defined by **n**. A scalar order parameter is defined as

$$S = \frac{3}{2}\left[\langle(\mathbf{n} \cdot \mathbf{m})^2\rangle - \frac{1}{3}\right]. \tag{1.2}$$

By increasing the temperature, the orientational order is lost and a phase transition to the isotropic phase occurs. In the isotropic phase $S = 0$ because all the orientations are distributed at random, and then $\langle(\mathbf{n} \cdot \mathbf{m})^2\rangle = 1/3$. The nematic phase is less symmetric than the isotropic phase. For this reason, it is possible to define, as a macroscopic order parameter, some anisotropic quantity Q such that $Q = 0$ for the symmetric phase, and $Q \neq 0$ for the less symmetric phase. The order parameter usually considered in liquid crystal's literature is connected with the anisotropy in diamagnetic susceptibility χ, represented by a second-rank Cartesian tensor. Since the molecules are anisotropic in shape, the physical quantities representing the phase are expected to be also anisotropic. If **M** is the magnetic moment due to the molecular diamagnetism, its connection with the magnetic field can be expressed as

$$M_i = \sum_j \chi_{ij} H_j, \quad \text{for} \quad i, j = x, y, z, \tag{1.3}$$

in which χ_{ij} are the elements of the susceptibility tensor. In the isotropic phase the tensor is diagonal, i.e.,

$$\chi_{ij} = \chi \delta_{ij}, \tag{1.4}$$

whereas in the nematic phase it can be put in a diagonal form (as we will show later)

$$\begin{pmatrix} \chi_\perp & 0 & 0 \\ 0 & \chi_\perp & 0 \\ 0 & 0 & \chi_\parallel \end{pmatrix}, \tag{1.5}$$

where $\chi_\perp$ and $\chi_\parallel$ refer, respectively, to the directions perpendicular and parallel to the symmetry axis (the director **n**). In the form introduced above, χ is not yet a good order parameter because it cannot vanish in the isotropic phase. In fact, in that phase we have

$$\chi = \frac{1}{3}\left(2\chi_\perp + \chi_\parallel\right) = \frac{1}{3}\sum_k \chi_{kk}. \tag{1.6}$$

However, we can obtain from the susceptibility its anisotropic part

$$Q_{ij} = Q_0 \left(\chi_{ij} - \frac{1}{3}\delta_{ij} \sum_k \chi_{kk} \right), \tag{1.7}$$

which can be used as an order parameter. The constant Q_0 is the normalization and can be chosen as the inverse of the maximum observed anisotropy (for a perfectly aligned nematic phase) [5]. It is possible to establish a connection between S and Q because the diamagnetic anisotropy

$$\chi_a = \chi_{\|} - \chi_{\perp} \propto S, \tag{1.8}$$

and, therefore, a usual form of the macroscopic order parameter is

$$Q_{ij} = \frac{3}{2} S \left(n_i n_j - \frac{1}{3}\delta_{ij} \right). \tag{1.9}$$

This tensorial order parameter of quadrupolar symmetry contains all the elements of symmetry of the NLC phase. This means that all the tensors characterizing the NLC phase have to be decomposed in terms of the elements of the symmetry of the phase. They can be decomposed in terms of **n**, unity tensor (Kronecker delta), δ_{ij}, and Levi-Civita tensor, ϵ_{ijk}; or, alternatively, they can be decomposed in terms of Q, unit tensor and Levi-Civita tensor [8]. This second way permits the investigation of the temperature dependence of the order parameter by means of the Landau-de Gennes theory [1].

1.2 Maier-Saupe model for the nematic-isotropic phase transition

The intermolecular forces giving rise to the nematic phase are such that they depend not only on the spatial separation between the molecules, but also on the relative orientation of the molecules. A successful model to describe the nematic-isotropic phase transition is due to Maier and Saupe [12]. In the model, they consider that the molecules interact through the induced dipole-dipole term of the Coulomb interaction (van der Waals forces). A brief account of this approach will be given here just to present the main lines of a molecular approach for the nematic-isotropic phase transition. * In Chapter 4, the Maier-Saupe model will be employed to investigate the temperature dependence of the surface energy.

*The presentation here follows closely that of Plischke and Bergersen, *Equilibrium Statistical Mechanics* (World Scientific, Singapore, 1999, 2nd edition).

We consider a molecular system (N rod-like molecules) such that the center of the $i-$ molecule is located in $\mathbf{r}_i$ whereas the vector pointing along the symmetry direction is $\mathbf{n}_i$. The interaction among the molecules will be denoted by $V(\mathbf{r}_{ij}, \mathbf{n}_i, \mathbf{n}_j)$, where $\mathbf{r}_{ij} = \mathbf{r}_j - \mathbf{r}_i$. Furthermore, $g(\mathbf{r}_{ij}, \mathbf{n}_i, \mathbf{n}_j)$ will denote the pair distribution function, i.e., the conditional probability that a molecule is found in $\mathbf{r}_i$, with orientation $\mathbf{n}_i$, when in $\mathbf{r}_j$ another molecule is localized with orientation $\mathbf{n}_j$. The orientational dependent part of the interaction energy (written as an average on the spatial and angular variables) will be

$$E(\mathbf{n}_i) = c + \rho \int d^3\mathbf{r}_{ij} \int d\Omega_j V(\mathbf{r}_{ij}, \mathbf{n}_i, \mathbf{n}_j) f(\mathbf{n}_j) g(\mathbf{r}_{ij}, \mathbf{n}_i, \mathbf{n}_j), \quad (1.10)$$

where ρ is the number of molecules per unit volume (which will be assumed as constant), c is a term independent of the orientation, $f(\mathbf{n}_j)$ is the orientational distribution function and Ω_j denotes the solid angle of $\mathbf{n}_j$ [13]. There are two important approximations for the Maier-Saupe model. The first one is to consider that the pair distribution function does not depend on the orientation of the molecules, i.e.,

$$g(\mathbf{r}_{ij}, \mathbf{n}_i, \mathbf{n}_j) = g(\mathbf{r}_{ij}). \quad (1.11)$$

The second approximation is to consider that the potential has the form

$$V(\mathbf{r}_{ij}, \mathbf{n}_i, \mathbf{n}_j) = -U(\mathbf{r}_{ij}) P_2(\mathbf{n}_i \cdot \mathbf{n}_j), \quad (1.12)$$

which is obtained by considering that the potential can be developed as a multipole expansion where only the first terms are taken into account. More precisely, one considers that

$$V(\mathbf{r}_{ij}, \mathbf{n}_i, \mathbf{n}_j) = \sum_m U_m P_m(\mathbf{n}_i \cdot \mathbf{n}_j), \quad (1.13)$$

with

$$U_m = \sum_k \langle U_{mk}(\mathbf{r}) \rangle. \quad (1.14)$$

In the above expressions, $P_m(\mathbf{n}_i \cdot \mathbf{n}_j)$ are the Legendre polynomials. In particular $P_2(x) = (3x^2 - 1)/2$. We have, furthermore, that

$$\langle U_{mk}(\mathbf{r}) \rangle = \int U_{mk}(\mathbf{r}) g(\mathbf{r}) d^3\mathbf{r}. \quad (1.15)$$

Therefore, apart the constant term, the orientational dependent part of the energy (1.10) becomes

$$E(\mathbf{n}_i) = -\rho U \int d\Omega_j f(\mathbf{n}_j) P_2(\mathbf{n}_i \cdot \mathbf{n}_j). \quad (1.16)$$

Associated with each particle i is a symmetric traceless tensor:

$$\sigma^i_{jk} = \frac{1}{2}\left(3n^i_j n^i_k - \delta_{jk}\right), \tag{1.17}$$

where $\mathbf{n}^i$ is a unit vector pointing along the long axis of the molecule specified by its solid angle on the unit sphere. In terms of (1.17) the tensor order parameter can be defined as

$$Q^i_{jk} = \langle\sigma^i_{jk}\rangle = \langle\frac{1}{2}\left(3n^i_j n^i_k - \delta_{jk}\right)\rangle. \tag{1.18}$$

We can now write

$$P_2(\mathbf{n}_i \cdot \mathbf{n}_j) = \frac{2}{3}\sum_{k,l}\sigma^i_{kl}\sigma^j_{kl}, \tag{1.19}$$

where the labels i, j are for the $i-$ and $j-$molecule, respectively. By using (1.19) in (1.16) one obtains

$$E(\mathbf{n}_i) = -\frac{2}{3}\rho U\sum_{k,l} Q^i_{kl}\sigma^i_{kl} \tag{1.20}$$

because

$$Q_{kl} = \langle\sigma_{kl}\rangle = \int d\Omega\, \sigma_{kl}\, f(\mathbf{n}). \tag{1.21}$$

The total energy of the system of N particles is

$$E = \frac{1}{2}N\langle E(\mathbf{n}_i)\rangle, \tag{1.22}$$

which, in view of (1.20), can be rewritten as

$$\begin{aligned} E &= -\frac{1}{3}\rho U\sum_{j,k} Q_{jk}\int d\Omega_i f(\mathbf{n}_i)\sigma^i_{jk} \\ &= -\frac{1}{3}\rho U\sum_{j,k} Q_{jk}Q_{jk}. \end{aligned} \tag{1.23}$$

The orientational entropy of the system is

$$\Sigma = -Nk_{\mathrm{B}}\int d\Omega f(\mathbf{n})\ln\left[f(\mathbf{n})\right], \tag{1.24}$$

with the distribution function given by

$$f(\mathbf{n}) = \frac{e^{-\beta E(\mathbf{n})}}{Z}, \tag{1.25}$$

where

$$Z = \int d\Omega e^{-\beta E(\mathbf{n})} \tag{1.26}$$

is the single molecule partition function. In the above expressions, $\beta = 1/k_{\rm B}T$, where $k_{\rm B}$ is the Boltzmann constant and T the absolute temperature. From these expressions we can investigate the thermodynamical behavior of the system. The order parameter is obtained in a self-consistent manner by means of Eq. (1.21).

The case $Q = 0$ corresponds to the isotropic phase (high temperature); for low temperatures, the above equation can present solutions $Q \neq 0$ which can correspond to a preferential direction for the orientation of the molecules, i.e., to the nematic phase.

The elements of the order parameter in an arbitrary reference system are given by (1.18). The order parameter is then given by a real 3×3 symmetric matrix. Therefore, it is always possible to find a reference system in which Q is diagonal. If we consider, for instance

$$n_x = \sin\theta\cos\phi, \quad n_y = \sin\theta\sin\phi, \quad \text{and} \quad n_z = \cos\theta, \tag{1.27}$$

where θ and ϕ are, respectively, the polar and azimuthal angles, one obtains

$$Q_{jk} = \begin{pmatrix} -\frac{1}{2}\left[\langle P_2\rangle - \langle q\rangle\right] & 0 & 0 \\ 0 & -\frac{1}{2}\left[\langle P_2\rangle - \langle q\rangle\right] & 0 \\ 0 & 0 & \langle P_2\rangle \end{pmatrix}, \tag{1.28}$$

where

$$\langle P_2\rangle = S = \frac{1}{2}\langle 3\cos^2\theta - 1\rangle \quad \text{and} \quad \langle q\rangle = \langle \frac{3}{2}\sin^2\theta\cos 2\phi\rangle. \tag{1.29}$$

If we now choose the $z-$axis as the preferential direction, i.e., $\mathbf{n} \parallel \mathbf{z}$, we have $\langle q\rangle = 0$ and (1.28) becomes

$$Q_{jk} = \begin{pmatrix} -\frac{1}{2}\langle P_2\rangle & 0 & 0 \\ 0 & -\frac{1}{2}\langle P_2\rangle & 0 \\ 0 & 0 & \langle P_2\rangle \end{pmatrix}. \tag{1.30}$$

In this case we immediately obtain

$$\sum_{jk} Q_{jk}Q_{jk} = \frac{3}{2}S^2. \tag{1.31}$$

The above result gives for (1.23)

$$E = -\frac{1}{2}\rho U S^2 \tag{1.32}$$

and, for (1.20)

$$E(\mathbf{n}) = -\rho U S P_2(\cos\theta). \tag{1.33}$$

The single molecule partition function can be rewritten in the form

$$Z = \int d\Omega e^{\beta\rho U S P_2(\cos\theta)}. \tag{1.34}$$

With this choice of preferential axis the self-consistency condition can be expressed as

$$S = \langle P_2(\cos\theta)\rangle = \frac{1}{Z}\int d\Omega P_2(\cos\theta) e^{\beta\rho U S P_2(\cos\theta)}. \tag{1.35}$$

The above equation has to be numerically solved to furnish the behavior of $S(T)$. The procedure is the standard one, namely, one finds a solution for (1.35) and then verifies whether it corresponds to an absolute minimum of the free energy

$$F = E - T\Sigma = -\frac{1}{2}N\rho\beta U S^2 - N k_B T \ln Z, \tag{1.36}$$

which can be explicitly written as

$$\frac{F}{N} = -k_B T \ln\left[4\pi \int_0^1 dx e^{\frac{1}{2}\rho U\beta\left[(3x^2-1)S - S^2\right]}\right]. \tag{1.37}$$

Equation (1.35), with the help of (1.34), can be rewritten in the form

$$S(T) = \frac{1}{2}\frac{\int_0^1 dx\ (3x^2-1)\ e^{aS(T)x^2}}{\int_0^1 dx\ e^{aS(T)x^2}}, \tag{1.38}$$

where $a = 3\rho U/(2k_B T)$. In Figure 1.1, a numerical solution of Eq. (1.38) is depicted for illustrative purposes. Notice the universal behavior of $S(T)$, starting from $S = 1$ for $T = 0\,\mathrm{K}$, and going to $S = 0$, at $T_R^* = 0.2202\rho U/k_B$, where the isotropic phase starts. The order parameter at the transition is $S_c = 0.4289$.

1.3 Extension of the Maier-Saupe theory: Influence of the dimerization process on the nematic ordering

One of the commonly used approximations for the investigation of the liquid crystal macroscopical properties is to assume the constituent molecule as a rigid rod. This approximation, together with the Maier-Saupe mean field analysis, presented earlier, interprets most of the phenomena experimentally

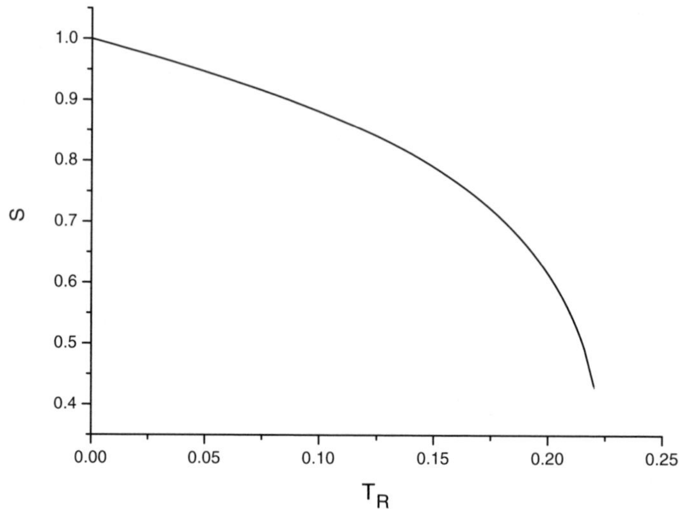

FIGURE 1.1
Scalar order parameter, S, versus reduced temperature $T_R = 2k_{\rm B}T/(3\rho U)$ as predicted by the self-consistent relation (1.38) of the Maier-Saupe approach.

observed in the conventional (classical) liquid crystals like MBBA and PAA. The investigations [14, 15] of the 4,n-alkyloxybenzoic acids, displaying nematic (N) and smectic S (S_C) phases like HOBA, OOBA and NOBA (heptyl, octyl and nonyloxybenzoic acids), however indicated that the rigid rod approximation cannot be used for explanation of the observed anomalies in the nematic state of these substances. The anomalies are in the temperature dependence of the elastic constants [16], the electroconductivity [17] and the dielectric permittivity [18]. The temperature trends of these parameters strongly deviate from those of the classical nematics. The unusual thermal behavior of the nematic phase of the 4,n-alkyloxybenzoic acids was found and analyzed by some optical methods. The experimental results demonstrate a strong textural transition and a dramatic increasing of the depolarized Rayleigh scattering at a given temperature T^* in the nematic range [14]. It is reasonable to assume that this phenomenon is connected to the temperature variable form of the constituent molecules of the 4,n-alkyloxybenzoic acids. These substances change the molecular structure with the temperature between three states: closed dimers, open dimers and monomers [19, 20].

In this section we present a theoretical model [21], which is also a further development of the classical mean field theories, but takes into account the internal degrees of freedom of the molecules due to the variations of the three molecular forms: closed dimers, open dimers and monomers. In the model the energy of the hydrogen bonding is one of the quantities connected with the influence of the molecular structure (dimerization process) on the macroscopical properties. This energy was measured in Ref. [19]. The model could explain a part of the experimental results. A consideration of the short range forces and fluctuations, leading to a smectic short range order in a coexistence with the nematic long range order, must be taken into account as well.

Let us fix the following notation for this section: $\mathcal{N}$ is the total number of closed dimers, at $T = 0$; N^*, $N_m = 2N^*$, N_c and N_o are, respectively, the total number of dissociated dimers, monomers, closed dimers and open dimers at a temperature T. Furthermore, $E > 0$ is the energy necessary to break a hydrogen bond of the closed dimer, the activation energy of the chemical reaction 1 *closed dimer* $\rightarrow$ 1 *open dimer* is then E; the activation energy of the chemical reactions 1 *closed dimer* $\rightarrow$ 2 *monomers* is $2E$ and 1 *open dimer* $\rightarrow$ 2 *monomers* is E.

First we determine the equilibrium concentrations of closed dimers (n_c), open dimers (n_o) and monomers (n_m) by means of simple statistical mechanics methods. Let μ be the chemical potential of the mixture under consideration. At a given temperature T, the equilibrium distributions of open dimers, closed dimers and monomers are respectively given by

$$n_o = e^{-(\mu+E)/k_{\mathrm{B}}T} \quad n_c = e^{-\mu/k_{\mathrm{B}}T} \quad \text{and} \quad n_m = 2e^{-(\mu+2E)/k_{\mathrm{B}}T},$$

subject to the condition $n_c + n_o + n_m/2 = 1$, where the concentrations are $n_o = N_0/\mathcal{N}$, $n_c = N_c/\mathcal{N}$ and $n_m = N_m/\mathcal{N}$, which implies that

$$e^{\mu/k_{\mathrm{B}}T} = 1 + e^{-1/x} + e^{-2/x}, \tag{1.39}$$

where $x = k_{\mathrm{B}}T/E$. This relation defines the chemical potential in terms of the energy of the hydrogen bond and of the temperature. The fundamental limits for the above expressions are the following: $T \rightarrow 0$ gives $n_o \rightarrow 0$, $n_c \rightarrow 1$ and $n_m \rightarrow 0$. On the other hand, for $T \rightarrow \infty$, $n_o = n_c = n_m/2 = 1/3$, which imply also that $N_m = 2\mathcal{N}/3$. The temperature dependencies of n_c, n_o and n_m are shown in Figure 1.2. These results are in agreement with the experimental data of Ref. [20], according to which the closed dimer concentration decreases, whereas the open dimers and monomers concentration increases, for increasing temperature.

Let us consider now the nematic order in the mixture formed by closed dimers, open dimers and monomers. We assume that the monomers do not contribute to the nematic order. This is equivalent to suppose that the monomers are nearly of spherical shape, which is not too far from the reality. In this framework the nematic order is due only to the interactions among the molecules of the closed and open dimers. Furthermore, we suppose that the closed and open dimers are formed by rod-like molecules. Let $\mathbf{n}$ be the nematic director of the mixture. The angle formed by the molecular long axis of

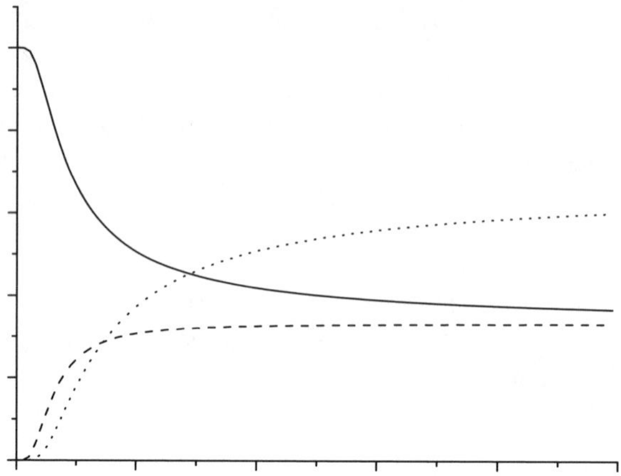

FIGURE 1.2
Equilibrium concentrations of closed dimers, n_c (solid line), open dimers, n_o (dotted line), and monomers, n_m (dashed line), versus $x = k_B T/E$. Reprinted from Ref. [21] with permission from Elsevier.

the closed and open dimer with **n** will be indicated by θ_c and θ_o, respectively. In the mean field approximation the total nematic potential is given by

$$V(\theta_o, \theta_c) = \sum_{i,j} V_{ij} = V_{oo} + V_{cc} + V_{oc} + V_{co}, \tag{1.40}$$

where V_{oo} and V_{oc} are the mean field potentials acting on a molecule of open dimer due to the other open dimers and closed dimers, respectively. A similar meaning has V_{cc}, connected to the closed dimer–closed dimer interaction, and V_{co}, due to the interaction of the closed dimer in the nematic field created by the open dimers. The partial nematic mean field potentials V_{ij}, in the Maier-Saupe approximations [see Eq. (1.12)] are given by [5]

$$V_{ij} = -\alpha_{ij} P_2(\theta_i) \langle P_2(\theta_j) \rangle, \tag{1.41}$$

where the coupling constants α_{ij} depend on the distance between the center of mass of the molecules and on a molecular property, which is roughly the anisotropy of the polarizability.

The above expressions allow one to write the potential in the form

$$V(\theta_o, \theta_c) = V_o(\theta_o) + V_c(\theta_c), \tag{1.42}$$

where

$$V_i(\theta_i) = -\left[\alpha_{ii} S_i + \alpha_{ij} S_j\right] P_2(\theta_i), \tag{1.43}$$

if we introduce the notation $S_i = \langle P_2(\theta_i) \rangle$ for the scalar order parameters of closed (S_c) and open (S_o) dimers, respectively. The partition function can be then factorized as follows

$$\begin{aligned} Z &= \int_{-1}^{1} d(\cos\theta_o) \int_{-1}^{1} d(\cos\theta_c) e^{-(V_o+V_c)/k_{\mathrm{B}}T} \\ &= \int_{-1}^{1} d(\cos\theta_o) e^{-V_o/k_{\mathrm{B}}T} \int_{-1}^{1} d(\cos\theta_c) e^{-V_c/k_{\mathrm{B}}T} \\ &= Z_o \cdot Z_c, \end{aligned} \tag{1.44}$$

and the free energy can be written as

$$F = -\mathcal{N} k_{\mathrm{B}} T \ln Z + \frac{1}{2} \mathcal{N} \sum_{i,j} \alpha_{ij} S_i S_j. \tag{1.45}$$

The self-consistency has to be imposed now, leading to the following set of coupled equations

$$S_i = \frac{\int_{-1}^{1} d(\cos\theta_i) P_2(\theta_i) e^{-\beta V_i}}{Z_i} \tag{1.46}$$

for $i = o, c$, which will be numerically solved. Equation (1.46) is the simple generalization of (1.35) to the case of two component systems. We have to take into account that the above set of coupled equations depends on the following quantities: (a) the temperature T, (b) the activation energy E, and (c) the quantities α_{ij} entering in the potential, which contain the strengths of the (mean-field) interaction closed dimer–closed dimer, closed dimer–open dimer and open dimer–open dimer. The concentration dependence of these quantities is determined by assuming that

$$\alpha_{ij} = \alpha'_{ij} = u_{ij} n_j. \tag{1.47}$$

This assumption for the coupling constants α_{ij} can be justified with the following arguments. At this stage we suppose that only one kind of nematic molecules is present. Let us consider a nematic molecule interacting with a volume element $d\tau$ containing $dN = \rho d\tau$ molecules, where ρ is the density of particles. The interaction is $dV = J(r) P_2 \langle P_2 \rangle dN$, where r is the distance between the considered molecule and $d\tau$, and $J(r) = -c/r^6$ in the Maier-Saupe approximation [5]. The total nematic mean field is obtained by integrating dV over r from a lower cut-off, of molecular dimension r_0, to infinity. A simple calculation gives $V = -(4\pi/3r_0^3) c \rho P_2 \langle P_2 \rangle$, in the Fowler approximation in which ρ is constant. Hence, for fixed volume, α is proportional to the concentration of particles. An extension of this analysis to the case in which two types of nematic molecules are present allows, easily, to deduce Eqs. (1.47). Note that α_{ij} is not expected to be symmetric in the exchange between i and j, as we can see by considering the following example. Let us consider the special case of a system where there is only one molecule of the i-type in the whole sample and all the other molecules are of the j-type. In such a case the i-molecule is completely surrounded by j-molecules and its mean field energy, in the field of the j-molecules, will be relatively high. On the contrary, a generic j-molecule has a high probability to be at large distance from the i-molecule. Then, the mean field energy of the j-molecules in the field of the i-molecule is small.

However, previous analysis is not the only one possible. An alternative analysis is the following. α_{ij} have to depend on the average distance R among the interacting molecules. This distance is proportional to $\rho^{-1/3}$. Hence, if we assume that the interaction energy depends on R^{-6}, we find that $\alpha_{ij} \propto \rho^2$, i.e., instead of Eqs. (1.47) we have

$$\alpha_{ij} = \alpha''_{ij} = u_{ij} n_j^2. \tag{1.48}$$

Using Eqs. (1.43), (1.46) and (1.47) (linear case) or Eqs. (1.43), (1.46) and (1.48) (quadratic case) we can numerically determine the nematic scalar order parameters for the dimers in the closed, S_c, or open, S_o, configurations. When these quantities are known, the nematic order parameter of the mixture is evaluated by means of the relation

$$S = \frac{n_c S_c + n_o S_o}{n_c + n_o}. \tag{1.49}$$

In a similar manner we can define the macroscopic anisotropy $\Delta X = X_{\parallel} - X_{\perp}$, where, as usual, $\parallel$ and $\perp$ refer to the nematic director, of a second order tensor X_{ij}. As is well known, in the case of a pure compound, if $\Delta X^{(m)}$ is the molecular anisotropy and S the scalar order parameter, the macroscopic anisotropy is $\Delta X = S \Delta X^{(m)}$ [1]. In the case under consideration, where the nematic is a mixture of two nematics, we have

$$\Delta X = \frac{n_c S_c \Delta X_c^{(m)} + n_o S_o \Delta X_o^{(m)}}{n_c + n_o}, \tag{1.50}$$

where $\Delta X_c^{(m)}$ and $\Delta X_o^{(m)}$ are the molecular anisotropies of the closed and open dimers, respectively.

Figure 1.3 shows the nematic scalar order parameter of the mixture S when S_c and S_o are obtained for the linear case and for the quadratic case. In the same figure the nematic order parameter of a simple nematic liquid crystal, formed only with one type of molecule (Maier-Saupe approach), is shown for comparative purposes.

In Figure 1.4, the theoretical predictions are compared with the experimental data of Deloche and Cabane [19]. The best fit is obtained by assuming $\alpha_{ij} \propto n_i n_j$, $u_{cc} \simeq 1$ eV [5], $u_{oo} = u_{co} = u_{oc} = 2u_{cc}$, and $E = 4.8$

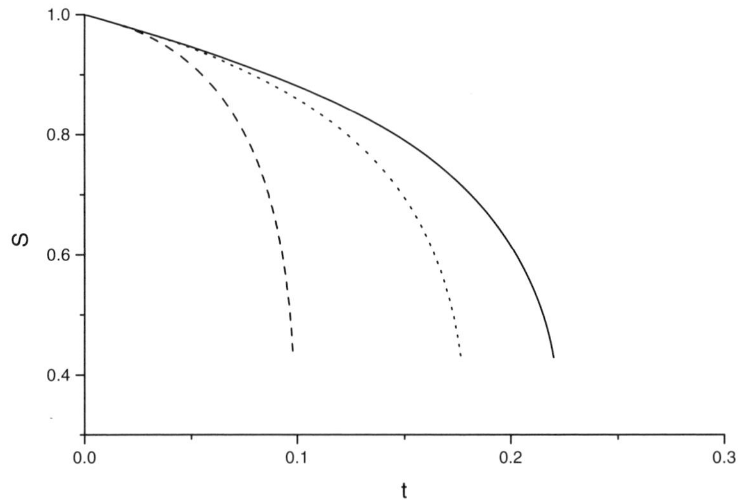

FIGURE 1.3
Nematic order parameter S of the mixture formed by closed dimers, open dimers and monomers in thermodynamical equilibrium. Dotted lines: $\alpha_{ij} \propto \sqrt{n_i n_j}$. Dashed lines: $\alpha_{ij} \propto n_i n_j$. Solid lines: classical Maier-Saupe theory. We have assumed that $u_{oo} = u_{co} = u_{oc} = 2u_{cc}$, with $u_{cc} \simeq 1$ eV [5] and $E = 4.8$ kcal/mole [19]. The reduced temperature is $t = k_B T/2u_{cc}$. Reprinted from Ref. [21] with permission from Elsevier.

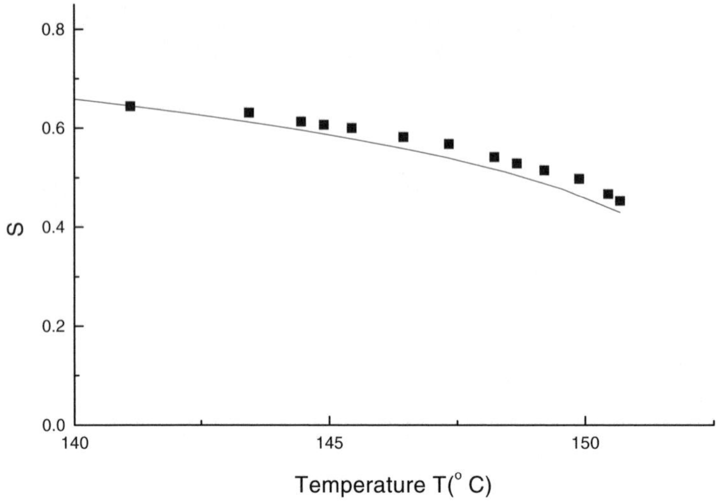

FIGURE 1.4
Nematic order parameter S of a mixture of closed dimers, open dimers and monomers in thermodynamical equilibrium, in the quadratic approximation for $u_{cc} \simeq 1$ eV, $u_{oo} = u_{cc} = u_{co} = 2u_{cc}$ [5] and $E = 4.8$ kcal/mole [19]. The squares are the experimental data of Ref. [19]. Reprinted from Ref. [21] with permission from Elsevier.

kcal/mole [19]. The agreement is fairly good. However, new experimental data are necessary in order to decide whether the coupling constants α_{ij} are given by Eqs. (1.47) or Eqs. (1.48).

In the model the molecules representing the closed and open dimers are supposed to be rod-like. In this framework their scalar order parameters are positive, as well as the macroscopic anisotropies.

1.4 Continuum description of the nematic phase

If the director $\mathbf{n}$ is independent of the position, the NLC medium is undistorted and its elastic energy density is supposed to be minimum. Let f_0 be this quantity. If we consider now $\mathbf{n} = \mathbf{n}(\mathbf{r})$ the NLC is distorted, and its elastic energy density will be indicated by f. In this case its first spatial derivatives, $n_{i,j} = \partial n_i/\partial x_j$ are different from zero. Let us assume that the first derivatives of $\mathbf{n}$ are sufficient to describe the distorted state. Then

$$f = f(n_{i,j}). \tag{1.51}$$

If the first derivatives of $\mathbf{n}$ are small quantities, it is possible to expand f in power series of $n_{i,j}$, as is usually done in the theory of elasticity. From (1.51) we then have

$$f = f_0 + L_{ij}n_{i,j} + \frac{1}{2}K_{ijkl}n_{i,j}n_{k,l} \geq f_0, \tag{1.52}$$

where the tensors L_{ij} and $K_{ijkl} = K_{klij}$ are defined as

$$L_{ij} = \left(\frac{\partial f}{\partial n_{i,j}}\right)_0 \quad \text{and} \quad K_{ijkl} = \left(\frac{\partial^2 f}{\partial n_{i,j}\partial n_{k,l}}\right)_0,$$

with the subscript 0 indicating that the derivatives are calculated with respect to the undistorted state. In the above expressions and henceforth the summation convention is adopted. The phenomenological tensors L_{ij} and K_{ijkl} can be decomposed by taking into account the elements of symmetry characterizing the medium, i.e., $\mathbf{n}$, the Kronecker delta, δ_{ij}, and the completely antisymmetric tensor, ϵ_{ijk}. Moreover, since in an NLC medium $\mathbf{n}$ and $-\mathbf{n}$ are equivalent, each term of (1.52) has to be even in $\mathbf{n}$.

Let us consider the tensor L_{ij}. It can be decomposed in the following way

$$L_{ij} = L_1 n_i n_j + L_2\delta_{ij} + L_3 n_k \epsilon_{kij},$$

where $L_1 = L_2 = 0$ since the NLC medium is not polar. Therefore the linear term in the first derivatives can be written as

$$L_{ij}n_{i,j} = L_3 n_k \epsilon_{kij} n_{i,j} = -L(\mathbf{n}.\nabla \times \mathbf{n}), \tag{1.53}$$

where, for simplicity, $L = L_3$. This coefficient is a pseudo-scalar like $\mathbf{n}\cdot\nabla \times \mathbf{n}$, since the energy is a true scalar. It is different from zero only for cholesteric liquid crystals, because they present a spontaneous deformation even in the ground state. For what concerns the tensor $K_{ijkl} = K_{klij}$ we can write

$$\begin{aligned} K_{ijkl} = K_1 n_i n_j n_k n_l &+ \frac{1}{2}K_2(n_i n_j \delta_{kl} + n_k n_l \delta_{ij}) + K_3 n_i n_k \delta_{jl} \\ &+ \frac{1}{2}K_4(n_i n_l \delta_{jk} + n_j n_k \delta_{il}) + K_5 n_j n_l \delta_{ik} \\ &+ K_6 \delta_{ij}\delta_{kl} + K_7 \delta_{ik}\delta_{jl} + K_8 \delta_{il}\delta_{jk}. \end{aligned}$$

In principle, K_{ijkl} is characterized by *eight* numbers. However, by taking into account that $n_i n_i = 1$, we deduce that K_i $(i = 1, ..., 4)$ do not contribute to f. Moreover,

$$K_5 n_j n_l \delta_{ik} n_{i,j} n_{k,l} = K_5(\mathbf{n}\times\nabla\times\mathbf{n})^2, \qquad K_6 \delta_{ij}\delta_{kl} n_{i,j} n_{k,l} = K_6(\nabla\cdot\mathbf{n})^2,$$
$$K_7 \delta_{ik}\delta_{jl} n_{i,j} n_{k,l} = K_7 n_{k,j} n_{k,j}, \qquad K_8 \delta_{il}\delta_{jk} n_{i,j} n_{k,l} = K_8 n_{l,j} n_{j,l}.$$

Then, the quadratic contribution in the first derivative to the elastic energy is

$$\frac{1}{2}\left[K_5(\mathbf{n}\times\nabla\times\mathbf{n})^2 + K_6(\nabla\cdot\mathbf{n})^2 + K_7 n_{k,j} n_{k,j} + K_8 n_{l,j} n_{j,l}\right]. \tag{1.54}$$

Since

$$n_{k,j} n_{k,j} = n_{k,j} n_{j,k} + (\mathbf{n}.\nabla\times\mathbf{n})^2 + (\mathbf{n}\times\nabla\times\mathbf{n})^2,$$

and

$$n_{k,j} n_{j,k} = (\nabla\cdot\mathbf{n})^2 - \nabla\cdot(\mathbf{n}\nabla\cdot\mathbf{n} + \mathbf{n}\times\nabla\times\mathbf{n}),$$

Equation (1.54) can be rewritten as

$$\frac{1}{2}(K_6 + K_7 + K_8)(\nabla\cdot\mathbf{n})^2 + \frac{1}{2}K_7(\mathbf{n}.\nabla\times\mathbf{n})^2 + \frac{1}{2}(K_5 + K_7)(\mathbf{n}\times\nabla\times\mathbf{n})^2$$
$$- (K_7 + K_8)\nabla\cdot(\mathbf{n}\nabla\cdot\mathbf{n} + \mathbf{n}\times\nabla\times\mathbf{n}). \tag{1.55}$$

Usually Eq. (1.55) is rewritten by defining

$$K_6 + K_7 + K_8 = K_{11}, \qquad K_7 = K_{22}, \qquad K_5 + K_7 = K_{33}, \qquad K_8 = K_{24},$$

and assumes the form

$$f_{\text{Frank}} = \frac{1}{2}K_{11}(\nabla\cdot\mathbf{n})^2 + \frac{1}{2}K_{22}(\mathbf{n}.\nabla\times\mathbf{n})^2 + \frac{1}{2}K_{33}(\mathbf{n}\times\nabla\times\mathbf{n})^2$$
$$- (K_{22} + K_{24})\nabla\cdot(\mathbf{n}\nabla\cdot\mathbf{n} + \mathbf{n}\times\nabla\times\mathbf{n}). \tag{1.56}$$

Equation (1.56) is the Frank expression for the elastic energy density of a deformed NLC, proposed in 1958. K_{11}, K_{22}, K_{33}, and $(K_{22}+K_{24})$ are known as splay, twist, bend and saddle-splay elastic constant, respectively.

The last term in (1.56), by means of Gauss theorem, gives only a surface contribution. Then the elastic energy density, proportional to the square of the spatial derivatives of the director, depends only on *three* elastic constants. These elastic constants are positive. By means of (1.53) and (1.56) the elastic energy density given by Eq. (1.52) is written as

$$f = f_0 - L(\mathbf{n} \cdot \nabla \times \mathbf{n}) + \frac{1}{2}K_{11}(\nabla \cdot \mathbf{n})^2 + \frac{1}{2}K_{22}(\mathbf{n} \cdot \nabla \times \mathbf{n})^2 + \frac{1}{2}K_{33}(\mathbf{n} \times \nabla \times \mathbf{n})^2 - (K_{22} + K_{24})\, \nabla \cdot\, (\mathbf{n}\, \nabla \cdot\, \mathbf{n} + \mathbf{n} \times \nabla \times \mathbf{n}). \tag{1.57}$$

In this book our attention will be devoted only to NLC, i.e., $L = 0$. In this case one can rewrite (1.57) in the compact form

$$f = f_0 + \frac{1}{2}K_{ijkl}n_{i,j}n_{k,l}, \tag{1.58}$$

to be used subsequently in this book. A possible source for the spatial dependence of K_{ijkl} will be discussed in Section 2.3.2.

1.5 External field contribution to the bulk elastic energy density

1.5.1 Flexoelectric and dielectric contributions

Let us now obtain an expression for the bulk elastic energy density in terms of first spatial derivatives of $\mathbf{n}$, when the system is submitted to an external electric field $\mathbf{E}$. In this case the energy density depends also on the applied electric field $\mathbf{E}$ of components E_i, namely $f = f(n_{i,j}, E_i)$. This quantity can be expanded in power series of the coordinates characterizing the thermodynamical state, which are $n_{i,j}$ and the components of the electric field E_i. It can be written in the form

$$f = f_0 + \frac{1}{2}K_{ijkl}n_{i,j}n_{k,l} - \frac{1}{2}\epsilon_{ij}E_iE_j - e_{ijk}n_{i,j}E_k. \tag{1.59}$$

The second addendum of (1.59) represents the usual dielectric energy. By decomposing ϵ_{ij} according to the general rule we obtain

$$\epsilon_{ij} = an_in_j + b\delta_{ij}, \tag{1.60}$$

where a and b are two constants to be determined. Simple calculations give

$$\epsilon_{ii} = a + 3b \quad \text{and} \quad n_in_j\epsilon_{ij} = a + b, \tag{1.61}$$

where ϵ_{ii} is the trace of ϵ_{ij}. In the local reference frame in which $\mathbf{n}$ is the z-axis, as we have considered in Section 1.2, the tensor ϵ_{ij} has the diagonal form

$$\begin{pmatrix} \epsilon_\perp & 0 & 0 \\ 0 & \epsilon_\perp & 0 \\ 0 & 0 & \epsilon_\parallel \end{pmatrix}, \tag{1.62}$$

because NLC materials are uniaxial crystals. It follows that $\epsilon_{ii} = 2\epsilon_\perp + \epsilon_\parallel$ and $n_i n_j \epsilon_{ij} = \epsilon_\parallel$. Consequently, from Eq. (1.61) we obtain

$$a = \epsilon_\parallel - \epsilon_\perp = \epsilon_a \quad \text{and} \quad b = \epsilon_\perp .$$

The quantity ϵ_a is called dielectric anisotropy. Hence ϵ_{ij}, from Eq. (1.60), is

$$\epsilon_{ij} = \epsilon_a n_i n_j + \epsilon_\perp \delta_{ij}, \tag{1.63}$$

and the dielectric energy appearing in Eq. (1.59) becomes

$$\frac{1}{2}\epsilon_{ij} E_i E_j = \frac{1}{2}\epsilon_a (\mathbf{n}\cdot\mathbf{E})^2 + \frac{1}{2}\epsilon_\perp E^2. \tag{1.64}$$

The last addendum of (1.59) can be rewritten as $-P_k E_k$, where

$$P_k = e_{ijk} n_{i,j}, \tag{1.65}$$

plays the role of a polarization induced by the deformation. Equation (1.65) is the equivalent of the electric polarization in a piezoelectric material. The quantities e_{ijk} are the components of the flexoelectric tensor $\mathbf{e}$. If the usual decomposition is performed we obtain

$$e_{ijk} = e_1 n_i n_j n_k + e_2 n_i \delta_{jk} + e_3 n_j \delta_{ik} + e_4 n_k \delta_{ij}, \tag{1.66}$$

where we have taken into account that e_{ijk} must be odd in $\mathbf{n}$. Since $n_i n_{i,j} = 0$, by substitution of (1.66) into (1.65) we obtain

$$P_k = e_3 n_j \delta_{ik} n_{i,j} + e_4 n_k \delta_{ij} n_{i,j}. \tag{1.67}$$

By taking into account that

$$n_j \delta_{ik} n_{i,j} = n_j n_{k,j} = -(\mathbf{n} \times \nabla \times \mathbf{n})_k,$$
$$n_k \delta_{ij} n_{i,j} = n_k n_{i,i} = n_k \nabla \cdot \mathbf{n},$$

Equation (1.67) can be written in a covariant form as

$$\mathbf{P} = e_4 \, \mathbf{n} \, \nabla \cdot \mathbf{n} - e_3 \, \mathbf{n} \times \nabla \times \mathbf{n}.$$

The phenomenological coefficients e_4 and e_3 are generally indicated as $e_4 = e_{11}$ and $e_3 = e_{33}$. Consequently, $\mathbf{P}$ writes

$$\mathbf{P} = e_{11} \mathbf{n} \nabla \cdot \mathbf{n} - e_{33} \mathbf{n} \times \nabla \times \mathbf{n}, \tag{1.68}$$

which has been proposed for the first time by Meyer in 1969 [22]. The molecular interpretation of the electrical polarization (1.68) is given in dipolar terms, i.e., by considering particular forms for the molecules constituting the NLC phase.

For practical applications we will consider that the elastic energy density, by taking into account the dielectric and the flexoelectric contribution, can be put in the form (apart from the terms that are not orientation dependent)

$$f = f_0 + \frac{1}{2} K_{ijkl} n_{i,j} n_{k,l} - \frac{1}{2} \epsilon_a \left(\mathbf{n} \cdot \mathbf{E} \right)^2 - \mathbf{P} \cdot \mathbf{E}. \tag{1.69}$$

1.5.2 Magnetic interaction

When an NLC is submitted to a magnetic field $\mathbf{H}$, the interaction energy writes

$$-\frac{1}{2} \chi_{ij} H_i H_j, \tag{1.70}$$

where χ_{ij} is the diamagnetic permittivity tensor introduced in (1.3). By decomposing χ_{ij} in the usual manner, as we have done for ϵ_{ij}, one obtains

$$\chi_{ij} = \chi_a n_i n_j + \chi_\perp \delta_{ij}. \tag{1.71}$$

Nematic liquid crystals usually have $\chi_a > 0$. Taking into account (1.71), the interaction energy Eq. (1.70) becomes

$$-\frac{1}{2} \chi_a (\mathbf{n} \cdot \mathbf{H})^2, \tag{1.72}$$

besides an inessential term, independent of the NLC orientation. The term describing the magnetic interaction has the same form of the one describing the dielectric interaction. Hence the equation written for NLC, submitted to an electric field, holds also when $\mathbf{E}$ is changed into $\mathbf{H}$, if ϵ_a is substituted by χ_a. However, since $\chi_a \approx 10^{-7}$ (cgs) [1], $\mathbf{H}$ may be considered constant across the sample.

1.6 Equilibrium configuration in strong and weak anchoring situations

The basic principle involved in the application of the continuum theory to the solution of real problems is that the equilibrium state of the director field $\mathbf{n}$ is always given by the director configuration that minimizes the total elastic energy of the system [8]–[10]. To illustrate this principle, let us particularize the analysis to the case of a sample of nematic liquid crystal limited by two uniform surfaces, placed in $z = \pm d/2$, i.e., a sample in the shape of a slab of thickness d. If the surface treatment, as supposed, is uniform, $\mathbf{n} = \mathbf{n}(z)$, where z, in this geometry, is the coordinate normal to the bounding walls. In this case, the elastic energy density f can be written in a simple way in terms of the angles defining $\mathbf{n}$. If we use the same geometry which gives (1.27) we will have now that $\theta = \theta(z)$ and $\phi = \phi(z)$. If we discard the uniform term f_0, and the surface-like term in (1.57), it can be written for the nematic phase as

$$\begin{aligned} f(\theta, \theta', \phi, \phi') &= \frac{1}{2}\left(K_{11}\sin^2\theta + K_{33}\cos^2\theta\right)\theta'^2 \\ &+ \frac{1}{2}\left(K_{22}\sin^2\theta + K_{33}\cos^2\theta\right)\sin^2\theta\phi'^2, \end{aligned} \tag{1.73}$$

where the prime indicates derivative with respect to z. Some particular cases are useful. If $\mathbf{n}$ is always parallel to the (x, z) plane, $\phi = 0$ and Eq. (1.73) gives

$$\begin{aligned} f(\theta, \theta') &= \frac{1}{2}\left(K_{11}\sin^2\theta + K_{33}\cos^2\theta\right)\theta'^2 \\ &= \frac{1}{2}K(\theta)\theta'^2, \end{aligned} \tag{1.74}$$

where $K(\theta) = K_{11}\sin^2\theta + K_{33}\cos^2\theta$. This case is relevant to a splay-bend deformation, i.e., no twist is considered. However, if $\mathbf{n}$ is always in the (x, y) plane, i.e., $\theta = \pi/2$, Eq. (1.73) gives

$$f(\phi, \phi') = \frac{1}{2}K_{22}\phi'^2, \tag{1.75}$$

which is relevant to a pure twist deformation. Finally, the one-constant approximation ($K_{11} = K_{22} = K_{33} = K$) is commonly employed and, in this case, Eq. (1.73) is written as

$$f(\theta, \theta', \phi, \phi') = \frac{1}{2}K\left(\theta'^2 + \sin^2\theta\,\phi'^2\right). \tag{1.76}$$

The total energy of an NLC of volume τ bounded by a surface $\mathcal{S}$, in the slab approximation, is given by

$$\mathcal{F} = \iiint_\tau f(\theta, \theta', \phi, \phi')\mathrm{d}\tau + \iint_\mathcal{S} f_S(\theta, \phi)\mathrm{d}\mathcal{S}, \tag{1.77}$$

where $f_S(\theta, \phi)$ takes into account the surface contribution to the total energy and will be considered in detail in the next chapter. The integration over x and y can be easily performed, and from Eq. (1.77) we obtain that the total energy per unit surface is given by

$$F = \frac{\mathcal{F}}{A} = \int_{-d/2}^{d/2} f(\theta, \theta', \phi, \phi')\mathrm{d}z + f_S(\theta_1, \phi_1) + f_S(\theta_2, \phi_2), \tag{1.78}$$

where A is the area of the bounding surfaces, $\theta_1 = \theta(-d/2)$, $\phi_1 = \phi(-d/2)$, $\theta_2 = \theta(d/2)$ and $\phi_2 = \phi(d/2)$.

In the slab approximation, $\theta(z)$ and $\phi(z)$ are then found by putting $\delta F = 0$. By using standard variational techniques [8], from Eq. (1.78) we obtain that $\theta(z)$ and $\phi(z)$ are solutions of the differential equations (Euler-Lagrange equations)

$$\frac{\partial f}{\partial \theta} - \frac{d}{dz}\frac{\partial f}{\partial \theta'} = 0, \quad \text{and} \quad \frac{\partial f}{\partial \phi} - \frac{d}{dz}\frac{\partial f}{\partial \phi'} = 0, \tag{1.79}$$

satisfying the boundary conditions

$$\begin{aligned} -\frac{\partial f}{\partial \theta'} + \frac{\partial f_S}{\partial \theta_1} &= 0, \qquad -\frac{\partial f}{\partial \phi'} + \frac{\partial f_S}{\partial \phi_1} = 0, \\ \frac{\partial f}{\partial \theta'} + \frac{\partial f_S}{\partial \theta_2} &= 0, \qquad \frac{\partial f}{\partial \phi'} + \frac{\partial f_S}{\partial \phi_2} = 0, \end{aligned} \tag{1.80}$$

for $z = -d/2$ and $z = d/2$, respectively. This situation is known as the *weak anchoring* situation. It corresponds to the case in which the angles θ_i and ϕ_i ($i = 1, 2$), which are the actual angles at the border of the sample, are not known, and have to be determined to establish the equilibrium configuration of the director. The surface energy f_S is comparable with the bulk term in (1.78). This case will be considered in more details in the next chapter, dedicated to analyzing the origin and the possible forms for the surface energy.

Another situation occurs when the surface term is much greater than the bulk term, i.e., if in (1.78) one has to assume that

$$f_S \gg \int_{-d/2}^{d/2} f(\theta, \theta', \phi, \phi')\mathrm{d}z. \tag{1.81}$$

In this case, called the *strong anchoring* situation, the profile of the director field has to be determined by satisfying boundary conditions of the kind

$$\theta_i = \Theta_i, \quad \text{and} \quad \phi_i = \Phi_i, \tag{1.82}$$

for $i = 1, 2$, where the values of Θ_i and Φ_i $(i = 1, 2)$ are known.

1.6.1 Example 1: Hybrid aligned nematic cell (HAN)

As a simple and illustrative application of the above formalism, we determine the equilibrium configuration of the director field in the case of a hybrid aligned nematic cell (HAN) [8] in the hypothesis of strong anchoring. We suppose that the treatment of the surface imposes the director to be perpendicular to the surface at $z = -d/2$ (homeotropic alignment) and parallel to that at $z = d/2$ (planar alignment). In this case the director is everywhere parallel to the (x, z) plane and only splay-bend distortions are expected, as discussed above. The total elastic energy of the sample is

$$F[\theta] = \int_{-d/2}^{d/2} \frac{1}{2} K(\theta) \theta'^2 \mathrm{d}z. \tag{1.83}$$

Now $\theta(z)$ must satisfy the boundary conditions

$$\theta_1 = 0 \quad \text{and} \quad \theta_2 = \pi/2. \tag{1.84}$$

Equation (1.79) for the present problem is

$$K(\theta)\theta'' + \frac{1}{2}\frac{dK(\theta)}{d\theta}\theta'^2 = 0. \tag{1.85}$$

In the considered case f does not depend explicitly on z. Hence, the quantity

$$p = \theta' \frac{\partial f}{\partial \theta'} - f \tag{1.86}$$

is independent of z. In fact, if $f = f(\theta, \theta')$ as in the present case, from Eq. (1.86) it follows that

$$\begin{aligned}
\frac{dp}{dz} &= \theta'' \frac{\partial f}{\partial \theta'} + \theta' \frac{d}{dz}\frac{\partial f}{\partial \theta'} - \frac{df}{dz} \\
&= \theta'' \frac{\partial f}{\partial \theta'} + \theta' \frac{d}{dz}\frac{\partial f}{\partial \theta'} - \frac{\partial f}{\partial \theta}\theta' - \frac{\partial f}{\partial \theta'}\theta'' \\
&= \theta' \left[\frac{d}{dz}\frac{\partial f}{\partial \theta'} - \frac{\partial f}{\partial \theta} \right] = 0,
\end{aligned}$$

because $\theta(z)$ has to satisfy Eq. (1.79). Therefore, by observing that

$$p = \theta' \left(\frac{\partial f}{\partial \theta'} \right) - f = \frac{1}{2} K(\theta) \theta'^2,$$

we can use to solve the proposed problem, instead of (1.85), the following equivalent equation:

$$K(\theta)\theta'^2 = C^2, \tag{1.87}$$

where $C^2 (= 2p)$ is an integration constant determined by boundary conditions (1.84). From (1.87) we have

$$C = \frac{1}{d} I(0, \pi/2). \tag{1.88}$$

The tilt angle is then given by

$$\frac{I[0, \theta(z)]}{I(0, \pi/2)} = \frac{z}{d}, \tag{1.89}$$

where $I(\alpha, \beta) = \int_\alpha^\beta \sqrt{K(x)} \mathrm{d}x$. As follows from (1.74), in the one-constant approximation $K(\theta) = K$. In this case, (1.88) gives

$$C = \frac{\pi}{2d} \sqrt{K},$$

and the tilt angle is found to be

$$\theta(z) = \frac{\pi}{2d} \left(z + \frac{d}{2} \right). \tag{1.90}$$

Note that in the one-constant approximation, from (1.90) and (1.83) it follows that F is given by

$$F = \pi^2 \frac{K}{8d}. \tag{1.91}$$

This equation shows that in the strong anchoring case F diverges in the limit $d \to 0$. The same result is found in the general case where $K_{11} \neq K_{33}$. In the strong anchoring situation, the NLC is distorted for any sample thickness.

1.6.2 Example 2: Fréedericksz transition–strong anchoring

The Fréedericksz transition in a uniform sample is a well-known effect described in many textbooks [1, 3]. It is a transition of orientation in NLC induced by an external field. For a particular situation of symmetry this transition of order is a second order phase transition, for which the control parameter is the applied field and the order parameter the maximum value of the tilt angle [1, 2, 3, 5]. In the case of compensated NLC, where hyperpolarizabilities need to be taken into account, the transition can be of first order at the inversion point of the main anisotropy [23]–[25].

In order to review the main aspects of this phenomenon in the strong- and weak-anchoring hypothesis a nematic slab of thickness d is again considered. We consider first the case of strong anchoring. The case of weak anchoring will be treated in Section 2.2.1. The easy direction is supposed to be parallel to the z axis (homeotropic alignment). It is also assumed that the director lies always in the plane (x, z) in such a way that $\mathbf{n} = \sin\theta(z)\mathbf{i} + \cos\theta(z)\mathbf{k}$, where $\theta(z)$ is the tilt angle, $\mathbf{i}$ and $\mathbf{k}$ are the unit vectors parallel to the x and z axes, respectively.

In the one-constant approximation, the bulk energy density due to elastic distortion and to the field $\mathbf{E}$, applied along the $z-$direction, apart a constant term, is

$$f = \frac{1}{2} K \left(\frac{d\theta}{dz}\right)^2 + \frac{1}{2}\epsilon_a\, E^2 \sin^2\theta. \tag{1.92}$$

In this framework, the total energy of the nematic sample per unit area is given by the functional

$$F[\theta(z)] = \int_{-d/2}^{d/2} \left[\frac{1}{2} K \left(\frac{d\theta}{dz}\right)^2 + \frac{1}{2}\epsilon_a\, E^2 \sin^2\theta\right] dz. \tag{1.93}$$

The actual $\theta(z)$ profile is obtained by minimizing the total energy given by (1.93). According to whether the strong- or weak-anchoring situation is considered, different boundary conditions must be fulfilled by $\theta(z)$.

In the hypothesis of strong anchoring $\theta(\pm d/2) = \Theta = 0$, and one can achieve normalization by imposing that $d\theta/dz = 0$ at $z = 0$ and $\theta(0) = \theta_M$. From the above considerations, one can deduce that $\theta(z)$ minimizing (1.93) is the solution of the differential equation

$$\left(\frac{d\theta}{dz}\right)^2 = \frac{1}{\xi^2}\left(\sin^2\theta_M - \sin^2\theta\right), \tag{1.94}$$

where

$$\xi = \sqrt{\frac{K}{-\epsilon_a\, E^2}} \tag{1.95}$$

is the dielectric coherence length. From (1.94), in the strong homeotropic case, one obtains

$$\left[\frac{d\theta(\pm d/2)}{d\,z}\right]^2 = \mp\frac{\epsilon_a\,E^2}{K}\sin^2\theta_M. \tag{1.96}$$

If $\epsilon_a > 0$, ξ is imaginary and $\theta(0) = 0$ for any applied field. On the contrary, if $\epsilon_a < 0$, ξ and $d\theta/dz$ are real and the homeotropic pattern can be destabilized ($\theta \neq 0$ at $z = 0$ is possible). There is a Fréedericksz transition when

$$E = E_{c\,\infty} = \frac{\pi}{d}\sqrt{\frac{K}{-\epsilon_a}}. \tag{1.97}$$

This critical field can be determined by rewriting (1.94) in the form

$$\int_0^{\theta_M}\frac{d\theta}{\sqrt{\sin^2\theta_M - \sin^2\theta}} = \frac{d}{2\xi}. \tag{1.98}$$

If one introduces $\sin\psi = \sin\theta/\sin\theta_M$, Eq. (1.98) can be rewritten as

$$\frac{d}{2\xi} = \int_0^{\pi/2}\frac{d\psi}{\sqrt{1-\sin^2\theta_M\sin^2\psi}} = K(\sin\theta_M), \tag{1.99}$$

where $K(\sin\theta_M)$ is the complete elliptic integral of the first kind. This equation can be analyzed in the limit of small θ_M. One has

$$\frac{d}{2\xi} = \frac{\pi}{2} + \frac{\pi}{8}{\theta_M}^2 + \mathcal{O}(\theta_M)^3, \tag{1.100}$$

which can be solved for θ_M to give

$$\theta_M = \sqrt{4(\frac{d}{\pi\xi} - 1)}. \tag{1.101}$$

Then, if $\xi < d/\pi$ the solution is real and $\theta_M \neq 0$, otherwise $\theta_M = 0$. It follows that $\xi_c = d/\pi$, which defines the critical field introduced above and denoted by $E_{c\infty}$. Since in the case we are considering $V = Ed$, the threshold voltage for the Freédericksz transition is

$$V_{c\infty} = \pi\sqrt{\frac{K}{-\epsilon_a}}, \tag{1.102}$$

i.e., it is independent of the thickness of the sample and is of the order of a few volts for typical NLC systems.

[1] de Gennes PG and Prost J. *The Physics of Liquid Crystals.* Clarendon Press, Oxford, 1994.

[2] Priestley EB, Wojtowicz PJ, and Sheng P (Editors). *Introduction to Liquid Crystals.* Plenum Press, New York, 1975.

[3] Chandrasekhar S. *Liquid Crystals.* Cambridge University Press, Cambridge, 1977.

[4] de Jeu WH. *Physical Properties of Liquid Crystalline Materials.* Gordon and Breach, Philadelphia, 1982.

[5] Vertogen G and de Jeu WH. *Thermotropic Liquid Crystals, Fundamentals.* Springer Verlag, Berlin, 1988.

[6] Leslie FM. *Theory and Applications of Liquid Crystals.* Edited by Ericksen JL and Kinder Leher D. Springer Verlag, Berlin, 1987.

[7] Coolings PJ. *Liquid Crystals.* Adam Hilger, Bristol, 1990. Chapters 6 and 7.

[8] Barbero G and Evangelista LR. *An Elementary Course on the Continuum Theory for Nematic Liquid Crystals.* World Scientific, Singapore, 2001.

[9] Stewart IW. *The Static and Dynamic Continuum Theory of Liquid Crystals.* Taylor & Francis, London, 2004.

[10] Virga EG. *Variational Theories for Liquid Crystals.* Chapman and Hall, London, 1994.

[11] Sonin AA. *The Surface Physics of Liquid Crystals.* Gordon and Breach, Luxembourg, 1995.

[12] Maier W and Saupe A. Eine einfache molekulare theorie des nematischen kristallinflussigen zustandes. *Zeitschrift fur Naturforschung A*, **13** 564 (1958); Eine einfache molekular-statistiche theorie der nematischen kristallinflussigen phase I. *Zeitschrift fur Naturforschung A*, **14**, 882 (1959); Eine einfache molekular-statistiche theorie der nematischen kristallinflussigen phase II . *Zeitschrift fur Naturforschung A*, **15**, 287 (1960).

[13] Plischke M and Bergersen B. *Equilibrium Statistical Mechanics.* World Scientific, Singapore, 1999, 2nd edition.

[14] Petrov MP and Simova PD. Depolarized Rayleigh scattering and textures in the nematic phase of some 4,n-alkyloxybenzoic acids. *Journal of Physics D–Applied Physics*, **18**, 293 (1985).

[15] Petrov MP, Braslau A, Levelut AM, and Durand G. Surface–induced transition in the nematic phase of 4-n octyloxybenzoic acid. *Journal de Physique II*, **2**, 1159 (1992).

[16] Gruler H and Meier G. Investigations on elastic constants of nematic homologous series of 4,4'-di(alkoxy)azoxybenzene. *Molecular Crystals and Liquid Crystals*, **23**, 261 (1973).

[17] Rondelez F. Conductance measurements above a smectic-c[–] nematic transition. *Solid State Communications*, **11**, 1675 (1972).

[18] Carr EF. Conductivity anisotropy in para-normal-nonyloxybenzoic acid. *Physical Review A*, **12**, 327 (1975).

[19] Deloche B and Cabane B. Coupling of hydrogen-bonding to orientational fluctuation modes in liquid-crystal PHBA. *Molecular Crystals and Liquid Crystals*, **19**, 25 (1972).

[20] Petrov MP, Anachkova E, Kirov N, Ratajczak H, and Baran J. IR spectroscopic investigations of peculiar behavior of 4,n-alkoxybenzoic acids in their mesomorphic state–nonligned samples. *Journal of Molecular Liquids*, **61**, 221 (1994).

[21] Barbero G, Evangelista LR, and Petrov MP. Influence of the dimerization process on the nematic ordering. *Physics Letters A*, **256**, 399 (1999).

[22] Meyer RB. Piezoelectric effects in liquid crystals. *Physical Review Letters*, **29**, 918 (1969).

[23] Barberi R, Barbero G, and Evangelista LR. Order phase transition near the inversion point of main anisotropy. *Modern Physics Letters B*, **4**, 751 (1990).

[24] Barbero G, Evangelista LR, and Gabbasova Z. Fréedericksz-like transitions in compensated nematic liquid crystals. *International Journal of Modern Physics B*, **5**, 697 (1991).

[25] Barbero G, Evangelista LR, and Krekhov AP. Tricritical point in order-disorder phase transitions for nematics. *International Journal of Modern Physics B*, **6**, 2521 (1992).

[26] Jerome B. Surface effects and anchoring in liquid crystals. *Reports on Progress in Physics*, **54**, 391 (1991).

2

NEMATIC LIQUID CRYSTALS: SURFACE PROPERTIES

In this chapter, we analyze the main effects of the presence of a surface limiting the sample on the anchoring properties of nematic liquid crystals. We introduce the concept of anchoring and discuss some proposed forms for the surface energy in a macroscopic perspective. Subsequently, some sources for the surface energy are investigated. The stochastic contribution to the anchoring energy is shown to be responsible for the deviation from the Rapini-Papoular expression for the surface energy. The effect of the spatial variation of the scalar order parameter and of the elastic constant on the surface free energy is discussed in a pseudo-molecular approach. At the end of the chapter, we present an estimation of the contribution of the smectic-nematic interface to the surface energy.

2.1 Anchoring

While the bulk properties of liquid crystals depend strongly on the molecular structure and on the molecular interactions, the surface properties depend, in addition, on the surface-liquid crystal interactions. When the nematic phase is limited by a surface created by contact with another phase (solid, liquid, gas) its orientation may change in a drastic manner. In particular, the characteristic features of the *anchoring* of the liquid crystal are very important from the fundamental point of view for a correct physical description of the properties of the sample, and are very important for the performance of liquid crystal devices. The anchoring may be defined as the phenomenon of orientation of a liquid crystal by a surface. It is a result of a delicate balance between a number of interactions and was discovered by Mauguin at the beginning of the last century [1]. The surface normally imposes some preferred directions, called anchoring directions, or, simply, *easy direction.* The *easy axis* is then the direction of spontaneous orientation of $\mathbf{n}$ on the surface, in the absence of an external torque.

The energy of the interfacial region, formed near the limiting surface, depends also on the orientation of the molecules in the nematic phase. Typical problematic situation regarding the anchoring requires determination of the orientation of the director at the interface between the NLC and other media. The origin of the surface alignment can be discussed in terms of anisotropic torques of either physico-chemical or geometrical nature. This implies the existence of an anisotropic surface energy, f_s. To find the molecular orientation of an NLC at the equilibrium, one must then minimize the volume free energy (describing the bulk distortion), plus the surface free energy, f_s. This is the situation stated in Chapter 1 for a slab [see Eq. (1.77)]. However, the bulk part of the total energy is well known. As established in Chapter 1, it is a quadratic function of the curvature, i.e., of the spatial derivatives of $\mathbf{n}$. The surface energy, f_s, on the other hand, has not a well defined form and is still subject of many discussions.

In practice, many methods have been used to impose a preferred direction of $\mathbf{n}$ on a solid surface. A detailed explanation of them can be found in Ref. [1]. Here we mention only one method in order to establish the main features of the alignment by a surface and its consequences on the molecular orientation of the NLC phase. The most common procedure consists in rubbing a surface. In this manner one can make grooves on it, and one obtains usually an alignment parallel to the grooves. This can be explained in terms of a surface localized additional bulk curvature energy, as shown by Berreman [2]. Molecules which are parallel to the grooves are not perturbed. Molecules normal to the grooves undergo a curvature distortion. This curvature extends in the bulk on a thickness comparable to the groove size. For thin enough grooves, the additional curvature energy can be considered as a surface energy. However, the real surface energy, f_s, is assumed to be infinite, because it is assumed to fix the molecules parallel to the surface. The real origin of f_s is related to the local chemical interaction of the molecules with the surface and the highly delocalized anisotropic van der Waals dispersion forces. This gives rise to a surface energy which can be considered as having a geometric origin.

Despite the richness and the complexity of the subject, in this book we will focus our attention on the phenomenological description of the surface energy, f_s, and its commonly accepted forms.

2.2 The anchoring energy function

In the simplest case of glass plate boundaries, which are locally isotropic along their surface, the only defined direction is the normal to the plates, which we denote by $\mathbf{k}$. The surface energy, f_s, is then expected to depend only on $\mathbf{n} \cdot \mathbf{k}$ and possible on its gradient along the surface which, in the case of uniform treatment of flat surface, can be ignored. The possibility of f_s depending on the normal derivatives of $\mathbf{n}$ has been shown to be irrelevant.

As discussed in Chapter 1, in the bulk $\mathbf{n}$ and $-\mathbf{n}$ are physically equivalent. However, on a polar surface (for instance, on a ferroelectric crystal) this degeneracy could be removed, but the commonly used glass plates behave as apolar surfaces. Therefore, f_s must be an even function of $\mathbf{n} \cdot \mathbf{k}$.

The director orientation at the interface between an NLC and another medium is defined in terms of a polar surface angle, θ_s, and of an azimuthal angle, ϕ_s. Since the anisotropic part of the surface energy has to depend on the director orientation at the interface, one can write [3]

$$f_s = f_s(\Theta, \Phi) + W(\theta_s - \Theta, \phi_s - \Phi), \tag{2.1}$$

where Θ and Φ correspond to the values of the polar and azimuthal angles which minimizes the surface energy (they characterize the easy direction). The positive function W is called *anchoring energy function.* As follows from the definition of easy direction, at the equilibrium and in the absence of external torque, $\theta_s = \Theta$ and $\phi_s = \Phi$. The anchoring energy can be physically interpreted as the work which must be done to rotate the director from the easy direction to the actual one. It is then possible to define the polar and azimuthal torques, per unit area, respectively, in the following forms:

$$\tau_P = -\frac{\partial f_s}{\partial \theta_s} \quad \text{and} \quad \tau_A = -\frac{\partial f_s}{\partial \phi_s}. \tag{2.2}$$

Near the equilibrium position these torques can be approximately given by

$$\tau_P = -2W_P(\theta_s - \Theta) \quad \text{and} \quad \tau_A = -2W_A(\phi_s - \Phi), \tag{2.3}$$

where we have introduced two important quantities:

$$W_P = \frac{1}{2}\left(\frac{\partial^2 f_s}{\partial \theta_s^2}\right)_{\Theta,\Phi}, \tag{2.4}$$

called *polar anchoring coefficient* or polar anchoring energy and

$$W_A = \frac{1}{2}\left(\frac{\partial^2 f_s}{\partial \phi_s^2}\right)_{\Theta,\Phi}, \tag{2.5}$$

called *azimuthal anchoring coefficient* or azimuthal anchoring energy. These anchoring energies can be measured by several different experimental techniques. Typical experimental values of W_P and W_A are of the order of 10^{-4} to 10^{-1} erg/cm^2.

Let us particularize the analysis to the case of isotropic substrates. In this case, the surface free energy, f_s, does not depend on the azimuthal angle, and therefore $W_A = 0$. The function f_s depends only on the scalar product $\mathbf{n} \cdot \mathbf{k} = \cos\theta_s$ and can be expressed as a Taylor expansion in terms of $\cos\theta_s$, in the form

$$f_s(\theta_s) = W_0 + W_1 \cos\theta_s + W_2 \cos^2\theta_s + ... + W_n \cos^n\theta_s + ..., \tag{2.6}$$

which can be shown to be not useful from the practical point of view. In fact, as stated, the expansion terms are not orthogonal functions, which implies that the coefficients obtained from experiments, by truncating the expansion to a given order, depend on the truncation order. An alternative expression, dealing with orthogonal functions, can be proposed in the form

$$f_s(\theta_s) = W_0 + W_1 \cos\theta_s + W_2 \cos 2\theta_s + ... + W_n \cos n\theta_s + ... \tag{2.7}$$

In Eq. (2.7) the odd coefficients (W_{2n+1}), for $n = 0, 1, 2, ...$ must vanish for non-ferroelectric systems as NLC. However, this is true only for what concerns the bulk properties. Near the surface the translational symmetry of the nematic phase is broken and a surface ferroelectricity may occur. In general, however, the system can be considered as non-polar and the above expressions can be very useful in describing–from the macroscopic point of view–several properties of a nematic in contact with a solid, isotropic, surface.

The most common expression for the surface free energy is the one due to Rapini and Papoular (1969) which can be obtained from (2.6) (up to the second term) as [4]

$$f_s(\theta_s) = W_0 + W_P \cos^2\theta_s. \tag{2.8}$$

This simplified expression permits a simple interpretation for the surface free energy. In this case W_P is the *anchoring energy* which, as defined before, corresponds to the work that has to be done to rotate the director from the stable equilibrium position to the unstable one. In general, it is possible to start with the Rapini-Papoular expression by operating in the following manner. If a substrate is characterized by a surface easy direction given by $\mathbf{n}_0$, then the surface free energy is simply defined as

$$f_s = -\frac{1}{2} W (\mathbf{n} \cdot \mathbf{n}_0)^2, \tag{2.9}$$

in such a manner to underline the fact that the easy direction is the one which, in the absence of external torques, minimizes the surface free energy. This simple expression will be employed throughout this book to discuss several important surface effects in the nematic phase of liquid crystalline materials. In Section 2.3.1, we present a calculation of the stochastic contribution to the anchoring. In this case, it is possible to detect a deviation from the Rapini-Papoular expression stated above.

2.2.1 Example 1: Fréedericksz transition–weak anchoring

A common experimental determination of the anchoring energy is often performed by measuring the threshold field, E_c, for the Fréedericksz transition, in a thin nematic layer of thickness d. This kind of measurement only furnishes the anchoring energy coefficient W and, therefore, does not give useful information about the angular dependence of f_s.

Let us analyze the influence of the anchoring energy on the critical field by supposing the surface anchoring energy of the Rapini-Papoular type, as given by (2.9). We consider a slab of thickness d and assume that the surfaces are of the same kind, with the $z-$axis being perpendicular to them. In this case the surface free energy (2.9) can be written as

$$f_s = \frac{1}{2} W \sin^2 \theta_s, \tag{2.10}$$

where $\theta_s = \theta(\pm d/2)$ and a constant term, independent of the orientation at the surface, was dropped. In this case, the bulk equation is still the Euler-Lagrange equation (1.79) but the boundary conditions are

$$-K\theta' + \frac{W}{2}\sin(2\theta_s) = 0 \qquad \text{and} \qquad K\theta' + \frac{W}{2}\sin(2\theta_s) = 0, \tag{2.11}$$

for $z = -d/2$ and $z = d/2$ respectively, as follows from (1.80). Taking into account that

$$\theta(\mp d/2) = \pm\frac{1}{\xi}(\sin^2 \theta_m - \sin^2 \theta s)^{1/2},$$

where θ_m is such that $d\theta/dz = 0$ for $\theta = \theta_m$, boundary conditions (2.11) give

$$(\sin^2 \theta_m - \sin^2 \theta_s)^{1/2} = \frac{\xi}{2L}\sin(2\theta_s), \tag{2.12}$$

where $L = K/W$ is the extrapolation length. By operating as in Section 1.6.2, from (2.12) we have now

$$\int_{\theta_s}^{\theta_m} \frac{1}{\sqrt{\sin^2 \theta_m - \sin^2 \theta}} \mathrm{d}\theta = \frac{d}{2\,\xi}, \tag{2.13}$$

since $\theta_s \neq 0$ for $E > E_c$, and

$$\int_{\theta_s}^{\theta(z)} \frac{1}{\sqrt{\sin^2 \theta_m - \sin^2 \mu}} \mathrm{d}\mu = \frac{z}{\xi}, \quad \text{for} \quad -d/2 \le z \le 0. \tag{2.14}$$

In order to evaluate the critical field, as in Section 1.6.2, we use again $\sin\psi = \sin\theta/\sin\theta_m$. In this way as well, $\theta = \theta_s$ corresponds to the substitution $\psi = \psi_s = \sin^{-1}(\sin\theta_s/\sin\theta_m)$, and to $\theta = \theta_m$, $\psi = \pi/2$. Thus, (2.13) can be written now in the form

$$\int_{\psi_s}^{\pi/2} \frac{1}{\sqrt{1-\sin^2\theta_m \sin^2\psi}} \mathrm{d}\psi = \frac{d}{2\xi}, \tag{2.15}$$

from which, in the limit $\theta_m \to 0$, we obtain

$$\lim_{\theta_m \to 0} \int_{\psi_s}^{\pi/2} \frac{1}{\sqrt{1-\sin^2\theta_m \sin^2 X}} \mathrm{d}X = \frac{\pi}{2} - \lim_{\theta_m \to 0} \psi_s. \tag{2.16}$$

In the considered limit (2.12) gives

$$\theta_m^2 - \theta_s^2 = \left(\frac{\xi}{L}\right)^2 \theta_s^2.$$

Hence,

$$\theta_s = \frac{\theta_m}{[1+(\xi/L)^2]^{1/2}}.$$

Consequently,

$$\lim_{\theta_m \to 0} \psi_s = \arctan\left(\frac{L}{\xi_c}\right). \tag{2.17}$$

Equation (2.15), taking into account (2.16) and (2.17), gives finally

$$\frac{\pi}{2} - \arctan\left(\frac{L}{\xi_c}\right) = \frac{d}{2\xi_c}, \tag{2.18}$$

which is usually written as

$$\cot\left(\frac{d}{2\xi_c}\right) = \frac{L}{\xi_c}, \tag{2.19}$$

and is known as the Rapini-Papoular relation. Equation (2.19) determines the critical field E_c. In the limit of large d, (2.19) again gives (1.97) in the case of strong anchoring. Note that in the case of weak anchoring we have a *threshold field* E_c, whereas in the strong anchoring case we have a *threshold voltage* $V_{c\infty}$.

The coherence length given by (1.95), taking into account Eq. (1.97), can be written in the form

$$\frac{1}{\xi} = \frac{\pi}{d}\frac{E}{E_{c\infty}} = \frac{\pi}{d}h, \tag{2.20}$$

where $h = E/E_{c\infty}$ is the field measured in units of $E_{c\infty}$. By means of (2.20), Eq. (2.19) writes

$$\frac{d}{2L} = \frac{(\pi/2)h_c}{\cot[(\pi/2)h_c]}, \tag{2.21}$$

where h_c is the reduced critical field in the weak anchoring case. Note that for $d/2L \to 0$, $h_c \to 0$, i.e., there is no threshold, whereas in the limit of large d, $h_c \to 1$ as in the case of strong anchoring. For $h < h_c$ the stable pattern is the homeotropic one ($\theta = 0$). The reverse applies to $h > h_c$; the NLC is distorted ($\theta \neq 0$).

Equation (2.18) shows that the extrapolation length L has a simple meaning. In fact, if the considered sample has a relatively strong anchoring, $L = K/W$ is a small quantity. Consequently, in (2.18), $\arctan(L/\xi_c) \approx L/\xi_c$. For this approximation, (2.18) can be rewritten as

$$\pi = \frac{d + 2L}{\xi_c}.$$

This relation shows that in the case of finite anchoring energy the effective sample thickness is $d+2L$ instead of d. In other words, $\theta(\pm d/2) = \theta_s$, whereas $\theta[\pm(d/2 + L)] = 0$.

2.2.2 Example 2: Fréedericksz transition–nonhomogeneous anchoring energy

In this example a generalization of the preceding results for the case of nonhomogeneous anchoring energy is proposed [5]. It will be supposed that the anchoring strength $W = W(x)$ or, equivalently $L = L(x)$, where x is the coordinate parallel to the bounding plates. In the case in which the NLC slab is submitted to an electric field $\mathbf{E}$ parallel to z, the total energy per unit length along the y axis is given by

$$\begin{aligned} F = &\int_{-\infty}^{\infty}\int_{-d/2}^{d/2}\left[\frac{1}{2}K(\nabla\theta)^2 + \frac{\epsilon_a}{2}E^2\theta^2\right]dxdz \\ &+ \int_{-\infty}^{\infty}\frac{1}{2}W_+(x)\theta^2\left(x, \frac{d}{2}\right)dx + \int_{-\infty}^{\infty}\frac{1}{2}W_-(x)\theta^2\left(x, -\frac{d}{2}\right)dx, \end{aligned} \tag{2.22}$$

written in the limit of small θ. As in the preceding section, note that for the experimental arrangement considered here the Fréedericksz effect exists only for $\epsilon_a < 0$. The novelty of the above expression is that the surfaces are assumed non-homogeneous with respect to the anchoring energy. Thus, in order to obtain the $\theta(x,z)$-profile one minimizes (2.22), obtaining

$$\frac{\partial^2\theta}{\partial x^2}+\frac{\partial^2\theta}{\partial z^2}+\lambda^2\theta=0,\quad -\frac{d}{2}\le z\le\frac{d}{2},\quad -\infty<x<\infty, \tag{2.23}$$

where $\lambda^2 = -\epsilon_a E^2/K$. The boundary conditions to be satisfied are similar to (2.11), for $\Theta = 0$, but now rewritten for non-homogeneous L and small θ, namely

$$\pm L_\pm(x)\left[\frac{\partial\theta}{\partial z}\right]_{z=\pm d/2}+\theta(x,\pm d/2)=0. \tag{2.24}$$

The solution of (2.23) can be written in general form as

$$\theta(x,z)=\int_{-\infty}^{\infty}h(k,z)e^{ikx}dk, \tag{2.25}$$

where $h(k,z) = h(-k,z)$, since $\theta(x,z)$ is a real quantity. When (2.25) is placed in (2.23) one obtains

$$h(k,z)=\alpha(k)e^{\gamma z}+\beta(k)e^{-\gamma z}, \tag{2.26}$$

where

$$\gamma\equiv\sqrt{k^2-\lambda^2}, \tag{2.27}$$

and $\alpha(k)$ and $\beta(k)$ have to be determined by means of the boundary conditions (2.24). It is then possible to obtain for the tilt angle $\theta(x,z)$ the general expression

$$\theta(x,z)=\frac{1}{2\pi}\int_{-\infty}^{\infty}\left[\theta\left(x',\frac{d}{2}\right)p^{(+)}(x'-x,z)-\theta\left(x',-\frac{d}{2}\right)p^{(-)}(x'-x,z)\right]dx', \tag{2.28}$$

given in terms of new propagators $p^{(\pm)}(x-x',z)$. They are defined as

$$p^{(\pm)}(x-x',z)=\int_{-\infty}^{\infty}\frac{\sinh\left[\gamma\left(z\pm\frac{d}{2}\right)\right]}{\sinh(\gamma d)}e^{ik(x-x')}dk. \tag{2.29}$$

The general solution (2.28) is placed in the boundary condition (2.24) giving a system of two new coupled integral equations

$$\pm\frac{L_\pm(x)}{2\pi}\times\int_{-\infty}^{\infty}\left[\theta(x',\frac{d}{2})q^{(\pm)}(x-x')-\theta\left(x',-\frac{d}{2}\right)q^{(\mp)}(x-x')\right]dx'$$
$$+\theta\left(x,\pm\frac{d}{2}\right)=0, \tag{2.30}$$

with

$$q^{(+)}(x-x') = \int_{-\infty}^{\infty} e^{ik(x-x')}\gamma\frac{\cosh(\gamma d)}{\sinh(\gamma d)}dk,$$

$$q^{(-)}(x-x') = \int_{-\infty}^{\infty} e^{ik(x-x')}\frac{\gamma}{\sinh(\gamma d)}dk. \tag{2.31}$$

Before proceeding it is interesting to underline that the set of equations (2.23) through (2.31) constitutes the general theoretical framework to investigate the Fréedericksz effect in the weak-anchoring situation. Without loss of generality, it is possible to restrict the considerations to the case of symmetric surfaces, for which, obviously, it is expected that $L_+(x) = L_-(x) = L(x)$. This implies that $\theta(x, d/2) = \theta(x, -d/2)$ and, in general, $\theta(x, z) = \theta(x, -z)$. From the above hypothesis, Eq. (2.28) is rewritten as

$$\theta(x,z) = \frac{1}{2\pi}\int_{-\infty}^{\infty} \theta\left(x', \frac{d}{2}\right)\left[p^{(+)}(x-x',z) - p^{(-)}(x-x',z)\right]dx', \tag{2.32}$$

whereas the boundary conditions assume the form

$$\frac{L(x)}{2\pi}\int_{-\infty}^{\infty} \theta(x', \frac{d}{2})\left[q^{(+)}(x-x') - q^{(-)}(x-x')\right]dx' + \theta\left(x, \frac{d}{2}\right) = 0. \tag{2.33}$$

If one uses (2.31), it is possible to obtain

$$q^{(+)}(x-x') - q^{(-)}(x-x') = -\int_{-\infty}^{\infty} e^{ik(x-x')}\gamma\tanh\left(\frac{\gamma d}{2}\right)dk. \tag{2.34}$$

For the particular case in which $L(x)$ is position independent, also the actual tilt angle at the surface $\theta(x, d/2)$ is expected to be x independent. In fact, since

$$\int_{-\infty}^{\infty} e^{ik(x-x')}dk = 2\pi\delta(x-x'), \tag{2.35}$$

Equation (2.33) can be easily written as

$$L\lambda\tan\left(\frac{\lambda d}{2}\right) = 1, \tag{2.36}$$

which is the Rapini-Papoular relation already introduced [Eq. (2.19)]. The general result can be found by rewriting Eq. (2.33) as

$$\theta\left(x,\frac{d}{2}\right)=\frac{L(x)}{2\pi}\int_{-\infty}^{\infty}\theta\left(x',\frac{d}{2}\right)\left[q^{(-)}(x-x')-q^{(+)}(x-x')\right]dx', \quad (2.37)$$

which clearly implies that

$$q^{(-)}(x-x')-q^{(+)}(x-x')=\frac{2\pi}{L(x')}\delta(x-x'). \quad (2.38)$$

Now one can equate (2.34) and (2.38) in order to obtain

$$\frac{2\pi}{L(x')}\delta(x-x')=-\int_{-\infty}^{\infty}e^{ik(x-x')}\gamma\tanh\left(\frac{\gamma d}{2}\right)dk. \quad (2.39)$$

Finally, one can integrate (2.39) over x' and by means of (2.35), easily deduce that

$$L(x)\lambda\tan\left(\frac{\lambda d}{2}\right)=1. \quad (2.40)$$

Equation (2.40) states that the threshold field is fixed by the minimum value of the anchoring energy, no matter what the spatial dependence of W is. This result is *a posteriori* physically consistent with the definition of threshold in a sample characterized by weak-anchoring energy . Thus, the true theoretical threshold for the Fréedericksz effect in an NLC sample having non-homogeneous anchoring energy is fixed by the lowest value of this parameter.

2.3 Sources for the surface free energy

In order to understand the behavior and the main dependencies of the interfacial energy in NLC, many phenomenological [6]–[10] or molecular models [11]–[16] have been proposed. The crucial question regarding these models is to explain not only planar easy alignment, but also homeotropic and tilted alignments at the interface. The major part of the proposed models is able to predict the planar easy alignment, but only few of them are successful to explain the other two situations. Among the proposed theories, the one presented by Mada and Kobayashi [7] was able to explain tilted alignment by introducing terms in the surface free energy which are dependent on the surface derivatives of the director. But these surface derivatives make unsolvable the corresponding variational problem [9, 17]. On the other hand, Parsons [18] was able to explain the tilted alignment as being due to the competition of polar and non-polar interactions. The phenomenological expression for the surface energy has the form

$$fs = w_0 + w_1 \cos\theta_s + w_2 \cos^2\theta_s, \tag{2.41}$$

in such a manner that homeotropic, planar or tilted easy alignment could be obtained according to the values of w_1 and w_2. Furthermore, a second order phase transition from tilted to homeotropic alignment was predicted for $w_1 = -2w_2$. A different development was proposed by Sluckin and Poniewierski [10] who showed that the assumption of polar interactions is not necessary to predict the director tilt. They have proposed a power expansion of the surface free energy in terms of the order parameter Q_{ij} evaluated at the surface, in analogy with the Landau-de Gennes model. They have proposed an expression in the form

$$fs = w_0(Q) + w_2(Q)\cos^2\theta_s + w_4(Q)\cos^4\theta_s, \tag{2.42}$$

which is strictly connected with the Landau-de Gennes description of the bulk free energy. Through this expression it is possible to describe the orientational properties of the interface and also some surface order phenomena [19, 20] like wetting transitions.

Another approach consists of calculating the surface free energy by considering the microscopic interactions between molecules. The calculations are very complex and drastic simplifications have to be made in order to obtain analytical expressions. Van der Waals interactions lead to a parallel alignment of molecules at the surface [6]. This works well for the case of PAA. But experimental results dealing, for instance, with MBBA indicate that the molecules are tilted from the surface through an angle which changes with temperature [21]. Other attempts have been made. As pointed out by Faetti [14], the mean field surface free energy was calculated by expressing the generic interaction between axially-symmetric molecules as a power expansion in spherical harmonics (see Chapter 4). It was shown that van der Waals forces cannot explain the tilted alignment, while quadrupolar interactions and short range interactions produce a surface free energy similar to (2.41). These results indicate that short-range and quadrupolar interactions can explain the experimental results. The role of short range hard rod was investigated by Kimura and Nakano [15] and by Holyst and Poniewierski [16] by means of an extension of the Onsager approach to the interface. They found that short-range interactions favor the homeotropic alignment at the free surface; in the case of nematic-isotropic interface in Ref. [15] an easy planar alignment is predicted, while a tilted alignment at an angle of $\pi/3$ is predicted in Ref. [16].

A different approach [22] demonstrated that an electric polarization of the medium takes place, if the spatial variation of the director field (ferroelectricity) and of the scalar order parameter (ordoelectricity) are taken into account. This polarization is related to the presence of a non-vanishing average electric quadrupole proportional to the order parameter tensor [22, 23]. Near the interface the order parameter exhibits a strong variation and, thus, a surface polarization occurs. The free energy is minimized when the director angle at the surface is equal to the strange number $\theta_s = 54.7°$, which agrees very well with the experimental results for the nematic-isotropic interface of 7CB. By assuming that ordoelectric coefficients are of the same order as the flexoelectric ones, a satisfactory quantitative agreement with experimental measurements of the anchoring energy coefficient at the nematic-isotropic interface of 7CB was obtained.

In the following, we will analyze this problem from three different but not mutually exclusive points of view.

2.3.1 Stochastic contribution to the anchoring energy

The effect of a stochastic contribution to the surface energy, coming from the direct interaction between an orienting film and a solid substrate, indicates that, in the hypothesis in which the NLC orientation coincides with the film orientation, the Rapini-Papoular expression for the anisotropic part of the surface energy has to be modified. There are two important modifications: (1) the first one is a renormalization of the anchoring strength coefficient, connected with the square of the sine of the deformation angle; (2) the second one is the presence of an additional contribution, proportional to the fourth power of the same quantity.

To show these modifications we consider a simple model in which the orienting film (like a Langmuir-Blodgett (LB) film) is monomolecular and formed of rodlike molecules [24]. The main orientation of the film is due to the interactions between the molecules forming the film itself and between the molecules of the film and the substrate over which the film is deposited. It is supposed, furthermore, that the NLC orientation follows that of the orienting film, which is equivalent to suppose that the film is not compact. Therefore, the molecules of the NLC may enter the holes (free places) present in the structure of the film. In this manner the NLC molecules in "contact" with the film are oriented by steric interaction (with the molecules of the film). Then, this first layer of nematic molecules orients the bulk NLC by means of the anisotropic intermolecular interaction characterizing the nematic phase. This model of the interface LB film NLC was proposed some years ago by Hiltrop and Stegemeyer [25] and was reconsidered in connection with temperature-induced surface transitions in NLC [26, 27].

The monomolecular film is formed by rigid rod–like molecules of length l, having the direction of their long molecular axis characterized by $\mathbf{m}$. Due to the interaction of the molecules and the substrate a film with two–dimensional (2D) order is formed. One end of each rod is attached to the solid substrate in a quasiregular pattern, whereas the other end is free. In this model, a molecule of the film may bend with an angle θ under the action of the direct interaction with the substrate and the interaction connected with the other molecules of the film. The reference frame is the same used in other systems in this book: the x and $y-$ axes are parallel to the substrate and the $z-$ axis normal to it. The starting point for a molecular approach to evaluate the elastic energy of the monomolecular film is the assumption of a two-body interaction of the kind

$$g(\mathbf{m}, \mathbf{m}', \mathbf{r}) dS dS', \tag{2.43}$$

between the surface elements dS and dS', where $\mathbf{m}$ and $\mathbf{m}'$ are the directions of the molecular major axes at $\mathbf{R}$ and $\mathbf{R}'$, respectively, and $\mathbf{r} = \mathbf{R}' - \mathbf{R}''$ [28]. The order of the film is supposed to be perfect, which implies that $\mathbf{m}$ coincides with the statistical average of the molecular major axis of the molecules in the film.

In a continuum description the vector $\mathbf{m}$ ($|\mathbf{m}|^2 = 1$) depends on the coordinates $\mathbf{R}$ of the considered point on the solid substrate, i.e., $\mathbf{m} = \mathbf{m}(\mathbf{R})$ and $g(\mathbf{m}, \mathbf{m}', \mathbf{r})$ is the intermolecular interaction energy. In the present calculations this energy is supposed to be short-range, i.e.,

$$g(\mathbf{m}, \mathbf{m}', \mathbf{r}) = 0, \quad \text{for} \quad |\mathbf{r}| > \rho \approx \frac{1}{\sqrt{\sigma}}, \tag{2.44}$$

where σ is the surface density of the molecules of the film. This assumption is equivalent to stating that long range electrostatic contributions in the interaction energy between the molecules of the film can be neglected. To express the excess energy associated with a surface distribution of $\mathbf{m}$, we assume that

$$\mathbf{m}' = \mathbf{m}(\mathbf{R}') = \mathbf{m}(\mathbf{R}) + \delta\mathbf{m}, \tag{2.45}$$

where $|\delta\mathbf{m}| \ll 1$. This hypothesis means that $\mathbf{m}$ is considered a macroscopic quantity and hence changes over macroscopic distances $d \gg \rho$. Only in this case it is possible to build an elastic theory for the orienting film. In this framework $g(\mathbf{m}, \mathbf{m}', \mathbf{r})$ may be expanded in power series of $\delta\mathbf{m}$, in the form

$$\begin{aligned} g(\mathbf{m}, \mathbf{m}', \mathbf{r}) &= g(\mathbf{m}, \mathbf{m} + \delta\mathbf{m}, \mathbf{r}) \\ &= g(\mathbf{m}, \mathbf{m}, \mathbf{r}) + \lambda_i \delta m_i + \frac{1}{2}\mu_{ij}\delta m_i \delta m_j, \end{aligned} \tag{2.46}$$

where

$$\lambda_i = \left[\frac{\partial g}{\partial m_i'}\right]_{\delta m = 0} \quad \text{and} \quad \mu_{ij} = \left[\frac{\partial^2 g}{\partial m_i' \partial m_j'}\right]_{\delta m = 0}, \tag{2.47}$$

and $g(\mathbf{m}, \mathbf{m}, \mathbf{r})$ is the uniform part of the interaction energy. Equation (2.46) holds because $g(\mathbf{m}, \mathbf{m}', \mathbf{r})$ is supposed to be an analytical function of the scalar quantities (rotational invariants) $\mathbf{m} \cdot \mathbf{m}'$, $\mathbf{m} \cdot \mathbf{u}$ and $\mathbf{m}' \cdot \mathbf{u}$, where $\mathbf{u} = \mathbf{r}/|\mathbf{r}|$, in the form

$$g(\mathbf{m}, \mathbf{m}', \mathbf{r}) = \sum_{a,b,c} C_{a,b,c}(r)(\mathbf{m} \cdot \mathbf{m}')^a (\mathbf{m} \cdot \mathbf{u})^b (\mathbf{m}' \cdot \mathbf{u})^c, \tag{2.48}$$

where the expansion coefficient $C_{a,b,c}(r)$ depends only on the modulus of $\mathbf{r}$. In the hypothesis that the interaction range of the intermolecular forces is small, δm_i may be expanded in power series of $\mathbf{r}$ as

$$\delta m_i = m_{i,\alpha}(\mathbf{R})x_\alpha + \frac{1}{2}m_{i,\alpha\beta}(\mathbf{R})x_\alpha x_\beta, \tag{2.49}$$

where

$$m_{i,j}(\mathbf{R}) = \left(\frac{\partial m_i}{\partial x_j}\right)_{\mathbf{R}}, \tag{2.50}$$

and x_α are the Cartesian components of $\mathbf{r}$. We assume that the elastic approximation works, i.e., that the vector $\mathbf{m}$ changes smoothly with $\mathbf{R}$. This means that the following inequalities hold:

$$|m_{i,\alpha}| \ll \frac{1}{\rho}, \quad \text{and} \quad |M_{i,\alpha\beta}| \ll \frac{1}{\rho^2}. \tag{2.51}$$

Substitution of (2.49) into (2.46) yields

$$\begin{aligned} g(\mathbf{m},\mathbf{m}',\mathbf{r}) = g(\mathbf{m},\mathbf{m},\mathbf{r}) &+ \lambda_i(\mathbf{r})x_\alpha m_{i,\alpha}(\mathbf{R}) \\ &+ \frac{1}{2}\left[\lambda_i(\mathbf{r})m_{i,\alpha\beta}(\mathbf{R}) + \mu_{ij}(\mathbf{r})m_{i,\alpha}(\mathbf{R})m_{j,\beta}(\mathbf{R})\right]x_\alpha x_\beta. \end{aligned} \tag{2.52}$$

The above expression gives the interaction energy between two elements characterized by the orientations $\mathbf{m} = \mathbf{m}(\mathbf{R})$ and $\mathbf{m}' = \mathbf{m}(\mathbf{R}' = \mathbf{R} + \mathbf{r})$, whose relative position is $\mathbf{r}$, in terms of the spatial derivatives of $\mathbf{m}$. It is a rapidly decreasing function of $\mathbf{r}$, like λ_i and μ_{ij}. In the elastic limit

$$|\lambda_i x_\alpha m_{i,\alpha}| \gg |(\lambda_i m_{i,\alpha\beta} + \mu_{ij}m_{i,\alpha}m_{j,\beta})x_\alpha x_\beta|, \tag{2.53}$$

as follows from the above discussion concerning $|m_{i,\alpha}|$ and $m_{i,\alpha\beta}$.

In the mean field approximation, the energy of the film $f(\mathbf{R})$ at the point $\mathbf{R}$ is obtained by integrating $g/2$ over $\mathbf{r} \in (x, y)$. By using (2.52) we obtain

$$\begin{aligned} f(\mathbf{R}) &= \int_{\mathcal{S}_\infty} g(\mathbf{m},\mathbf{m}',\mathbf{r})dS' \\ &= f_0(\mathbf{m}) + L_{i\alpha}m_{i,\alpha}(\mathbf{R}) + \frac{1}{2}N_{i\alpha\beta}m_{i,\alpha\beta}(\mathbf{R}) + \frac{1}{2}M_{ij\alpha\beta}m_{i,\alpha}(\mathbf{R})m_{j,\beta}(\mathbf{R}), \end{aligned} \tag{2.54}$$

where $\mathcal{S}_\infty$ means that the integration over r is extended over the range of the intermolecular forces giving rise to the film. In Eq. (2.54) we have introduced the quantities

$$f_0(\mathbf{m}) = \frac{1}{2}\int_{\mathcal{S}_\infty} g(\mathbf{m}, \mathbf{m}, \mathbf{r})d\mathcal{S}', \tag{2.55}$$

$$L_{i\alpha} = \frac{1}{2}\int_{\mathcal{S}_\infty} \lambda_i(\mathbf{r})x_\alpha d\mathcal{S}', \tag{2.56}$$

and, furthermore,

$$N_{i\alpha\beta} = \frac{1}{2}\int_{\mathcal{S}_\infty} \lambda_i(\mathbf{r})x_\alpha x_\beta d\mathcal{S}', \tag{2.57}$$

$$M_{ij\alpha\beta} = \frac{1}{2}\int_{\mathcal{S}_\infty} \mu_{ij}(\mathbf{r})x_\alpha x_\beta d\mathcal{S}'. \tag{2.58}$$

From (2.57) one derives

$$N_{i\alpha\beta} = N_{i\beta\alpha}. \tag{2.59}$$

Furthermore, from (2.58) and (2.47), it follows that

$$M_{ij\alpha\beta} = M_{ji\alpha\beta} = M_{ij\beta\alpha} = M_{ji\beta\alpha}. \tag{2.60}$$

The meaning of the different terms introduced before is very simple. $f_0(\mathbf{m})$ is the surface energy density of a uniformly oriented film ($\mathbf{m}$ position independent), whereas $\mathcal{L}$, $\mathcal{N}$, and $\mathcal{M}$ play the role of elastic constants. Tensors $\mathcal{L}$, $\mathcal{N}$, and $\mathcal{M}$ have to be decomposed in terms of the elements of symmetry of the film. In the present case, in which the film is assumed to be flat, the elements of symmetry are the geometrical normal $\mathbf{k}$ (parallel to the $z-$axis) and the vector $\mathbf{m}$ (see details in [29]).

The term $f_0(\mathbf{m})$ can be expanded in power series of $\mathbf{m} \cdot \mathbf{k}$, or in terms of Legendre polynomials. If $\mathbf{m}$ is uniform, the energy of the film reduces to $f_0(\mathbf{m})$. Since the film is in contact with the substrate, we have to take into account also the direct interaction between the molecules of the film and the molecules of the substrate [30].

If we suppose that the solid substrate is isotropic, the surface free energy due to the direct interaction is of the kind $\psi(\mathbf{m} \cdot \mathbf{k})$. This energy depends on the van der Waals interaction, dielectric interaction, and so on. It depends on the physical properties of the solid substrate. In the ideal case of a homogeneous surface, ψ is position independent. But real surfaces are never homogenous! More precisely, on average they have approximately the same properties, but from point to point they change in a more or less stochastic manner. Hence, we can write

$$\psi(\mathbf{m} \cdot \mathbf{k}, \mathbf{r}) = \psi_{\rm av}(\mathbf{m} \cdot \mathbf{k}) + \delta\psi(\mathbf{m} \cdot \mathbf{k}, \mathbf{r}), \tag{2.61}$$

where $\delta\psi(\mathbf{m} \cdot \mathbf{k}, \mathbf{r})$ takes into account the stochastic part of the direct interaction between the film and the solid substrate (due, for instance, to free ions or local irregularities). The total surface energy (due to the intrinsic part and the direct interaction) is then

$$F_0 = f_0(\mathbf{m}\cdot\mathbf{k}) + \psi(\mathbf{m}\cdot\mathbf{k},\mathbf{r}) = G_0(\mathbf{m}\cdot\mathbf{k}) + \delta\psi(\mathbf{m}\cdot\mathbf{k},\mathbf{r}), \tag{2.62}$$

where

$$G_0(\mathbf{m}\cdot\mathbf{k}) = f_0(\mathbf{m}\cdot\mathbf{k}) + \psi_{\rm av}(\mathbf{m}\cdot\mathbf{k}) \tag{2.63}$$

is the uniform part of the total surface energy. We assume now that the uniform film is homeotropically aligned on the considered surface, such that $\mathbf{m} \parallel \mathbf{k}$. In a first approximation we use for $G_0(\mathbf{m}\cdot\mathbf{k})$ the expression

$$G_0(\mathbf{m}\cdot\mathbf{k}) = -\frac{1}{2}W(\mathbf{m}\cdot\mathbf{k})^2. \tag{2.64}$$

The total surface energy F_0 is then given by

$$F_0 = -\frac{1}{2}W(\mathbf{r})(\mathbf{m}\cdot\mathbf{k})^2, \tag{2.65}$$

where

$$W(\mathbf{r}) = W + \Delta W(\mathbf{r}), \tag{2.66}$$

in which $\Delta W(\mathbf{r})$ is the stochastic contribution to W. We suppose that for $\mathbf{m}$ position independent, $f(\mathbf{R})$ is minimum. This implies that the tensor $\mathcal{L}$, which is connected with spontaneous distortion, has to be identically zero. If we suppose also that $\mathbf{m}$ remains always in the plane (x,z), by indicating by θ the angle between $\mathbf{m}$ and $\mathbf{k}$, the total energy of the film is

$$F = \int_{\mathcal{S}} \frac{1}{2}\left[K(\nabla\theta)^2 + W(\mathbf{r})\sin^2\theta\right] d\mathcal{S}, \tag{2.67}$$

where $\mathcal{S}$ is the surface of the sample and $\nabla = \mathbf{i}\,\partial/\partial x + \mathbf{j}\,\partial/\partial y$. In (2.67) K accounts for the elastic properties of the film, coming from the tensor $\mathcal{M}$ introduced in (2.58). The contribution coming from the tensor $\mathcal{N}$ can be neglected since it can be integrated over $\mathcal{S}$, giving only a line contribution. To obtain the $\mathbf{m}$ equilibrium distribution, we have to minimize (2.67), thus obtaining

$$K\left[\frac{\partial^2\theta}{\partial x^2} + \frac{\partial^2\theta}{\partial y^2}\right] - W(\mathbf{r})\sin\theta\cos\theta = 0. \tag{2.68}$$

By extracting the fluctuating part of $W(\mathbf{r})$, as was done in (2.67), and by putting

$$\theta(\mathbf{r}) = \Theta + \delta\theta(\mathbf{r}), \tag{2.69}$$

where Θ is position independent, we can linearize Eq. (2.69). In this limit Eq. (2.68) is written as

$$\frac{\partial^2 \delta\theta}{\partial x^2} + \frac{\partial^2 \delta\theta}{\partial y^2} - \frac{W}{K}\cos(2\Theta) = \frac{\Delta W(\mathbf{r})}{2\,K}\sin(2\Theta). \tag{2.70}$$

The problem can be solved by means of a Green function $G(\mathbf{r},\mathbf{r}')$, in the form

$$\delta\theta(\mathbf{r}) = \int G(\mathbf{r},\mathbf{r}')\frac{\Delta W(\mathbf{r}')}{2\,K}\sin(2\Theta)d\mathbf{r}'. \tag{2.71}$$

Let us suppose now that

$$\langle\gamma(\mathbf{r})\gamma(\mathbf{r}')\rangle = D_\gamma e^{-|\mathbf{r}-\mathbf{r}'|/R_\gamma}, \tag{2.72}$$

where

$$\gamma(r) = \frac{\Delta W(r)}{K}, \tag{2.73}$$

D_γ is the dispersion, and R_γ is the correlation length of the random distribution $\gamma(\mathbf{r})$. This kind of correlation satisfies the fundamental property of stochastic systems, i.e.,

$$\lim_{r\to\infty}\langle\gamma(\mathbf{r})\gamma(\mathbf{r}')\rangle = 0. \tag{2.74}$$

By taking into account (2.72), the effective surface energy, defined as

$$F_{\text{eff}} = \frac{1}{\mathcal{S}}\int F(\mathbf{r})d\mathbf{r}, \tag{2.75}$$

is found to be

$$F_{\text{eff}} = \frac{1}{2}W\sin^2\Theta + (2\pi)^2 D_\gamma R_\gamma^2 K\,I(X)\sin^2(2\Theta), \tag{2.76}$$

where

$$I(X) = \frac{1}{1-X} + \frac{\ln X}{2(1-X)^{3/2}} - \frac{\ln(1+\sqrt{1-X})}{(1-X)^{3/2}}, \tag{2.77}$$

and

$$X = \alpha R_\gamma^2, \quad \text{with} \quad \alpha(\Theta) = \frac{W}{K}\sin(2\Theta). \tag{2.78}$$

The Green function of (2.70) is

$$G(K) = -\frac{1}{K^2+\alpha(\Theta)}, \tag{2.79}$$

and, therefore,

$$\langle\gamma(\mathbf{r})\delta\theta(\mathbf{r})\rangle = -2\pi\int_0^\infty r'dr'\int_0^\infty KdK\int_0^{2\pi}\frac{e^{iKr'\cos Y}}{K^2+\alpha(\Theta)}e^{-r'/R_\gamma}dY, \tag{2.80}$$

which can be written as

$$\langle\gamma(\mathbf{r})\delta\theta(\mathbf{r})\rangle = -(2\pi)^2\int_0^\infty r'dr'\int_0^\infty KdK\frac{J_0(Kr')}{K^2+\alpha(\Theta)}e^{-r'/R_\gamma}dY, \tag{2.81}$$

if the Bessel function of zeroth order

$$J_0(a) = \frac{1}{2\pi}\int_0^{2\pi} e^{ia\cos Y}dY, \tag{2.82}$$

is introduced. By observing that

$$\int_0^\infty \frac{KdK}{K^2+\alpha(\Theta)}J_0(Kr) = K_0(\sqrt{\alpha(\Theta)}r), \tag{2.83}$$

where $K_0(\sqrt{\alpha(\Theta)}r)$ is the modified Bessel function, by substituting this function into the expression of $\langle\gamma(\mathbf{r})\delta\theta(\mathbf{r})\rangle$ and integrating over r, we obtain [31] Eq. (2.76) and (2.77) reported above.

The fluctuation additive term present in (2.76) arises from the expression

$$\Delta F = \frac{K}{2\mathcal{S}}\int_{\mathcal{S}}\langle\gamma(\mathbf{r})\delta\theta(\mathbf{r}')\rangle\sin(2\Theta)d\mathcal{S}. \tag{2.84}$$

Equation (2.76) clearly shows that the inclusion of a stochastic spatial variation of the surface field, caused by the direct interaction between the film and the substrate, gives rise to a new functional form of the effective anchoring energy [32]. More precisely, the usual anchoring strength, in the Rapini-Papoular sense, is renormalized: W is substituted by

$$W_{\text{eff}} = W + \Delta W, \tag{2.85}$$

where

$$\Delta W = (2\pi)^2 D_\gamma R_\gamma^2 I(\alpha R_\gamma^2)K. \tag{2.86}$$

Furthermore, a new term appears, proportional to $\sin^4\Theta$ and characterized by a coefficient equal to ΔW, but with opposite sign. In the hypothesis that the film is not compact, the bulk orientation of the NLC is due to the steric interaction between the first nematic layer and the film. Consequently, the anchoring energy of the NLC coincides with the anchoring energy of the film. Of course the total surface energy of the NLC may contain, besides the steric term discussed above, other contributions due, for instance, to dispersion interactions. However, the steric one is usually the most important in the interface between the NLC and the film. Hence, the theory presented above is expected to work well, at least for this kind of interface. It is possible to estimate the order of magnitude of ΔW, given by (2.86). Let us assume that $K \simeq 10^{-13}$ erg, the anchoring energy $W \simeq 10^{-2}$ erg/cm^2. As follows from (2.72)

$$D_\gamma = \left[\frac{W}{K}\right]^2 \langle \frac{\Delta W(\mathbf{r})}{W} \frac{\Delta W(0)}{W} \rangle = \left[\frac{W}{K}\right]^2 \sigma. \tag{2.87}$$

By assuming $\sigma \simeq 10^{-1}$ and $I \simeq 1$, we obtain for $\Delta W \simeq 10^{-2}$ erg/cm^2, i.e., of the same order of W.

This simple estimation shows that the influence of the stochastic part may be very important. In particular, it has to be pointed out that a term proportional to $\sin^4 \Theta$ strongly modifies the phase diagram relevant to the surface transitions induced by temperature or by external fields [33].

2.3.2 Validity of the elastic model for nematic surface anchoring energy

According to the hypothesis of the elastic theory, briefly discussed in Section 1.4, the elastic energy density of an NLC sample, F, is a positive defined quadratic form in $n_{i,j}$. This means that

$$F = \frac{1}{2} K_{ijkl}\, n_{i,j}\, n_{k,l}, \tag{2.88}$$

where the elastic tensor $K_{ijkl} = K_{klij}$ is position independent in the bulk [see Eq. (1.58)]. It can be decomposed, as well known [29, 34], in terms of the elements of symmetry of the NLC phase. By means of pseudo-molecular models it is possible to evaluate the elastic tensor K_{ijkl} when the intermolecular interaction responsible for the nematic phase is known [35]–[38]. This kind of calculation is relatively simple in the bulk, where the elements of symmetry of the NLC phase reduce to $\mathbf{n}$. Near the surface the situation is more complicated for two different reasons. First, the elements of symmetry of the NLC phase are $\mathbf{n}$ and the geometrical normal to the bounding surface, $\mathbf{k}$. This implies that usually the number of surface elastic constants is larger than the one in the bulk [17], [36]–[38]. Second, the "elastic constants" are expected to be position dependent. Consequently, in a surface layer whose thickness is of the order of the range of the intermolecular forces giving rise to the NLC phase, the elastic energy density is

$$F = \frac{1}{2} K_{ijkl}(z)\, n_{i,j}\, n_{k,l} + \delta F(n_{i,j}), \tag{2.89}$$

where $K_{ijkl}(z)$ takes into account the incomplete NLC-NLC interaction and δF for the new elements of symmetry of the phase. By means of a Maier-Saupe [39] interaction the elastic description of an NLC close to a substrate was analyzed, by supposing an interaction volume of ellipsoidal shape [37]. In that context, the $K_{ijkl}(z)$ and δF have been evaluated. The analysis shows that $K_{ijkl}(z)$ and δF exist in a surface layer whose thickness is of the order of 20 to 30 molecular dimensions. In the same framework, it was shown that the uniform part [40] of the free energy density is intrinsically anisotropic with an easy axis [41] parallel to the geometrical normal (homeotropic orientation). The associated anisotropic anchoring energy is found to be of the order of $1\,\mathrm{erg/cm^2}$ using reasonable values for the physical parameters characterizing usual NLC [42]. Since this anisotropic anchoring energy comes only from the NLC itself, from now on it will be called intrinsic uniform and denoted by W_{iu} [40]. The extrapolation length connected to this energy is defined as $L_{\mathrm{iu}} = K_b/W_{\mathrm{iu}}$, with K_b being the bulk value of the average Frank elastic constant.

Let us now analyze all the possible contributions to the anisotropic part of the surface energy, and evaluate them in the framework of the Maier-Saupe model [43]. As shown in [40], if the interaction volume is of ellipsoidal shape, the uniform part of the surface energy is anisotropic with homeotropic easy direction. Let us consider a semi-infinite NLC sample and a Cartesian reference frame whose z axis is normal to the bounding wall placed at $z = 0$ and $z > 0$ corresponds to the NLC half-space. The NLC distortion is supposed planar in the (x, z) plane. The average NLC orientation is described by means of the tilt angle $\theta = \arccos(\mathbf{n} \cdot \mathbf{k})$. For the sake of simplicity, $\theta = \theta(z)$ only. Assuming the Maier-Saupe law for the intermolecular interaction, and considering that both the molecular and the effective interaction volumes are ellipsoids of revolution (of similar shape) around $\mathbf{n}$, it is possible to define the energy of the undistorted configuration, F_0, as:

$$F_0 = -\frac{1}{2}\int_{V_N} J(r)\, dV_N \tag{2.90}$$

where $J(r) = -C/r^6$, with C being a positive constant, and V_N the effective interaction volume. In the situation in which the interaction volume is incomplete, i.e., close to the surface, a surface free energy density can be defined as

$$G(e,\theta) = \int_{A_m}^{A_M} \Delta\, F_0(e,\theta; z)\, dz \tag{2.91}$$

where e is the eccentricity of the ellipsoidal interaction volume and it is supposed to be equal to the one of the NLC molecule. Furthermore, A_m and A_M are, respectively, the z coordinate of the lowest and highest point of the ellipsoid, and $\Delta\, F_0 = F_{0b} - F_0(z)$, where F_{0b} is the bulk value of the uniform part of the free energy density.

The physical meaning of Eq. (2.91) is that due to NLC-NLC interaction, there exists, in a surface layer whose thickness, b, is several molecular dimensions, an anisotropic surface field characterized by a well defined easy direction and anchoring energy. Of course, a direct NLC-substrate interaction can exist. It is delocalized over a surface layer of thickness ρ_{s} depending on the anisotropic part of the NLC-substrate interaction. This direct interaction, when integrated over this characteristic distance, gives an extrinsic contribution to the anchoring energy. It is characterized by another easy direction which depends on the physical symmetry of the surface and of the NLC. Furthermore, it is also characterized by an anisotropic anchoring strength, $W_{\rm eu}$. Since $b \simeq \rho_{\rm s}$, because the NLC-substrate and NLC-NLC interactions are of the same kind, it is possible to introduce a total uniform part of the surface energy, defined as the sum of the above mentioned terms. In principle the easy axes are different, and the actual one will be in between them. For simplicity we will suppose that they coincide with the z axis, and hence

$$W_{\rm u} = W_{\rm iu} + W_{\rm eu}. \tag{2.92}$$

However, since we are mainly interested in the analysis of the intrinsic part of the surface energy coming from the spatial variation of the elastic constants, this hypothesis is not very restrictive. Long ago, Yokoyama [44] has suggested that a spatial variation of the elastic constant is equivalent to a surface energy. This problem has been more recently reconsidered by other authors [3, 36, 45]. The main conclusions of these kinds of investigations are the following. If close to a boundary, $K = K(z)$, the associated anisotropic part of the anchoring energy is given by [36]

$$\frac{1}{W_{\rm elast}} = \int_0^\infty \frac{K_b - K(z)}{K_b\, K(z)}\, dz, \tag{2.93}$$

where, as before, K_b is the bulk value of the elastic constant, whereas $K(z)$ is its actual value at a distance z from the wall. It changes over a surface layer whose thickness will be denoted by b. Equation (2.93) holds only when b is very small with respect to the thickness of the real sample. This implies that long range parts of the surface energy, like the ones connected to electrical effects, are not considered [46]–[48].

We are now in order to use the results obtained previously in Ref. [37] to estimate the order of magnitude of $W_{\rm elast}$. In the case we are analyzing here, in which the surface is supposed flat and $\theta = \theta(z)$, Eq. (2.90) can be rewritten as

$$F(\theta, z) = \frac{1}{2}\, K(z)\, \theta'^2, \tag{2.94}$$

where $K(z)$ takes into account for the positional dependence of Frank elastic constants, and for the new elastic constants connected with the reduction of symmetry introduced by the wall (the terms c_i, for $i = 1, 2, 3, 4$ of Ref. [37]). As it follows from [37], $K(z)$ is well approximated by the function

$$K(z) = K_b - \frac{1}{2}\, K_b\, e^{-z/b}. \tag{2.95}$$

By substituting (2.95) into (2.93) straightforward calculations give

$$\frac{1}{W_{\rm elast}} = \frac{b}{K_b} \ln 2. \tag{2.96}$$

According to these pseudo-molecular calculations, b is expected to be of the order of $20-30\ a_0$, with a_0 being the minor axis of the internal ellipsoid (i.e., the molecular volume). By assuming $a_0 \simeq 5\,\AA$ and $K_b \simeq 10^{-6}\,\mathrm{erg/cm}$ we find $W_{\mathrm{elast}} \simeq 1\,\mathrm{erg/cm}^2$. This situation corresponds to the so-called strong anchoring [1, 49]. From this result it follows that some other mechanism could be responsible for the weak anchoring energy experimentally detected [1, 49].

The pseudo-molecular calculations have been performed by assuming perfect nematic order (i.e., the scalar order parameter $S = 1$). Thus, we need to reconsider this hypothesis. As is well known, in the framework of the Landau-Ginzburg theory the actual scalar order parameter profile has to be deduced by minimizing the functional written in terms of S. This functional is of the kind

$$G = \int_0^\infty [\frac{1}{2}\,LS'^2 + f(S)]dz, \tag{2.97}$$

where L is an elastic constant parameter, $S' = dS/dz$ and $f(S)$ the free energy density of the uniform nematic phase [50]. Standard calculations show that near the boundary S is given by

$$S(z) = S_b + (S_0 - S_b)e^{-z/\xi}, \tag{2.98}$$

where the bulk value of the order parameter, S_b, is fixed by the temperature and the surface value of the order parameter, S_0, is fixed by the temperature and the surface NLC-substrate interaction. In Eq. (2.98), ξ is the coherence length of the medium in the NLC phase [51]. The Frank elastic constants are expected to be proportional to S according to the law

$$K(z) = L\,S^2(z), \tag{2.99}$$

as shown in Refs. [35] and [50]. It follows that a spatial variation of K is expected in a surface layer whose thickness is of the order of ξ. This is a new contribution which is usually neglected in the pseudo-molecular approach. However, Eq. (2.93) remains valid because it was obtained without any assumptions on the source of the $K(z)$ term [36]. Hence, instead of Eq. (2.95) we have to consider the $K(z)$ behavior given by

$$K(z) = [K_b - \frac{1}{2}\,K_b\,e^{-z/b}][1 + \Delta e^{-z/\xi}]^2 \tag{2.100}$$

obtained by Eq. (2.99) taking into account Eq. (2.95), and introducing $\Delta = (S_0 - S_b)/S_b$. Notice that in the hypothesis of perfect nematic order, the elastic constant parameter, L, coincides with Eq. (2.95). Since for T close to T_c, $b \ll \xi$, Eq. (2.100) is equivalent to

$$K(z) = K_b(1 + \Delta e^{-z/\xi})^2. \tag{2.101}$$

It follows that the elastic contribution to the anisotropic part of the anchoring energy is given by

$$\frac{1}{W_{\rm elast}} = \frac{1}{K_b} \int_0^\infty \frac{(1+\Delta e^{-z/\xi})^2 - 1}{(1+\Delta e^{-z/\xi})^2} \, dz. \tag{2.102}$$

In the limit of $\Delta \ll 1$, which implies S_0 not very different from S_b, from (2.102) we obtain

$$\frac{1}{W_{\rm elast}} = 2\,\Delta\,\frac{\xi}{K_b}. \tag{2.103}$$

By assuming $\Delta = 0.3$, and $S_b = 0.5$, which implies $S_0 - S_b = 0.15$, one obtains for the extrapolation length associated with this elastic term, $L_{\rm elast} \simeq \xi$, i.e., in the macroscopic range, as experimentally observed [1, 49].

The total extrapolation length is the sum of the different contributions analyzed above. Hence, it is given by

$$L_T = L_{\rm u} + L_{\rm elast}, \tag{2.104}$$

where

$$L_{\rm u} = \frac{K_b}{W_{\rm u}} = \frac{K_b}{W_{\rm eu} + W_{\rm iu}}, \tag{2.105}$$

and

$$L_{\rm elast} = \frac{K_b}{W_b} + \frac{K_b}{W_\xi}. \tag{2.106}$$

Consequently, the total extrapolation length can be rewritten as

$$L_T = \frac{K_b}{W_{\rm eu} + W_{\rm iu}} + \frac{K_b}{W_b} + \frac{K_b}{W_\xi} \simeq \frac{K_b}{W_\xi}. \tag{2.107}$$

This simple result shows that the weak anchoring is mainly due to the spatial variation of the scalar order parameter.

The main conclusions can be summarized as follows:

(1) Due to the spatial variation of the elastic constants, there is an intrinsic contribution to the anisotropic surface energy. The extrapolation length connected to this surface energy is found to be of the order of the thickness of the surface layer over which the elastic constants change, when perfect nematic order is supposed. It is of the order of 70Å for usual nematics;

(2) When the spatial variation of the scalar order parameter is taken into account, there is an intrinsic elastic contribution whose extrapolation length is found to be of the order of the coherence length in the NLC phase. It is in the macroscopic range and it is of the same order of the one experimentally detected ($\approx 0.1 - 0.5\,\mu$m).

2.3.3 Contribution of the smectic-nematic interface to the surface energy

The contribution of the smectic-nematic interface to the surface energy of a nematic liquid crystal sample can be analyzed by means of a simple model [52]. This model shows that the surface energy depends on the thickness of the region over which the smectic-nematic transition takes place. For perfectly flat substrates this thickness is of the order of the correlation length entering in the transition. An estimate of this contribution shows that it is greater than the one arising from the nematic-substrate interaction. Moreover, it is possible to show that the surface energy determined in this way presents a non – monotonic behavior with the temperature.

When the liquid crystalline system is limited by a flat surface, the translational invariance of the phase is broken. The presence of a substrate can also induce a positional order in the molecules in the vicinity of the surface, as discussed before. In fact, near to the surface the centers of mass of the molecules tend to form layers parallel to the surface as indicated by several experiments [1]. Since these molecules are oriented perpendicularly or tilted with respect to the layers, they form smectic layers near the boundaries. As pointed out by Cognard, the energy confined in these layers is higher than the one that can be added to the LC film by other actions, like mechanical, thermal or electrical ones [53]. Therefore, it seems very important to analyze also the contribution to the surface energy coming from the smectic-nematic interface.

We investigate now the possibility that the main contribution to the surface energy of the system can be connected to the interaction between the smectic layer and the nematic. The smectic-nematic interface is supposed to have a thickness ε along which the system passes from one phase to the other. On the other hand, the system can be treated as a simple junction [29], then we explicitly evaluate the surface energy in the transition zone by introducing the smectic coherence length in the nematic phase. In this manner also the temperature dependence of the interfacial energy can be analyzed, in a first approximation, as is done for the nematic liquid crystal–wall interface [54]. In Chapter 4 this analysis is developed in more details.

We consider here a nematic slab of thickness d. The Cartesian reference frame has the (x, y)-plane coinciding with the surfaces of the slab. The problem will be supposed one-dimensional with all the quantities depending only on the z-coordinate. The director is everywhere parallel to the (x, z)-plane and $\theta(z) = \arccos(\mathbf{n} \cdot \mathbf{k})$ is the tilt angle formed by $\mathbf{n}$ and the z-axis. The tilt angle assumes the value $\theta_1(z)$ in the smectic layer, $\theta_2(z)$ in the nematic phase and $\theta_3(z)$ in the transition region smectic-nematic. By considering that the transition occurs in a layer of thickness ε the total elastic energy per unit surface can be written as

$$F = \int_0^b \frac{1}{2} K_1 {\theta_1'}^2 \mathrm{d}z + \int_b^{b+\varepsilon} \frac{1}{2} K(z) {\theta_3'}^2 \mathrm{d}z + \int_{b+\varepsilon}^d \frac{1}{2} K_2 {\theta_2'}^2 \mathrm{d}z, \quad (2.108)$$

where b is the thickness of the smectic region, K_1 and K_2 are the elastic constants of the smectic and nematic phases, respectively, and the prime denotes, as before, spatial derivative. The second addendum in (2.108) represents the contribution to the total elastic energy coming from the smectic-nematic interface. In this region, $K(z)$ can be written, in a first approximation, as

$$K(z) = K_1 + \frac{K_2 - K_1}{\varepsilon}(z - b). \quad (2.109)$$

In the strong anchoring hypothesis, the boundary conditions at the surface are $\theta(0) = 0$ and $\theta(d) = \theta$. The solutions of the Euler-Lagrange equations obtained from the minimization of the functional (2.108) are

$$\begin{aligned}
\theta_1(z) &= \frac{C}{K_1} z, \quad 0 \le z \le b, \\
\theta_3(z) &= \frac{C\varepsilon}{K_2 - K_1} \ln K(z) + C\left[\frac{b}{K_1} - \frac{\varepsilon}{K_2 - K_1} \ln K_1\right], \quad b \le z \le b + \varepsilon, \\
\theta_2(z) &= \theta + \frac{C}{K_2}(z - d), \quad b + \varepsilon \le z \le d,
\end{aligned} \quad (2.110)$$

where

$$C = \left[\frac{\varepsilon}{K_2 - K_1} \ln\left(\frac{K_2}{K_1}\right) + \frac{d - b - \varepsilon}{K_2} + \frac{b}{K_1}\right]^{-1} \theta. \quad (2.111)$$

Equations (2.110) and (2.111) give, for an arbitrary point $z = \tilde{b}$,

$$\theta_2(\tilde{b}) = \theta + \frac{C}{K_2}(\tilde{b} - d). \quad (2.112)$$

On the other hand, as already indicated, it is possible to treat the problem as a smectic-nematic junction where the total elastic energy per unit surface takes the form

$$\tilde{F} = \int_0^{\tilde{b}} \frac{1}{2} K_1 {\tilde{\theta}_1'}^2 \mathrm{d}z + \int_{\tilde{b}}^d \frac{1}{2} K_2 {\tilde{\theta}_2'}^2 \mathrm{d}z + \frac{1}{2}\beta\left[\tilde{\theta}_1(\tilde{b}) - \tilde{\theta}_2(\tilde{b})\right]^2. \quad (2.113)$$

In (2.113), the last term represents the contribution connecting the smectic-nematic junction to the total elastic energy, and β is the surface energy.

Again, by minimizing (2.113) subjected to the boundary conditions $\tilde{\theta}(0) = 0$ and $\tilde{\theta}(d) = \theta$, one obtains

$$\tilde{\theta}_1(z) = \frac{\tilde{C}}{K_1} z, \quad 0 \le z \le \tilde{b},$$
$$\tilde{\theta}_2(z) = \theta + \frac{\tilde{C}}{K_2}(z - d), \quad \tilde{b} \le z \le d, \tag{2.114}$$

where

$$\tilde{C} = \left[\frac{1}{\beta} + \frac{d - \tilde{b}}{K_2} + \frac{\tilde{b}}{K_1} \right]^{-1} \theta. \tag{2.115}$$

From Eq. (2.114) we obtain

$$\tilde{\theta}_2(\tilde{b}) = \theta + \frac{\tilde{C}}{K_2}(\tilde{b} - d). \tag{2.116}$$

The main measurements performed on a real nematic sample concern the bulk properties, like, for instance, the optical path difference. Since $d \gg b$ and $d \gg \varepsilon$, the physical situations described by F and $\tilde{F}$ must be the same in the bulk. Consequently, we will assume that $\theta_2(z) = \tilde{\theta}_2(z)$ and that the border of the nematic phase is localized in $z = b + \varepsilon$. From Eqs. (2.111), (2.112), (2.115), and (2.116), with $\tilde{b} = b + \varepsilon$, we obtain for the surface energy

$$\beta = \left[\frac{1}{K_2 - K_1} \ln \left(\frac{K_2}{K_1} \right) - \frac{1}{K_1} \right]^{-1} \frac{1}{\varepsilon}. \tag{2.117}$$

Moreover, if we consider $K_1 = \alpha K_2$ then

$$\beta = \lambda \frac{K_2}{\varepsilon}, \tag{2.118}$$

where

$$\lambda = \frac{\alpha(\alpha - 1)}{\alpha \ln \alpha - \alpha + 1}. \tag{2.119}$$

Notice that $\theta_2(z) = \tilde{\theta}_2(z)$ implies the equality of the total energies F and $\tilde{F}$.

In order to estimate β we remember that ε is essentially of the order of the correlation length ξ characterizing the region where the transition smectic-nematic occurs. Thus, ξ is expected to be of the order of several molecular lengths. It is also expected that ξ increases near the transition temperature. From these considerations a reasonable estimate is $\varepsilon \approx 1000\,\AA$. For a typical nematic like the PAA, $K_2 \approx 7 \times 10^{-7}$ dyn [51]. Moreover, it is expected that the elastic constant of the smectic phase, K_1 is greater than the elastic constant of the nematic phase K_2. If we assume that $\alpha \approx 3$, we obtain for the surface energy , $\beta \approx 0.3$ erg/cm^2. The surface energies measured on real samples are of the order of 10^{-1} and 10^{-2} erg/cm^2 [49]. Therefore, our results indicate that the surface energy is mainly connected to the smectic-nematic interface, in the hypothesis that ε is not too large. If we consider that $\tilde{b} < b + \varepsilon$, then, in general, β is a negative quantity for $K_1 > K_2$. This situation is not physically meaningful. It happens only because in this case we are extending the nematic phase to a region where the phase is not purely nematic. On the other hand, several recent measurements performed on lyotropic nematic samples [55] indicate an agreement with the present estimate. In these experiments, with discotic nematic liquid crystals, the surfaces of the substrates are with and without treatment. The measured values of the optical path difference are the same for both situations. This fact indicates that the lamellar phase formed between the glass plates and the NLC is responsible for a strong attenuation of the effect of the glass on the nematic phase. Another point which deserves mention is the dependence of the surface energy stored on the interface with the temperature. In this sense we have used a mean field approximation for the determination of the correlation length, namely $\xi \approx (T^* - T)^{-1/2}$, where T^* is a temperature for the structural phase transition in the smectic-nematic interface. From this observation and by considering that $\varepsilon \approx \xi$ and $\beta \approx K_2/\varepsilon$, we can conclude that $\beta \approx K_2(T^* - T)^{1/2}$. This result indicates that as the transition temperature is approximated, the surface energy of the interface becomes negligible. It indicates also that β has a non-monotonic behavior with the temperature, which agrees with the results obtained by di Lisi *et al.* [56] for a structural transition at a nematic-substrate interface.

For actual inhomogeneous surfaces, where the treatment hardly ensures a perfect uniform orientation of the director, geometrical effects can be present influencing the experimentally detected quantities [57]. Then, the above conclusions can be valid at least in the case in which perfectly flat substrates are supposed to form the slab.

[1] Jerome B. Surface effects and anchoring in liquid crystals. *Reports on Progress in Physics*, **54**, 391 (1991).

[2] Berreman DW. Solid surface shape and alignment of an adjacent nematic liquid crystal. *Physical Review Letters*, **28**, 1683 (1972).

[3] Faetti S. In *Physics of Liquid Crystalline Materials*. Edited by Khoo IC and Simoni F. Gordon and Breach, Philadelphia, 1991.

[4] Rapini A and Papoular MJ. Distortion d'une lamelle nématique

sous champ magnétique. Conditions d'ancrage aux parois. *Journal de Physique Colloque*, **30 (C4)**, 54 (1969).

[5] Evangelista LR and Barbero G. Walls of orientation induced in nematic liquid crystal samples by inhomogeneous surfaces. *Physical Review E*, **50**, 2120 (1993).

[6] Parsons JD. Strucutural critical-point at free-surface of a nematic liquid crystal. *Molecular Crystals and Liquid Crystals*, **31**, 79 (1975).

[7] Mada H. Study on the surface alignment of nematic liquid crystals–temperature dependence of pre-tilt angles. *Molecular Crystals and Liquid Crystals*, **51**, 43 (1979); Study on the surface alignment of nematic liquid crystals–determination of the easy axis and temperature dependence of its field energy. *Molecular Crystals and Liquid Crystals*, **53**, 127 (1979).

[8] Croxton CA. Statistical thermodynamics of the liquid crystal surface–smectic, nematic and isotropic systems. *Molecular Crystals and Liquid Crystals* **59**, 219 (1980); A Landau-de Gennes theory of the nematic liquid crystal surface. *Molecular Crystals and Liquid Crystals*, **66**, 223 (1981).

[9] Barbero G, Bartolino R, and Meuti M. A conjecture on the Mada theory for nematic liquid crystal–surface interaction. *Journal de Physique Lettres*, **45**, L-449 (1984).

[10] Sluckin TJ and Poniewierski A. In *Fluid and Interfacial Phenomena.* Edited by Croxton CA. John Wiley & Sons, Chichester, 1986.

[11] Bernasconi J, Strassler S, and Zeller HR. Van der Waals contribution to the surface and anchoring energies of nematic liquid crystals. *Physical Review A*, **22**, 276 (1980).

[12] Parsons JD. Molecular theory of surface tension in nematic liquid crystals. *Journal de Physique* **37**, 1187 (1976).

[13] Murakami J. Molecular theory of surface tension for liquid crystal. *Journal of the Physical Society of Japan*, **42**, 210 (1977).

[14] Faetti S. Anchoring at the interface between a nematic liquid crystal and an isotropic substrate. *Molecular Crystals and Liquid Crystals*, **179**, 217 (1990).

[15] Kimura H and Nakano H. Statistical theory of surface tension and molecular orientations in nematic liquid crystals II: The nematic–isotropic interface. *Journal of the Physical Society of Japan*, **55**, 4186 (1986).

[16] Holyst R and Poniewierski A. Orientation of liquid crystal molecules at the nematic-isotropic interface and the nematic free surface. *Molecular Crystals and Liquid Crystals*, **192**, 65 (1990).

[17] Oldano C and Barbero G. Possible boundary discontinuities of the tilt angle in nematic liquid crystals. *Journal de Physique Lettres*, **46**, L-451 (1985); An *ab initio* analysis of the 2nd order elasticity effect on nematic configurations. *Physics Letters A*, **110**, 213 (1985).

[18] Parsons JD. Structural critical point at free surface of a nematic liquid crystal. *Physical Review Letters*, **41**, 877 (1978).

[19] Yokoyama H, Kobayashi S, and Kamei H. Measurement of director orientation at the nematic isotropic interface using a substrate–nucleated nematic film. *Molecular Crystals and Liquid Crystals*, **107**, 331 (1984).

[20] Yokoyama H, Kobayashi S, and Kamei H. Deformations of a planar nematic-isotropic interface in uniform and non-uniform electric fields. *Molecular Crystals and Liquid Crystals*, **129**, 109 (1985).

[21] Bouchiat MA and Langevin-Cruchon D. Molecular order at free surface of a nematic liquid crystal from light reflectivity measurements. *Physics Letters A*, **34**, 331 (1971).

[22] Barbero G, Dozov I, Palierne JF, and Durand G. Order electricity and surface orientation in nematic liquid crystals. *Physical Review Letters*, **56**, 2056 (1986).

[23] Prost J and Marcerou JP. Microscopic interpretation of flexoelectricity. *Journal de Physique*, **38**, 315 (1977).

[24] Alexe-Ionescu AL, Barbero G, Gabbasova Z, Sayko G, and Zvezdin AK. Stochastic contribution to the anchoring energy: Deviation from the Rapini-Papoular expression. *Physical Review E*, **49**, 5354 (1994).

[25] Hiltrop K and Stegemeyer H. Contact angles and alignment of liquid crystals lecithin monolayers. *Molecular Crystals and Liquid Crystals*, **49**, 61 (1978).

[26] Komitov L, Lagerwall ST, Sparavigna A, Stebler B, and Strigazzi A. Surface transition in a nematic layer with reverse pretilt. *Molecular Crystals and Liquid Crystals*, **223**, 197 (1992).

[27] Alexe-Ionescu AL, Barbero G, Miraldi E, and Ignatov A. Surface transitions in nematic liquid crystals oriented with Langmuir-Blodgett films. *Applied Physics A*, **56**, 453 (1993).

[28] Nehring J and Saupe A. Elastic theory of uniaxial liquid crystals. *Journal of Chemical Physics*, **54**, 337 (1971); Calculation of elastic constants of nematic liquid crystals. *Journal of Chemical Physics*, **56**, 5527 (1972).

[29] Barbero G and Evangelista LR. *An Elementary Course on the Continuum Theory for Nematic Liquid Crystals*. World Scientific, Singapore, 2001.

[30] Osipov MA, Sluckin TJ, and Cox SJ. Influence of permanent molecular dipoles on surface anchoring of nematic liquid crystals. *Physical Review E*, **55**, 464 (1997).

[31] Gradshtein IS and Ryzhik IM. *Table of Integrals, Series and Products.* Academic Press, New York, 1965.

[32] Sayko GV, Utochkin SN, and Zvezdin AK. Spin-reorientation phase transitions in thin films of RE-TM amorphous alloys. *Journal of Magnetism and Magnetic Materials*, **113**, 194 (1993).

[33] Barberi R, Barbero G, Gabbasova Z, and Zvezdin AK. Flexoelectricity and alignment phase transitions in nematic liquid crystals. *Journal de Physique II*, **3**, 147 (1993).

[34] Barbero G and Barberi R. In *Physics of Liquid Crystalline Materials.* Edited by Khoo IC and Simoni F. Gordon and Breach, Philadelphia, 1991.

[35] Teixeira PIC, Pergamenshchik VM, and Sluckin TJ. A model calculation of the surface elastic constants of a nematic liquid crystal. *Molecular Physics*, **80**, 1339 (1993).

[36] Alexe-Ionescu AL, Barberi R, Barbero G, and Giocondo M. Surface elastic properties of nematic liquid crystals. *Physics Letters A*, **190**, 109 (1994).

[37] Barbero G, Evangelista LR, Giocondo M, and S. Ponti S. Interfacial energy for nematic liquid crystals: Beyond the spherical approximation. *Journal de Physique II* **4**, 1519 (1994).

[38] Faetti S and Nobili M. Elastic anomalies at the interface between a nematic liquid cystals and its vapor–a microscopic approach. *Journal de Physique II*, **4**, 1617 (1994).

[39] Maier W and Saupe A. Eine einfache molekulare theorie des nematischen kristallinflussigen zustandes. *Zeitschrift fur Naturforschung A*, **13** 564 (1958); Eine einfache molekular-statistiche theorie der nematischen kristallinflussigen phase I. *Zeitschrift fur Naturforschung A*, **14**, 882 (1959); Eine einfache molekular-statistiche theorie der nematischen kristallinflussigen phase II . *Zeitschrift fur Naturforschung A*, **15**, 287 (1960).

[40] Evangelista LR and Ponti S. The intrinsic part of the surface energy for nematics in a pseudo-molecular approach: Comparison with experimental results. *Physics Letters A*, **197**, 55 (1995).

[41] Barbero G, Madhusudana NV, and Durand G. Weak anchoring energy and pretilt of a nematic liquid crystal. *Journal de Physique Lettres*, **45**, L613 (1984).

[42] Kelker H and Hatz R. *Handbook of Liquid Crystals.* Verlag Chemie, Basel, 1980.

[43] Ponti S and Evangelista LR. On the validity of the elastic model for the nematic surface anchoring energy. *Liquid Crystals*, **20**, 105 (1996).

[44] Yokoyama H, Kobayashi S, and Kamei H. Temperature dependence of the anchoring strength at a nematic liquid crystal evaporated SiO interface. *Journal of Applied Physics*, **61**, 4501 (1987).

[45] Barbero G and Durand G. Anchoring energy or surface melting in nematic liquid crystals. *Molecular Crystals and Liquid Crystals*, **203**, 33 (1991).

[46] Alexe-Ionescu AL, Barbero G, and Petrov AG. Gradient flexoelectric effect and thickness dependence of anchoring energy. *Physical Review E*, **48**, R1631 (1993).

[47] Barbero G and Durand G. Order parameter spatial variation and anchoring energy for nematic liquid crystals. *Journal of Applied Physics*, **69**, 6968 (1991).

[48] Alexe-Ionescu AL, Barbero G, and Evangelista LR. Local and non-local terms in the elastic theory for piezoelectric materials. *Molecular Materials* **3**, 31 (1993).

[49] Blinov LM, Kabayenkov AY, and Sonin AA. Experimental studies of the anchoring energy of nematic liquid crystals: Invited lecture. *Liquid Crystals*, **5**, 645 (1989).

[50] Priestley EB, Wojtowicz PJ, and Sheng P (Editors). *Introduction to Liquid Crystals.* Plenum Press, New York, 1975.

[51] de Gennes PG and Prost J. *The Physics of Liquid Crystals.* Clarendon Press, Oxford, 1994.

[52] Evangelista LR, Fontanini S, Malacarne LC, and Mendes RS. Contribution of the smectic-nematic interface to the surface energy. *Physical Review E*, **55**, 1279 (1997).

[53] Cognard J. Alignment of nematic liquid crystals and their mixtures. *Molecular Crystals and Liquid Crystals*: 1, 1, Suppl. (1982).

[54] Rosenblatt C. Temperature dependence of the anchoring strength coefficient at a nematic liquid crystal-wall interface. *Journal de Physique*, **45**, 1087 (1984).

[55] Fontanini S, Barbero G, and Neto AMF. Determination of the effective splay-bend elastic constant of a lyotropic nematic liquid crystal. *Physical Review E*, **53**, 2454 (1996).

[56] di Lisi GA, Rosenblatt C, Akins RB, Griffin AC, and Hari U. Anchoring strength coefficient of a monomer and its dimer at a polymer-coated interface. *Liquid Crystals*, **11**, 63 (1992).

[57] Evangelista LR and Barbero G. Theoretical analysis of actual surfaces: The effect on the nematic orientation. *Physical Review E*, **48**, 1163 (1993).

3

BULK AND SURFACE ELASTIC EFFECTS OF LONG RANGE QUADRUPOLAR INTERACTIONS

In this chapter, we discuss the role of the quadrupolar interaction in nematic liquid crystal samples in the shape of a slab, limiting the study to planar deformations. We show that this interaction gives rise to a bulk energy density that, in the elastic approximation, depends linearly on the second spatial derivative and quadratically on the first spatial derivative of the nematic orientation. This bulk energy density can be separated in a surface-like term, which gives rise only to a surface contribution, plus a term having the usual form. Both terms depend on the first derivative of the tilt angle and are proportional to the square of the electrical quadrupolar density. The bulk term quadratic in the first derivative of the tilt angle renormalizes the usual elastic energy density connected to the short range forces. The surface-like term is proportional to the first derivative of the tilt angle. It recalls the splay-bend elastic term, although the tilt angle dependence is more complicated. The energy density in the surface layers, where the quadrupolar interaction is incomplete, is also evaluated. The solution of the variational problem by means of a simple version of the density functional theory is presented.

3.1 Introduction

As we have seen in Section 1.4, the elastic behavior of NLCs is described by the Frank elastic constants [1]. They are phenomenological parameters introduced by means of symmetry considerations [2]. From a molecular point of view it is possible to connect the elastic constants with the interparticles interaction responsible for the nematic phase [3]–[5]. However, as has been pointed out by Sluckin *et al.* [6]–[9], special attention has to be devoted to the long range interactions, and in particular to the quadrupole-quadrupole interaction. Since NLC are quadrupolar media, it is important to analyze, in some detail, the characteristics of this type of interaction. The electrostatic quadrupole-quadrupole interaction energy decays as R^{-5}, where R is the intermolecular distance. This slow decay of the quadrupolar interaction makes the quadrupolar free energy density intrinsically non-local. In particular, it has been shown [10] that this free energy density cannot be reduced to the Frank elastic form in the general three-dimensional case. In this case, a more complex non-local approach should be used to take into account the quadrupolar interaction. The literature on the macroscopic behavior of NLC is usually devoted to the study of planar director distortions in a nematic slab. In such a case, the director field lies everywhere in a plane orthogonal to the slab plane and depends only on the distance from the surfaces of the slab. In this chapter, we will show that, in this special case of planar director distortions, the quadrupolar free energy density can be reduced to an elastic form [11]. For this reason the analysis is limited, again, to a nematic sample of slab shape, with perfect nematic order. The nematic deformation is supposed to be planar and one-dimensional; it is also assumed that the director is fully described by the tilt angle formed by it with the geometrical normal to the walls of the sample. In this framework, it is shown that the quadrupolar interaction gives rise, in the elastic approximation, to an energy density which can be separated in a bulk and in a surface contribution. The bulk contribution is proportional to the square of the first spatial derivative of the tilt angle. It is characterized by an elastic constant which depends on the nematic orientation, proportional to the square of the nematic quadrupolar density. The surface contribution contains two terms. One is connected to a surface-like linear term in the first spatial derivative of the tilt angle. Due to this dependence, it recalls the splay-bend elastic term introduced long ago by Nehring and Saupe [3] and discussed recently by several authors [12]–[17]. The other term comes from the incomplete quadrupolar interaction in the surface layer, whose thickness is of the order of the molecular dimension. It depends only on the surface nematic orientation, and can be considered as an intrinsic anchoring energy.

3.2 Quadrupolar interaction

We shall evaluate the interaction energy due to the quadrupole-quadrupole interaction, by considering that the nematic order is due only to the attractive long range dispersion forces. Few assumptions have to be made: 1) the nematic molecules are hard spheres of radius $r_0/2$ and the single particle density is everywhere uniform ($\rho(r_1) = N =$ constant); 2) the quadrupole is in the center of the sphere; 3) the two-particle density $\rho(r_1, r_2)$ is assumed to be given by $\rho(r_1, r_2) = \rho(r_1)h(|\mathbf{r}_1 - \mathbf{r}_2|/r_0)$, where $h(X) = 0$ for $X < 1$ and $h(X) = 1$ for $X > 1$; 4) the impenetrability of the molecules is taken into account by means of another interaction (of contact) responsible for the Frank elastic constants of the nematic liquid crystal.

Since we assume perfect nematic order, i.e., $S = 1$, the molecular major axis $\mathbf{m}$ coincides with the nematic director $\mathbf{n}$. Hence, the molecular quadrupolar tensor is given by (1.9), that can be rewritten here as:

$$q_{ij} = q\left[n_i n_j - \frac{1}{3}\delta_{ij}\right],$$

where q is the molecular quadrupole. The bulk density of quadrupolar tensor is

$$Q_{ij} = Nq_{ij} = e\left(n_i n_j - \frac{1}{3}\delta_{ij}\right). \tag{3.1}$$

The quantity $e = Nq$ is the quadrupolar density. Its value, for usual NLCs, is of the order of the flexoelectric coefficient [18].

The electrical interaction between the quadrupoles $q_{ij} = q_{ij}(\mathbf{r})$ and $q'_{ij} = q_{ij}(\mathbf{r}')$, respectively located in $\mathbf{r}$ and in $\mathbf{r}'$, whose relative position is $\mathbf{R} = \mathbf{r}' - \mathbf{r}$, is given by [19]

$$g = \frac{1}{12}kq_{ij}\frac{\partial^2}{\partial x_i \partial x_j}\left\{\frac{(x_m - x'_m)q'_{mn}(x_n - x'_n)}{R^5}\right\}, \tag{3.2}$$

where k depends on the system of units used. By substituting the molecular quadrupolar tensor q_{ij} into Eq. (3.2) the quadrupolar interaction between two nematic molecules in $\mathbf{r}$ and in $\mathbf{r}'$, whose directors are $\mathbf{n} = \mathbf{n}(\mathbf{r})$ and $\mathbf{n}' = \mathbf{n}(\mathbf{r}')$ respectively, is given by [20]

$$\begin{aligned} g(\mathbf{n}, \mathbf{n}', \mathbf{R}) = \frac{kq^2}{12R^5} & \{1 + 2(\mathbf{n}\cdot\mathbf{n}')^2 - 20(\mathbf{n}\cdot\mathbf{u})(\mathbf{n}'\cdot\mathbf{u})(\mathbf{n}\cdot\mathbf{n}') \\ & -5[(\mathbf{n}\cdot\mathbf{u})^2 + (\mathbf{n}'\cdot\mathbf{u})^2] + 35(\mathbf{n}\cdot\mathbf{u})^2(\mathbf{n}'\cdot\mathbf{u})^2\}. \end{aligned} \tag{3.3}$$

The centers of the spherical molecules of radius $r_0/2$, indicated by $\mathbf{r}$ and $\mathbf{r}'$, are assumed to lie in a slab limited by two plane surfaces at $z = \pm d/2$ of a Cartesian reference frame. Then, the actual thickness of the nematic layer is $D = d + r_0$. However, from now on in this chapter, we will call d the thickness of the nematic layer. This parameter corresponds to the thickness of the distribution of quadrupoles. The problem is assumed as planar and one-dimensional with $\mathbf{n} = \mathbf{n}(z) = [\sin\theta(z), 0, \cos\theta(z)]$, where $\theta = \arccos(\mathbf{n}\cdot\mathbf{z})$, as before, is the tilt angle.

In this framework, the interaction energy between the quadrupole in $\mathbf{r} = (0, 0, z)$ and the one in $\mathbf{r}' = (x', y', z')$, for which $\mathbf{R} = (x', y', z' - z)$ can be easily evaluated by means of Eq. (3.3). It depends on the tilt angles of the interacting quadrupoles $\theta = \theta(z)$ and $\theta' = \theta(z')$ and on their relative position $\mathbf{R}$, viz $g = g(\theta, \theta'; x', y', z' - z)$. In the mean field approximation the energy density due to the quadrupolar interaction energy g is given by

$$f_q(\mathbf{r}) = \frac{1}{2}\int_\tau N^2 g(\mathbf{n},\mathbf{n}',\mathbf{R})d\tau', \tag{3.4}$$

where N is the molecular density that is assumed to be position independent. Due to assumption (3), the integration volume τ in Eq. (3.4) is the whole sample volume except for a sphere of radius r_0 around $\mathbf{r}$. In this case, Eq. (3.4) reads

$$f_q(z) = \frac{1}{2}\int_\tau N^2 g(\theta,\theta';x',y',z'-z)dx'dy'dz'. \tag{3.5}$$

It will be useful, for further considerations, to define the quantity

$$G(\theta,\theta';z'-z) = \frac{1}{2}\int_\Sigma N^2 g(\theta,\theta';x',y',z'-z)dx'dy', \tag{3.6}$$

that represents the contribution to the energy density in z due to a layer of thickness dz' at a distance $z'-z$. To evaluate $G(\theta,\theta';z'-z)$ we use a polar reference frame in the (x',y')-plane. Let be ρ and ϕ the polar coordinates of a point in the (x',y')-plane. A simple analysis shows that for $|z'-z|>r_0$, $0\le\rho<\infty$, and for $|z'-z|\le r_0$, $\sqrt{r_0^2-(z'-z)^2}\le\rho<\infty$. In the first case Σ is the whole (x',y')-plane, whereas in the second case Σ is the part of the (x',y')-plane outside the circle of radius $\rho_m=\sqrt{r_0^2-(z'-z)^2}$. In terms of $G(\theta,\theta';z'-z)$ the energy density given by Eq. (3.5) reads

$$f_q(z) = \int_{-d/2}^{d/2} G(\theta,\theta';z'-z)dz'. \tag{3.7}$$

A simple calculation shows that

$$G(\theta,\theta';z'-z) = \begin{cases} 0, & -|z'-z|\ge r_0, \\ \mathcal{H}(z'-z)T(\theta,\theta'), & |z'-z|\le r_0, \end{cases} \tag{3.8}$$

where the kernel $\mathcal{H}(z'-z)=\mathcal{H}(z-z')$ and the function $T(\theta,\theta')=T(\theta',\theta)$ are, respectively, defined as

$$\mathcal{H}(z'-z) = \frac{\pi k e^2}{96 r_0^3}\left[1-6\left(\frac{z'-z}{r_0}\right)^2+5\left(\frac{z'-z}{r_0}\right)^4\right], \tag{3.9}$$

and

$$T(\theta,\theta') = 3+19\cos^2\theta\cos^2\theta'-7(\cos^2\theta+\cos^2\theta')-4\sin(2\theta)\sin(2\theta'). \tag{3.10}$$

Equation (3.8) means that a compact layer of quadrupoles of thickness dz' does not produce any electrical effect outside the layer itself.

The total energy, per unit surface, of quadrupolar origin is given by

$$F_q = \int_{-d/2}^{d/2} f_q(z)dz. \tag{3.11}$$

The NLC sample is separated in three regions: the bulk, defined by $-d/2 + r_0 \le z \le d/2 - r_0$, and two surface layers, where $d/2 - r_0 \le |z| \le d/2$. Therefore, F_q can be decomposed as follows:

$$F_q = \int_{-d/2}^{-d/2+r_0} f_{qs}(z)dz + \int_{-d/2+r_0}^{d/2-r_0} f_{qb}(z)dz + \int_{d/2-r_0}^{d/2} f_{qs}(z)dz. \tag{3.12}$$

In (3.12), $f_{qs}(z)$ and $f_{qb}(z)$ are the bulk energy densities due to the quadrupolar interaction in the surface layers and in the bulk, respectively. Following a standard procedure, it is convenient to rewrite Eq. (3.12) as

$$F_q = \int_{-d/2}^{-d/2+r_0} [f_{qs}(z) - f_{qb}(z)]dz + \int_{-d/2}^{d/2} f_{qb}(z)dz + \int_{d/2-r_0}^{d/2} [f_{qs}(z) - f_{qb}(z)]dz, \tag{3.13}$$

i.e., as the sum of a bulk contribution plus two surface contributions, which are connected with the presence of the surface layers where the energy density is different from the bulk one. By introducing $\Delta f_q(z) = f_{qs}(z) - f_{qb}(z)$ and

$$\begin{aligned} F_s^{(-)} &= \int_{-d/2}^{-d/2+r_0} \Delta f_q(z)dz, \\ F_s^{(+)} &= \int_{d/2-r_0}^{d/2} \Delta f_q(z)dz, \quad \text{and} \\ F_{\text{bulk}} &= \int_{-d/2}^{d/2} f_{qb}(z)dz, \end{aligned} \tag{3.14}$$

it is possible to rewrite Eq. (3.13) as

$$F_q = F_s^{(-)} + F_{\text{bulk}} + F_s^{(+)}. \tag{3.15}$$

We shall consider in the following the energy of the bulk and of the surface layers separately.

In the bulk the energy density is given by Eq. (3.7) that, by taking into account Eqs. (3.8), can be rewritten as

$$f_{qb}(z) = \int_{z-r_0}^{z+r_0} \mathcal{H}(z' - z)T(\theta, \theta')dz', \tag{3.16}$$

that gives the energy density due to the quadrupolar interaction. If there are other interactions, the total energy density in the bulk is obtained, in a first approximation, adding all the contributions. The energy density given by Eq. (3.16) is *non-local*, because it depends on all the values of θ in the range $(z - r_0, z + r_0)$. To proceed further we have two possibilities. The first one is to reduce f_{qb} to a local form by means of some limit operation. The second one is to face the problem by taking into account the non-locality pointed out above. In Section 3.3, f_{qb} will be reduced to a local form, in the elastic approximation, whereas in Section 3.5 the non-local analysis will be presented.

We shall assume that, besides the quadrupolar interaction, there is also a short range interaction giving rise to an elastic energy density in the form (1.74), namely:

$$f_e = \frac{1}{2}(K_{11}\sin^2\theta + K_{33}\cos^2\theta)\left(\frac{d\theta}{dz}\right)^2 = \frac{1}{2}K_{33}(1 - \Delta\sin^2\theta)\left(\frac{d\theta}{dz}\right)^2, \quad (3.17)$$

where $\Delta = (K_{33} - K_{11})/K_{33}$ is the elastic anisotropy. In the one-constant approximation $K_{11} = K_{33} = K$ and $\Delta = 0$. In this framework, the Frank elastic energy density is given by

$$f_e = \frac{1}{2}K\left(\frac{d\theta}{dz}\right)^2, \quad (3.18)$$

that will be used in the following.

3.3 Quadrupolar energy density in the bulk: direct calculation

In this section we are interested in the calculation of the bulk free energy, $F_{\rm bulk}$, defined in Eq. (3.14). In the bulk, the energy density $f_{qb}(z)$, given by Eq. (3.16), depends only on the director angle in a very thin layer of molecular thickness $2r_0$. Then, if the director angle θ changes over a macroscopic length, the non-local energy density $f_{qb}(z)$ can be replaced by a local elastic expansion in the derivatives of θ at z. This local elastic description is only possible due to the planar character of the director distortion. Indeed, Eq. (3.16) is the direct consequence of the fact that two nematic layers of thickness dz and dz' having a uniform director orientation do not interact if $|z' - z| > r_0$ [see Eq. (3.8)]. To reduce f_{qb} to a local quantity we put $\theta' = \theta + \delta\theta(z, z')$. Since $|z' - z| \leq r_0$, which is a molecular dimension, $|\delta\theta(z, z')| \ll 1$. Consequently, $T(\theta, \theta')$ can be expanded in power series of $\delta\theta(z, z')$ as

$$T(\theta, \theta') = T_0(\theta) + T_1(\theta)\,\delta\theta + \frac{1}{2}T_2(\theta)\,(\delta\theta)^2, \quad (3.19)$$

where, as follows from Eq. (3.10),

$$T_0(\theta) = T(\theta, \theta) = (3 - 30\cos^2\theta + 35\cos^4\theta), \tag{3.20}$$

$$T_1(\theta) = \left(\frac{\partial T}{\partial \theta'}\right)_{\theta'=\theta} = 5\sin(2\theta)(3 - 7\cos^2\theta), \tag{3.21}$$

and

$$T_2(\theta) = \left(\frac{\partial^2 T}{\partial \theta'^2}\right)_{\theta'=\theta} = -2(7 - 65\cos^2\theta + 70\cos^4\theta). \tag{3.22}$$

In the elastic approximation $\delta\theta(z, z')$ can be written as

$$\delta\theta(z, z') = \frac{d\theta}{dz}(z' - z) + \frac{1}{2}\frac{d^2\theta}{dz^2}(z' - z)^2. \tag{3.23}$$

By means of Eqs. (3.19) and (3.23) we obtain for $T(\theta, \theta')$, when $z - r_0 \leq z' \leq z + r_0$, the approximate expression

$$T(\theta, \theta') = T_0(\theta) + (z' - z)\,T_1(\theta)\,\frac{d\theta}{dz} + \frac{1}{2}(z' - z)^2\left[T_1(\theta)\frac{d^2\theta}{dz^2} + T_2(\theta)\left(\frac{d\theta}{dz}\right)^2\right]. \tag{3.24}$$

The bulk energy density of quadrupolar origin, in the elastic approximation, is obtained by substituting Eq. (3.24) into Eq. (3.16). Simple calculations allow one to write

$$f_{qb} = f_0 + f_1 + f_{13} + f_2,$$

where

$$f_0 = T_0(\theta)\int_{z-r_0}^{z+r_0} \mathcal{H}(z' - z)dz' \tag{3.25}$$

is the homogeneous part of the energy density;

$$f_1 = T_1(\theta)\frac{d\theta}{dz}\int_{z-r_0}^{z+r_0}(z' - z)\mathcal{H}(z' - z)dz' \tag{3.26}$$

is the bulk term linear in the first spatial derivative of the tilt angle;

$$f_{13} = \frac{1}{2}T_1(\theta)\frac{d^2\theta}{dz^2}\int_{z-r_0}^{z+r_0}(z' - z)^2\mathcal{H}(z' - z)dz' \tag{3.27}$$

is the bulk term linear in the second order spatial derivative of the tilt angle, which recalls the splay-bend term [3], and finally

$$f_2 = \frac{1}{2}T_2(\theta)\left(\frac{d\theta}{dz}\right)^2\int_{z-r_0}^{z+r_0}(z' - z)^2\mathcal{H}(z' - z)dz', \tag{3.28}$$

which reminds the usual Frank elastic term, since it is quadratic in $d\theta/dz$. As follows from Eq. (3.9), which defines the kernel $\mathcal{H}(z'-z)$, in the bulk

$$\int_{z-r_0}^{z+r_0} \mathcal{H}(z'-z)dz' = \int_{z-r_0}^{z+r_0} (z'-z)\mathcal{H}(z'-z)dz' = 0, \tag{3.29}$$

and

$$\int_{z-r_0}^{z+r_0} (z'-z)^2\mathcal{H}(z'-z)dz' = -\frac{\pi k e^2}{315}. \tag{3.30}$$

Hence, in the bulk, $f_0 = f_1 = 0$, whereas, by taking into account Eqs. (3.20), (3.21), and (3.22), f_{13} and f_2 are, respectively,

$$f_{13} = \frac{1}{2}\left(-\frac{\pi k e^2}{315}\right) 5\sin(2\theta)(3-7\cos^2\theta)\frac{d^2\theta}{dz^2} \tag{3.31}$$

and

$$f_2 = -\frac{1}{2}\left(-\frac{\pi k e^2}{315}\right) 2(7-65\cos^2\theta+70\cos^4\theta)\left(\frac{d\theta}{dz}\right)^2. \tag{3.32}$$

The total bulk energy density is reduced to $f_{qb} = f_{13}+f_2$, that can be written in the form $f_{qb} = f_s + f_b$, where f_s and f_b are, respectively, the surface-like and bulk contributions, given by

$$f_s = \frac{1}{2}\left(-\frac{\pi k e^2}{315}\right)\frac{d}{dz}\left[5\sin(2\theta)(3-7\cos^2\theta)\frac{d\theta}{dz}\right], \tag{3.33}$$

and

$$f_b = \frac{1}{2}\left(-\frac{\pi k e^2}{315}\right)[16-35\sin^2(2\theta)]\left(\frac{d\theta}{dz}\right)^2. \tag{3.34}$$

Hence, the quadrupolar interaction gives rise to a bulk energy density that can be separated in surface-like and bulk contributions. In the International System of units, SI, where $k = 1/(4\pi\epsilon_0)$, Eqs. (3.33) and (3.34) become, respectively,

$$f_s = -\frac{e^2}{504\epsilon_0}\frac{d}{dz}\left[\sin(2\theta)(3-7\cos^2\theta)\frac{d\theta}{dz}\right] \tag{3.35}$$

and

$$f_b = -\frac{2e^2}{315\epsilon_0}\left(1-\frac{35}{16}\sin^2(2\theta)\right)\left(\frac{d\theta}{dz}\right)^2. \tag{3.36}$$

According to Eq. (3.15), the bulk contribution to the total energy F_q is obtained by integrating the bulk free energy density $f_{qb} = f_s + f_b$ over the whole interval $[-d/2, d/2]$. The first contribution, after integration, reduces to a surface energy density, which depends on $d\theta/dz$, of the kind

$$F_{sb}^{(-)} = \frac{e^2}{504\epsilon_0} \sin(2\theta)(3 - 7\cos^2\theta)\frac{d\theta}{dz}, \tag{3.37}$$

for $z = -d/2$, and to a similar contribution with opposite sign for $z = d/2$. The subscript sb means that this surface contribution to the surface energy density is, actually, a bulk term. By comparing expression (3.37) with the K_{13} elastic term [3], it is possible to define an "effective" K_{13}^e elastic constant as

$$K_{13}^e(\theta) = \frac{e^2}{252\epsilon_0}(3 - 7\cos^2\theta). \tag{3.38}$$

This splay-bend elastic constant is θ-dependent. Its value lies in the range

$$-e^2/63\epsilon_0 \leq K_{13}^e(\theta) \leq e^2/84\epsilon_0. \tag{3.39}$$

The bulk contribution, proportional to $(d\theta/dz)^2$, is given by Eq. (3.36). From this expression we deduce that the relevant elastic constant is θ-dependent and given by

$$K_b(\theta) = -\frac{4e^2}{315\epsilon_0}\left(1 - \frac{35}{16}\sin^2(2\theta)\right). \tag{3.40}$$

Its value lies in the range

$$-4e^2/315\epsilon_0 \leq K_b(\theta) \leq 19e^2/315\epsilon_0. \tag{3.41}$$

Note that in Eq. (3.40) the quadrupolar elastic constant depends on the tilt angle via $\sin^2(2\theta)$. This angular dependence is similar to the one already discussed for the influence of the flexoelectric polarization on the elastic properties of NLCs [21]. On the contrary, the usual elastic constant in the Frank expression, given by Eq. (3.17), depends on $\sin^2\theta$. This is a peculiarity of the quadrupolar interaction because the bulk energy density $f_{qb} = f_s + f_b$ does not follow from an expansion of the quadrupolar interaction energy of Eq. (3.3) in terms of $\partial n_i/\partial x_j$ and of $\partial^2 n_i/\partial x_j \partial x_k$, as usually done in the elastic theory. In fact this expansion does not converge, due to the R^{-5} of the quadrupolar interaction. A local expansion of the quadrupolar interaction is possible only for samples of slab shape.

For nematic orientations near to the homeotropic ($\theta \sim 0$) or planar ($\theta \sim \pi/2$) ones, $K_b(0) = K_b(\pi/2) = K_b(u)$ is negative and given by $K_b(u) = -4e^2/315\epsilon_0$. In the CGS system of units $\epsilon_0 = 1/4\pi$, and hence $K_b(u) = (16\pi/315)e^2 \sim 0.16e^2$. Since for usual NLCs $e^2 \sim K$, where K is the average Frank elastic constant, we can deduce that the quadrupolar contribution to the Frank elastic constant is rather small (nearly one order of magnitude smaller than the contribution due to the interactions of steric origin).

We note that $K_b(\theta)$ is negative near to the homeotropic and planar orientation, and positive for a homogeneous orientation close to $\theta = \pi/4$, as follows from Eq. (3.36). In fact, from Eq. (3.36) one can see that the quadrupolar interaction introduces a large elastic anisotropy given by $(e^2/36\epsilon_0)\sin^2(2\theta)$. It is similar to the one already discussed for the flexoelectric polarization [21]. From Eqs. (3.39) one has $-1 \leq K_{13}^e/K_b(u) \leq 1$. It follows that K_{13} is of the same order of magnitude of $K_b(u)$.

3.4 Quadrupolar energy density in the bulk: electrostatics approach

The aim of this section is to reobtain Eqs. (3.35) and (3.36) giving, respectively, the surface-like and bulk energy density due to the quadrupolar interaction by means of considerations based on the electrostatics theory. In this section we always use the SI units where $k = 1/(4\pi\epsilon_0)$. As is well known from elementary electrostatics, the electrostatic energy density in the bulk, connected with a continuous distribution of quadrupoles, is given by [19]

$$f_{qb} = -\frac{1}{12}\mathcal{Q}_{ij}\mathcal{E}_{ij}, \tag{3.42}$$

where, for convenience, $\mathcal{E}_{ij} = \partial E_i/\partial x_j$ denote the spatial derivatives of the total field acting on the quadrupole. The elements of the bulk density of quadrupolar tensor $\mathcal{Q}$ are given by Eq. (3.1), where $\mathbf{n} = [\sin\theta(z), 0, \cos\theta(z)]$. Consequently,

$$\mathcal{Q} = e\begin{pmatrix} \sin^2\theta - \frac{1}{3} & 0 & \sin\theta\cos\theta \\ 0 & -\frac{1}{3} & 0 \\ \sin\theta\cos\theta & 0 & \cos^2\theta - \frac{1}{3} \end{pmatrix}. \tag{3.43}$$

This implies that $\mathcal{Q}_{yy}$ is position independent. Furthermore, since $\mathbf{E}$ is an electrostatic field, $\nabla \times \mathbf{E} = 0$, and hence $\mathcal{E}_{ij} = \mathcal{E}_{ji}$. According to the model used in Section 3.2, the electric quadrupole is at the center of a sphere and, thus, the other quadrupoles are uniformly distributed out of a sphere of radius r_0. Consequently, the electric field acting on the quadrupole at the center of the sphere is only due to the uniform distribution outside the sphere. To evaluate $\mathcal{E}_{ij}$ we use the Clausius-Mossotti method [22]. According to this technique $\mathcal{E}_{ij}$ is given by

$$\mathcal{E}_{ij} = \mathcal{E}_{ij}^{(c)} - \mathcal{E}_{ij}^{(s)}, \tag{3.44}$$

where $\mathcal{E}_{ij}^{(c)}$ is due to the continuum distribution of quadrupoles (average field), and $\mathcal{E}_{ij}^{(s)}$ is due to the spherical quadrupole under consideration (spherical cavity field). To evaluate $\mathcal{E}_{ij}^{(c)}$ we have only to take into account that for a continuous quadrupolar material the electric displacement $\mathbf{D}^{(c)}$ is

$$D_i^{(c)} = \frac{1}{4\pi k} E_i^{(c)} - \frac{1}{6} \frac{\partial \mathcal{Q}_{ij}}{\partial x_j}, \tag{3.45}$$

if the medium is not polarizable and non–ferroelectric. By assuming the NLC as a perfect insulator we have $\nabla \cdot \mathbf{D}^{(c)} = 0$, that in the slab geometry reduces to $dD_z^{(c)}/dz = 0$. Since in the present case $\mathcal{Q}_{ij} = \mathcal{Q}_{ij}(z)$, from Eqs. (3.45) we obtain

$$\mathcal{E}_{zz}^{(c)} = \frac{2\pi k}{3} \frac{d^2 \mathcal{Q}_{zz}}{dz^2}. \tag{3.46}$$

Let us now consider the spatial gradients of the electric field $\mathbf{E}^{(s)}$, evaluated in the center of the sphere, created by the quadrupolar sphere. The electrical potential created in a point $\mathbf{r}$ by the sphere is

$$V^{(s)}(\mathbf{r}) = \frac{1}{6} k \int_{\tau_0} \mathcal{Q}_{ij}(\mathbf{r}') \frac{\partial^2}{\partial x_i' \partial x_j'} \left(\frac{1}{R} \right) d\tau', \tag{3.47}$$

where τ_0 is the volume of the sphere, $\mathbf{r}'$ a generic point of the sphere where the quadrupolar tensor density is $\mathcal{Q}_{ij}(\mathbf{r}')$, $d\tau'$ is a volume element around $\mathbf{r}'$ and, as before, $\mathbf{R} = \mathbf{r} - \mathbf{r}'$. By means of simple calculations Eq. (3.47) can be rewritten as $V^{(s)}(\mathbf{r}) = V_1(\mathbf{r}) + V_2(\mathbf{r}) + V_3(\mathbf{r})$, where

$$V_1(\mathbf{r}) = \frac{1}{6} k \int_{\tau_0} \frac{\partial^2 \mathcal{Q}_{ij}(\mathbf{r}')}{\partial x_i' \partial x_j'} \frac{1}{R} d\tau', \tag{3.48}$$

$$V_2(\mathbf{r}) = \frac{1}{6} k \int_{\Sigma_0} N_i \mathcal{Q}_{ij}(\mathbf{r}') \frac{\partial}{\partial x_j'} \left(\frac{1}{R} \right) d\Sigma', \tag{3.49}$$

and

$$V_3(\mathbf{r}) = \frac{1}{6} k \int_{\Sigma_0} -N_i \frac{\partial \mathcal{Q}_{ij}(\mathbf{r}')}{\partial x_j'} \frac{1}{R} d\Sigma'. \tag{3.50}$$

In Eqs. (3.49) and (3.50) Σ_0 is the surface of the sphere and $\mathbf{N}$ the geometrical normal to the surface of the sphere, directed outward. As is evident from Eq. (3.48), $V_1(\mathbf{r})$ is the electrical potential created in $\mathbf{r}$ by a bulk charge density given by

$$\rho(\mathbf{r}') = \frac{1}{6} \frac{\partial^2 \mathcal{Q}_{ij}(\mathbf{r}')}{\partial x_i' \partial x_j'}. \tag{3.51}$$

On the contrary, from Eq. (3.49) we deduce that $V_2(\mathbf{r})$ is the electrical potential created in $\mathbf{r}$ by a surface distribution of dipoles, whose surface density is

$$P_j(\mathbf{r}') = \frac{1}{6} N_i \mathcal{Q}_{ij}(\mathbf{r}'). \tag{3.52}$$

Finally, Eq. (3.50) shows that $V_3(\mathbf{r})$ is the electrical potential created in $\mathbf{r}$ by a surface charge density given by

$$\sigma(\mathbf{r}') = -\frac{1}{6} N_i \frac{\partial \mathcal{Q}_{ij}(\mathbf{r}')}{\partial x'_j}. \tag{3.53}$$

In Eqs. (3.52) and (3.53), $\mathbf{r}'$ is a point of Σ_0 where the geometrical normal is $\mathbf{N}$. Using a polar reference frame, with the z-axis as polar axis, we have that for a point on Σ_0, $\mathbf{r}' = r_0\mathbf{N}$, where $\mathbf{N} = (\sin\Theta\cos\Phi, \sin\Theta\sin\Phi, \cos\Theta)$, Θ and Φ represents the angular coordinates of a point on the sphere Σ_0. Furthermore, as usual in spherical coordinates, $d\Sigma' = r_0^2 \sin\Theta d\Theta d\Phi = -r_0^2 d(\cos\Theta)d\Phi$.

The analysis presented above shows that a body of volume τ_0 limited by a surface Σ_0, whose density of quadrupolar tensor is $\mathcal{Q}_{ij}(\mathbf{r}')$, is equivalent to a bulk density of charges and to a surface distribution of dipoles and of charges [23]. Because $\mathcal{Q}_{ij} = \mathcal{Q}_{ij}(z)$ the general equations giving $\rho(\mathbf{r}')$ and $\sigma(\mathbf{r}')$ become, respectively,

$$\rho(\mathbf{r}') = \frac{d^2\mathcal{Q}_{zz}(z')}{dz'^2} \quad \text{and} \quad \sigma(\mathbf{r}') = -N_i \frac{d\mathcal{Q}_{iz}(z')}{dz'}. \tag{3.54}$$

It follows that the electric field due to a sphere of quadrupolar material at an inner point $\mathbf{r} = (x, y, z)$ is given by $\mathbf{E}^{(s)}(\mathbf{r}) = \mathbf{E}_1(\mathbf{r}) + \mathbf{E}_2(\mathbf{r}) + \mathbf{E}_3(\mathbf{r})$, where, as follows from the discussion reported above,

$$\mathbf{E}_1(\mathbf{r}) = -k \int_{\tau_0} \rho(\mathbf{r}') \nabla \left(\frac{1}{R}\right) d\tau', \tag{3.55}$$

$$\mathbf{E}_2(\mathbf{r}) = -k \int_{\Sigma_0} \frac{R^2\mathbf{P}(\mathbf{r}') - 3[\mathbf{R}\cdot\mathbf{P}(\mathbf{r}')]\mathbf{R}}{R^5} d\Sigma', \tag{3.56}$$

and

$$\mathbf{E}_3(\mathbf{r}) = -k \int_{\Sigma_0} \sigma(\mathbf{r}') \nabla \left(\frac{1}{R}\right) d\Sigma', \tag{3.57}$$

where $\rho(\mathbf{r}')$, $\mathbf{P}(\mathbf{r}')$ and $\sigma(\mathbf{r}')$ are given by Eqs. (3.51), (3.52), and (3.53).

To evaluate f_{qb}, given by Eq. (3.42), we have now to calculate $\mathcal{E}_{ij}^{(s)}$ in the center of the sphere. After that, by means of Eq. (3.44), it is possible to calculate $\mathcal{E}_{ij}$ acting on the sphere of quadrupolar material, in the limit of $r_0 \to 0$. With this aim in mind, and using the same approximation as before [see Eq. (3.23)], $\mathcal{Q}_{ij}(z')$ can be expanded in power series of z' up to the second order. Hence

$$\mathcal{Q}_{ij}(z') = \mathcal{A}_{ij} + \mathcal{B}_{ij} z' + \frac{1}{2}\mathcal{C}_{ij} z'^2, \tag{3.58}$$

where

$$\mathcal{A}_{ij} = \mathcal{Q}_{ij}(0), \quad \mathcal{B}_{ij} = \left\{ \frac{d\mathcal{Q}_{ij}(z')}{dz'} \right\}_0, \quad \text{and} \quad \mathcal{C}_{ij} = \left\{ \frac{d^2\mathcal{Q}_{ij}(z')}{dz'^2} \right\}_0. \tag{3.59}$$

The matrices $\mathcal{A}$, $\mathcal{B}$ and $\mathcal{C}$ of elements $\mathcal{A}_{ij}$, $\mathcal{B}_{ij}$ and $\mathcal{C}_{ij}$–as follows from Eq. (3.43)–are given by

$$\mathcal{A} = e \begin{pmatrix} \sin^2\theta - \frac{1}{3} & 0 & \sin\theta\cos\theta \\ 0 & -\frac{1}{3} & 0 \\ \sin\theta\cos\theta & 0 & \cos^2\theta - \frac{1}{3} \end{pmatrix} \tag{3.60}$$

$$\mathcal{B} = e \begin{pmatrix} \sin(2\theta) & 0 & \cos(2\theta) \\ 0 & 0 & 0 \\ \cos(2\theta) & 0 & -\sin(2\theta) \end{pmatrix} \frac{d\theta}{dz}, \tag{3.61}$$

and

$$\mathcal{C} = 2e \begin{pmatrix} \cos(2\theta) & 0 & -\sin(2\theta) \\ 0 & 0 & 0 \\ -\sin(2\theta) & 0 & -\cos(2\theta) \end{pmatrix} \left(\frac{d\theta}{dz}\right)^2 + e \begin{pmatrix} \sin(2\theta) & 0 & \cos(2\theta) \\ 0 & 0 & 0 \\ \cos(2\theta) & 0 & -\sin(2\theta) \end{pmatrix} \frac{d^2\theta}{dz^2}, \tag{3.62}$$

where all the quantities are evaluated in the center of the sphere. Consequently, $\rho(\mathbf{r}')$ and $\sigma(\mathbf{r}')$, given by Eqs. (3.54), are

$$\rho(\mathbf{r}') = \frac{1}{6}\mathcal{C}_{33} = \text{ constant} \quad \text{and} \quad \sigma(\mathbf{r}') = -\frac{1}{6}N_\alpha \left(\mathcal{B}_{\alpha z} + \mathcal{C}_{\alpha z} z'\right). \tag{3.63}$$

Let us now consider $\mathbf{E}_1(\mathbf{r})$, given by Eq. (3.55). Since $\rho(z')$ in the elastic approximation is position independent, as follows from Eq. (3.63), $\mathbf{E}_1(\mathbf{r}) = (2\pi k/9)\mathcal{C}_{33}\mathbf{r}$. Therefore,

$$\mathcal{E}_1(0) = \frac{2\pi k}{9}\mathcal{C}_{33} \begin{pmatrix} 1 & 0 & 0 \\ 0 & 1 & 0 \\ 0 & 0 & 1 \end{pmatrix}. \tag{3.64}$$

To evaluate $\mathcal{E}_2(0)$ we have to take into account that, by Eqs. (3.43) and (3.52), it follows that

$$P_x(\mathbf{r}') = \frac{1}{6}[\mathcal{Q}_{xx}(z')\sin\Theta\cos\Phi + \mathcal{Q}_{xz}(z')\cos\Theta],$$
$$P_y(\mathbf{r}') = \frac{1}{6}\mathcal{Q}_{yy}\sin\Theta\sin\Phi,$$
$$P_z(\mathbf{r}') = \frac{1}{6}[\mathcal{Q}_{xz}(z')\sin\Theta\cos\Phi + \mathcal{Q}_{zz}(z')\cos\Theta], \tag{3.65}$$

where, on the surface of the sphere, $z' = r_0 \cos\Theta$. Furthermore, in Eq. (3.56), $\mathbf{R} = (x - r_0 \sin\Theta\cos\Phi, y - r_0\sin\Theta\sin\Phi, z - r_0\cos\Theta)$, as follows from the discussion reported above. By substituting Eqs. (3.65), with $\mathcal{Q}_{ij}(z')$ given by Eq. (3.58), into Eq. (3.56) after simple, but tedious calculations, we obtain

$$\mathcal{E}_2(0) = \frac{8\pi k}{35}\begin{pmatrix} -3(\mathcal{C}_{11}+\mathcal{C}_{33}) & 0 & \mathcal{C}_{13} \\ 0 & -(\mathcal{C}_{11}+3\mathcal{C}_{33}) & 0 \\ \mathcal{C}_{13} & 0 & 2(2\mathcal{C}_{11}+3\mathcal{C}_{33}) \end{pmatrix}. \tag{3.66}$$

From Eq. (3.66) it follows that $\nabla\cdot\mathbf{E}_2 = 0$, as expected.

Let us consider, finally, the spatial derivatives of the field $\mathbf{E}_3$ given by Eq. (3.57), in the center of the sphere, when $\sigma(\mathbf{r}')$ is defined by Eq. (3.63). We have

$$\mathcal{E}_3(0) = \frac{4\pi k}{15}\begin{pmatrix} 2\mathcal{C}_{33} & 0 & -3\mathcal{C}_{13} \\ 0 & 2\mathcal{C}_{33} & 0 \\ -3\mathcal{C}_{13} & 0 & -4\mathcal{C}_{33} \end{pmatrix}. \tag{3.67}$$

From Eq. (3.67) it follows again that $\nabla\cdot\mathbf{E}_3 = 0$. To obtain the tensor $\mathcal{E}$ entering into Eq. (3.42), we have just to remember that, in the elastic approximation, Eq. (3.46) reads

$$\mathcal{E}^{(c)}(0) = \frac{2\pi k}{3}\mathcal{C}_{33}\begin{pmatrix} 0 & 0 & 0 \\ 0 & 0 & 0 \\ 0 & 0 & 1 \end{pmatrix}. \tag{3.68}$$

By means of Eqs. (3.64), (3.66), and (3.67) we evaluate the spatial derivatives of the electric field due to the sphere of quadrupolar material in its center. After that, using Eq. (3.68) and Eq. (3.44), we obtain the tensor $\mathcal{E}$, of elements $\partial E_i/\partial x_j$. By substituting this result into Eq. (3.42), rewritten as $f_{qb} = -(1/12)\mathcal{A}_{ij}\mathcal{E}_{ij}$, and taking into account Eqs. (3.60), (3.61) and (3.62), we reobtain $f_{qb} = f_s + f_b$, where f_s and f_b are still given, respectively, by Eqs. (3.35) and (3.36).

In this calculation the field created by the sphere of quadrupolar material plays a fundamental role. If one identifies the field acting on the quadrupole with the field due to the continuum distribution of quadrupoles, given by

Eq. (3.46), one then obtains different results. In particular, one concludes that the quadrupolar interaction is responsible for an elastic anisotropy, which depends on the orientation, but one is unable to show that its contribution to the Frank elastic constant can be negative.

3.5 Non-local analysis

In the previous sections we have reduced the non-local quantity f_{qb}, given by Eq. (3.16), to a local bulk quantity f_b plus a surface-like contribution f_s, by means of a limit operation $r_0 \to 0$. Now, the problem can be analyzed by taking into account explicitly the non-locality of f_{qb}. As above we limit the analysis to the bulk, i.e., for $-d/2 + r_0 \leq z \leq d/2 + r_0$.

The total energy, per unit surface area, of quadrupolar origin is given by

$$F_q = \int_{-d/2}^{d/2} f_q dz = \int_{-d/2}^{d/2} \int_{-d/2}^{d/2} G(\theta, \theta', z' - z) dz' dz, \tag{3.69}$$

as follows from Eq. (3.7). The Frank elastic energy density, in the one-constant approximation, is given by Eq. (3.18). Hence, the total energy per unit surface area is

$$F = \int_{-d/2}^{d/2} [f_e + f_q] dz = \int_{-d/2}^{d/2} \left[\frac{1}{2} K \left(\frac{d\theta}{dz} \right)^2 + \int_{-d/2}^{d/2} G(\theta, \theta', z' - z) dz' \right]. \tag{3.70}$$

Functional (3.70) is of the kind

$$F = \int_{-d/2}^{d/2} \{ \mathcal{M}(d\theta/dz) + \int_{-d/2}^{d/2} \mathcal{N}[\theta(z), \theta(z'); |z' - z|] dz' \} dz, \tag{3.71}$$

where

$$\mathcal{N}[\theta(z), \theta(z'); |z' - z|] = \mathcal{N}[\theta(z'), \theta(z); |z' - z|].$$

The function minimizing F, given by Eq. (3.71), is the solution of the integro-differential equation [24]

$$-\frac{d}{dz} \frac{d\mathcal{M}}{d(d\theta/dz)} + 2 \int_{-d/2}^{d/2} \frac{\partial \mathcal{N}}{\partial \theta(z)} dz' = 0. \tag{3.72}$$

By taking into account Eqs. (3.8) and (3.18), Eq. (3.72) yields in the bulk

$$-K \frac{d^2\theta}{dz^2} + 2 \int_{z-r_0}^{z+r_0} \mathcal{H}(z' - z) \frac{\partial T(\theta, \theta')}{\partial \theta} dz' = 0, \tag{3.73}$$

which is the equilibrium equation for the non-local problem under consideration. It is an integro-differential equation, whose integral part is due to the non-local quadrupolar interaction. On the contrary, the differential part is due to the short range interaction, responsible for the usual elastic energy density, which admits a local description. Let us now assume that r_0 is a very small quantity. This implies that the nematic molecules may be assumed as practically dimensionless. Physically, this means that we are interested in spatial variations of the observables entering into the problem occurring over lengths very large with respect to r_0. We define

$$R(\theta,\theta') = \frac{\partial T(\theta,\theta')}{\partial \theta}, \tag{3.74}$$

and, as we have already done in Section 3.3, we assume that $|\delta\theta(z,z')| = |\theta' - \theta| \ll 1$, and furthermore that $\delta\theta(z,z')$ can be expanded in power series of $z' - z$, as shown in Eq. (3.23). In this framework

$$R(\theta,\theta') = R_0(\theta) + (z'-z)R_1(\theta)\frac{d\theta}{dz} + \frac{1}{2}(z'-z)^2\left[R_1(\theta)\frac{d^2\theta}{dz^2} + R_2(\theta)\left(\frac{d\theta}{dz}\right)^2\right], \tag{3.75}$$

where, as follows from Eqs. (3.10) and (3.74),

$$R_0(\theta) = R(\theta,\theta) = 5\sin(2\theta)(3 - 7\cos^2\theta) \tag{3.76}$$

and

$$R_1(\theta) = \left(\frac{\partial R}{\partial \theta'}\right)_{\theta'=\theta} = 35\sin^2(2\theta) - 16, \quad R_2(\theta) = \left(\frac{\partial^2 R}{\partial \theta'^2}\right)_{\theta'=\theta} = 35\sin(4\theta). \tag{3.77}$$

Substitution of Eq. (3.75) into Eq. (3.73), taking into account Eqs. (3.29) and (3.30), yields

$$\left(K - \frac{16}{315}\pi k e^2\right)\frac{d^2\theta}{dz^2} + \frac{\pi}{9}ke^2\left[\sin^2(2\theta)\frac{d^2\theta}{dz^2} + \sin(4\theta)\left(\frac{d\theta}{dz}\right)^2\right] = 0, \tag{3.78}$$

which is equivalent to

$$\frac{1}{2}\left[\left(K - \frac{16}{315}\pi k e^2\right) + \frac{\pi}{9}ke^2\sin^2(2\theta)\right]\left(\frac{d\theta}{dz}\right)^2 = \text{constant}. \tag{3.79}$$

This bulk equilibrium equation shows again that the quadrupolar interaction reduces the Frank elastic constant connected to short range interactions by a quantity of $(16/315)\pi k e^2 = 4e^2/315\epsilon_0$. It gives rise, furthermore, to an elastic anisotropy which depends on the nematic orientation, whose maximum value is $\pi k e^2/9 = e^2/36\epsilon_0$. These results agree with the ones obtained in Section 3.3. However this non-local bulk analysis does not give any information about the surface–like contribution, because it is based on the bulk integro-differential equilibrium equation, Eq. (3.73), in which surface contributions are absent.

3.6 Interfacial contributions to the surface free energy

We have shown that the bulk free energy density in Eq. (3.16) can be separated into a bulk elastic term, f_b, and a surface-like term, f_s, which is equivalent to a surface energy density linear in $d\theta/dz$ [Eq. (3.37)]. Close to the interfaces, in two thin interfacial layers of thickness r_0, the local free energy density is no longer given by $f_{qb}(z)$, of Eq. (3.16), as already underlined. Thus, there is an excess of interfacial free energy which gives a further contribution to the surface free energy, that we have indicated above by $F_s^{(-)}$ and $F_s^{(+)}$. Here we are interested in calculating these specific interfacial contributions to the surface free energy. Let us consider, for instance, the upper interfacial layer of the nematic slab $(d/2 - r_0 < z < d/2)$ and indicate simply by F_s the quantity $F_s^{(+)}$. Using Eqs. (3.7) to (3.8), we find that the local free energy density in the interfacial layer is

$$f_{qs}(z) = \int_{z-r_0}^{d/2} \mathcal{H}(z'-z)T(\theta,\theta')dz'. \tag{3.80}$$

The excess of interfacial free energy density Δf_q, as discussed in Section 3.2, is obtained by subtracting from $f_{qs}(z)$ the bulk free energy density $f_{qb}(z)$. We get, from Eq. (3.16) and Eq. (3.80),

$$\Delta f_q(z) = -\int_{d/2}^{z+r_0} \mathcal{H}(z'-z)T(\theta,\theta')dz'. \tag{3.81}$$

By operating as in Section 3.3, Δf_q can be written as

$$\Delta f_q(z) = \Delta f_0 + \Delta f_1 + \Delta f_{13} + \Delta f_2,$$

where

$$\Delta f_0 = -T_0(\theta)\int_{d/2}^{z+r_0} \mathcal{H}(z'-z)dz', \tag{3.82}$$

$$\Delta f_1 = -T_1(\theta)\frac{d\theta}{dz}\int_{d/2}^{z+r_0} (z'-z)\mathcal{H}(z'-z)dz', \tag{3.83}$$

$$\Delta f_{13} = -\frac{1}{2} T_1(\theta) \frac{d^2\theta}{dz^2} \int_{d/2}^{z+r_0} (z'-z)^2 \mathcal{H}(z'-z) dz', \tag{3.84}$$

and

$$\Delta f_2 = -\frac{1}{2} T_2(\theta) \left(\frac{d\theta}{dz}\right)^2 \int_{d/2}^{z+r_0} (z'-z)^2 \mathcal{H}(z'-z) dz'. \tag{3.85}$$

These are the direct equivalent of the bulk elastic terms in Eqs. (3.25) to (3.28). Note that, differently from the bulk case, Δf_0 and Δf_1 are now different from zero. According to the Gibbs theory of the interfacial phenomena, the surface free energy density F_s is obtained integrating Δf_q over the thin interfacial layer of thickness r_0 ($d/2 - r_0 < z < d/2$). Δf_{13} and Δf_2 are of the same order of magnitude as the bulk terms f_{13} and f_2. Then, the integral over the thin interfacial layer of thickness r_0 vanishes in the limit $r_0 \to 0$. Therefore, the surface free energy density due to the interfacial contributions is reduced to

$$F_s = F_0 + F_1, \tag{3.86}$$

where

$$F_0 = \int_{d/2-r_0}^{d/2} \Delta f_0 dz = -\int_{d/2-r_0}^{d/2} T_0(\theta) dz \int_{d/2}^{z+r_0} \mathcal{H}(z'-z) dz' \tag{3.87}$$

and

$$F_1 = \int_{d/2-r_0}^{d/2} \Delta f_1 dz = -\int_{d/2-r_0}^{d/2} T_1(\theta) \frac{d\theta}{dz} dz \int_{d/2}^{z+r_0} (z'-z) \mathcal{H}(z'-z) dz', \tag{3.88}$$

where θ is a function of z. In the thin interfacial layer, we can approximate the $\theta(z)$-function with the truncated expansion

$$\theta(z) = \theta_s + \left(\frac{d\theta}{dz}\right)_{z=d/2} [z - (d/2)], \tag{3.89}$$

where θ_s is the value of function $\theta(z)$ at $z = d/2$. Correspondingly, functions $T_0(\theta)$ and $T_1(\theta)$ can be approximated by

$$T_0(\theta) = T_0(\theta_s) + \left(\frac{dT_0}{d\theta}\right)_{\theta=\theta_s} \left(\frac{d\theta}{dz}\right)_{z=d/2} [z - (d/2)] \quad \text{and} \quad T_1(\theta) = T_1(\theta_s). \tag{3.90}$$

It can be easily verified that higher expansion terms can be disregarded in Eqs. (3.89) and (3.90) because they lead to surface energy contributions which vanish in the limit $r_0 \to 0$. By substituting Eqs. (3.90) into Eq. (3.87), we finally obtain the expression of the surface free energy in terms of the surface director angle and its first derivative. We find $F_s = F_0^0 + F_0^1 + F_1^0$, where

$$F_0^0 = -T_0(\theta_s) \int_{d/2-r_0}^{d/2} dz \int_{d/2}^{z+r_0} \mathcal{H}(z'-z)dz', \tag{3.91}$$

$$F_0^1 = -\left(\frac{dT_0}{d\theta}\right)_{\theta=\theta_s} \left(\frac{d\theta}{dz}\right)_{z=d/2} \int_{d/2-r_0}^{d/2} [z-(d/2)]dz \int_{d/2}^{z+r_0} \mathcal{H}(z'-z)dz', \tag{3.92}$$

and

$$F_1^0 = -T_1(\theta_s) \left(\frac{d\theta}{dz}\right)_{z=d/2} \int_{d/2-r_0}^{d/2} dz \int_{d/2}^{z+r_0} (z'-z)\mathcal{H}(z'-z)dz'. \tag{3.93}$$

In Eq. (3.91), F_0^0 represents the anchoring energy function that depends only on the surface director angle θ_s. In Eq. (3.92), F_0^1 is a linear elastic contribution coming from the homogeneous term in Eq. (3.87), whereas F_1^0, in Eq. (3.93), is another linear elastic term coming from F_1 in Eq. (3.88). By using the definition of $T_0(\theta)$ and $T_1(\theta)$, given in Eqs. (3.20) and (3.21), we get $T_1(\theta) = (1/2)dT_0/d\theta$. Thus, it is easy to show that $F_1^0 = -F_0^1$, as predicted by means of general considerations based on the symmetry of the nematic phase [25]. Therefore, the only non-vanishing contribution is the anchoring energy F_0^0. By taking into account the definitions of functions $T_0(\theta)$, given by Eq. (3.20), and $\mathcal{H}(z'-z)$, given by Eq. (3.9), Eq. (3.91) becomes

$$F_s = F_0^0 = \frac{ke^2\pi}{(24)^2 r_0}(3 - 30\cos^2\theta_s + 35\cos^4\theta_s). \tag{3.94}$$

This expression shows that the intrinsic anchoring energy of quadrupolar origin is characterized by an easy axis forming with the z-axis the angle $\theta_s = \arccos\sqrt{3/7}$. The extrapolation length connected with this anchoring energy is of the order of $10^2 \times r_0$. These results agree with the ones reported in Ref. [10].

In conclusion, the total surface energy due to the quadrupolar interaction is the sum of two contributions, that we have indicated by F_s and F_{sb}. The first one is a real interfacial term, having the form of a classical anchoring energy. It is given by Eq. (3.94). The second contribution is a surface-like elastic term, which comes from the bulk energy density. It is given by Eq. (3.37).

The analysis presented above shows that the quadrupolar energy density is a non-local quantity, in the sense that it depends on the nematic deformation in a layer, not only in a point. The reduction of this non-local energy density to a local one can be performed by means of a power expansion in terms of the spatial derivatives of the tilt angle.

Conclusions

The main results can be summarized as follows. In the bulk the quadrupolar energy density can be separated into two parts: a bulk and a surface-like contribution. The bulk contribution is characterized by an elastic constant $K_b(\theta)$ which depends on the nematic orientation. It is negative for nematic orientations close to the homeotropic or planar orientations, and positive for $\theta \sim \pi/4$. Its maximum value is of the order of $K/10$, where K is the detectable Frank elastic constant. The angular dependence of the bulk quadrupolar elastic constant is different from the usual elastic constant. This difference is connected with the circumstance that the elastic expression of the quadrupolar interaction obtained in this chapter is not simply a reduction of a three-dimensional expression. In fact such a general expression does not even exist. This different angular dependence takes into account a kind of compensation of the deformation "in plane". It is similar to the one discussed for the influence of the flexoelectric effect on the elastic properties of NLC. The elastic constant of the surface-like elastic term is of the same order of magnitude of the quadrupolar bulk elastic constant. We have also shown that the total surface energy due to the quadrupolar interaction contains two terms. The first term, having an elastic origin, is connected with a surface-like term, coming from the bulk energy density. This contribution depends on the surface nematic tilt angle and on its gradient. The other term is due to the incomplete bulk interaction in the surface layers. It depends only on the surface nematic orientation and is characterized by a tilted easy axis and by an extrapolation length of the order of $10^2 \times r_0$, where r_0 is of the order of the molecular dimension.

[1] de Gennes PG and Prost J. *The Physics of Liquid Crystals.* Clarendon Press, Oxford, 1994.

[2] Frank FC. On the theory of liquid crystals. *Discussions of the Faraday Society*, **25**, 19 (1958).

[3] Nehring J and Saupe A. Elastic theory of uniaxial liquid crystals. *Journal of Chemical Physics*, **54**, 337 (1971); Calculation of elastic constants of nematic liquid crystals. *Journal of Chemical Physics*, **56**, 5527 (1972).

[4] Vertogen G. Elastic constants and the continuum theory of liquid crystals. *Physica A*, **117**, 227 (1983).

[5] Teixeira PIC, Pergamenshchik VM, and Sluckin TJ. A model calculation of the surface elastic constants of a nematic liquid crystal. *Molecular Physics*, **80**, 1339 (1993).

[6] Evans R and Sluckin TJ. A density functional theory for inhomogeneous charged fluids–application to the sufaces of molten-salts. *Molecular Physics*, **40**, 413 (1980).

[7] Sluckin TJ. Density functional theory for simple molecular fluids. *Molecular Physics*, **43**, 817 (1981).

[8] Osipov MA and Sluckin TJ. Molecular theory of order electricity. *Journal de Physique II*, **3**, 793 (1993).

[9] Osipov MA, Sluckin TJ, and Cox SJ. Influence of permanent molecular dipoles on surface anchoring of nematic liquid crystals. *Physical Review E*, **55**, 464 (1997).

[10] Faetti S and Riccardi M. Elastic constants in the interfacial layer at the nematic liquid crystal vapor interface. *Nuovo Cimento D*, **17**, 1019 (1995).

[11] Barbero G, Evangelista LR, and Faetti S. Elastic effects of long-range quadrupolar interactions in nematic liquid crystals. *Physical Review E*, **58**, 7465 (1998).

[12] Oldano C and Barbero G. An *ab initio* analysis of the 2nd order elasticity effect on nematic configurations. *Physics Letters A* **110**, 213 (1985).

[13] Pergamenshchick VM. Phenomenological approach to the problem of the K_{13} surface-like elastic term in the free energy of a nematic liquid crystal. *Physical Review E*, **48**, 1254 (1993).

[14] Faetti S. Theory of surface-like elastic contributions in nematic liquid crystals. *Physical Review E*,**49**, 4192 (1994).

[15] Stallinga S and Vertogen G. Solution of the Oldano-Barbero paradox. *Physical Review E*, **53**, 1692 (1996).

[16] Dahl I and de Meyere A. On higher-order variational analysis in one-dimension and 3-dimenstions for soft boundaries. *Liquid Crystals*, **18**, 683 (1995).

[17] Yokoyama H. Density-functional theory of surface-like elasticity of nematic liquid crystals. *Physical Review E*, **55**, 2938 (1997).

[18] Prost J and Marcerou JP. Microscopic interpretation of flexoelectricity. *Journal de Physique*, **38**, 315 (1977).

[19] Landau LD and Lifchitz EI. *Theorie des Champs*. MIR, Moscow, 1972.

[20] Barbero G and Barberi R. In *Physics of Liquid Crystalline Materials*. Edited by Khoo IC and Simoni F. Gordon and Breach, Philadelphia, 1991.

[21] Barbero G and G. Durand. On the validity of the Rapini-Papoular surface anchoring energy form in nematic liquid crystals. *Journal de Physique*, **47**, 2129 (1986).

[22] Jackson JD. *Classical Electrodynamics*. John Wiley & Sons, Philadelphia, 1975.

[23] Durand E. *Electrostatique.* Masson, Paris, 1950.

[24] Alexe-Ionescu AL and Barbero G. Non-local description of nematic liquid crystals. *Liquid Crystals*, **25**, 189 (1998).

[25] Faetti M and Faetti S. Splay-bend surface elastic constant of nematic liquid crystals: A solution of the Somoza-Tarazona paradox. *Physical Review E*, **57**, 6741 (1998).

4

TEMPERATURE DEPENDENCE OF THE SURFACE FREE ENERGY

In this chapter, we analyze the temperature dependence of the surface energy of a nematic liquid crystal sample in three different perspectives. We start by discussing a model for the thermal renormalization of the anchoring energy, by expanding the surface energy in series of spherical harmonic functions. The coefficients of the expansion are the experimentally detectable anchoring coefficients. It is shown that all the anchoring coefficients of the same order depend on the temperature in the same manner. As a consequence, the alignment transitions driven by the surface (the so-called temperature surface transitions) are due to a surface anchoring energy which contains contributions of different orders. A local self-consistent model, that takes into account the presence of a surface field as well as an incomplete bulk nematic interaction, is presented in connection with the determination of the temperature dependence of the surface tension near the nematic–isotropic phase transition. It is shown that, according to the strength of the surface field, the surface tension presents a minimum just before the transition, followed by a strong variation at the transition. Finally, a microscopic approach to the surface gliding effect in lyotropic liquid crystals is discussed in terms of a thermally activated process. The problem is formulated in terms of a master equation describing the jumps between adjacent angular positions.

4.1 A model for the thermal renormalization of the anchoring energy

In Section 2.1, we have introduced the concept of *easy axis* or *easy direction* as being that direction along which a solid substrate is able to orient the nematic director $\mathbf{n}$. This is consequence of the direct interaction between the nematic medium and the substrate. There are experimental evidences that this direction, $\mathbf{n}_0$, depends on the temperature [1]–[10]. This dependence comes from the temperature dependence of the anisotropic part of the surface energy. This problem has been faced by means of Landau–type phenomenological theories [11]–[15], or with microscopic models based on Onsager's theory [16]–[20]. As pointed out in Ref. [21], the applicability of Landau-like approaches is questionable, whereas the proposed microscopic models contain a large number of unknown parameters. For this reason, the physical predictions of these microscopic models cannot be immediately tested. Following the exposition of Ref. [21], we discuss below a model for the thermal behavior of the anisotropic part of the surface energy based on a mean field analysis. This model can be useful for the understanding of the effect of the temperature on the anchoring coefficients, because it is valid for all values of the nematic scalar order parameter.

4.1.1 Symmetry considerations

We consider only uniform liquid crystals and suppose that the NLC occupies the semi-infinite space $z \geq 0$, bounded by a flat and homogeneous substrate at $z = 0$. In this framework, the presence of the surface does not introduce any biaxiality, and from the crystallographic point of view the nematic is characterized by the uniaxial tensor order parameter Q.

Let us consider the anisotropic part of the surface energy, W, characterizing the nematic-substrate interface. As we have seen in Chapter 2 [see Eq. (2.9) and below], when an undistorted NLC is in contact with a solid substrate or limited by another medium, the nematic director is oriented along the *easy axis*, $\mathbf{n}_0$. For deviations of the actual surface director $\mathbf{n}_s$ from the easy direction $\mathbf{n}_0$, $W = W(\mathbf{n}_0, \mathbf{n}_s)$. $W(\mathbf{n}_0, \mathbf{n}_0)$ corresponds to the minimum value of W, whereas $W(\mathbf{n}_0, \mathbf{n}_s) \geq W(\mathbf{n}_0, \mathbf{n}_0)$ is the surplus of surface energy due to the surface distortion $\delta\mathbf{n} = \mathbf{n}_s - \mathbf{n}_0$. For small deviations of $\mathbf{n}_s$ from $\mathbf{n}_0$, $W(\mathbf{n}_s, \mathbf{n}_0) = W_2 P_2(\mathbf{n}_s \cdot \mathbf{n}_0)$, where $W_2 < 0$ is the macroscopic anchoring strength and $P_2(\mathbf{n}_s \cdot \mathbf{n}_0) = (3/2)[(\mathbf{n}_s \cdot \mathbf{n}_0)^2 - (1/3)]$ the second order Legendre polynomial, as discussed in Chapter 1 [see Eq. (1.13) and below]. This simple expression does not hold true when the deviation of $\mathbf{n}_s$ from $\mathbf{n}_0$ is large. In this case, $W(\mathbf{n}_s, \mathbf{n}_0)$ is, usually, approximated by an expansion in terms of Legendre polynomials [22] of the kind $W(\mathbf{n}_s, \mathbf{n}_0) = \sum_l W_{2l} P_{2l}(\mathbf{n}_s \cdot \mathbf{n}_0)$, as was done in (1.13) with the interaction energy.

In the case of homogeneous substrates, $\mathbf{n}_0$ depends on the physical properties of the substrate and of the NLC. To find $\mathbf{n}_0$ in practice, it is necessary to write down the surface energy W in terms of the elements of symmetry of the surface and of the NLC, and to look for its minimum with respect to the nematic director. Experimental data show that $\mathbf{n}_0$ can be temperature dependent. This phenomenon has been termed *temperature surface transition* and discussed by several authors [5]–[9], [17], [23]–[25].

The temperature dependence of $\mathbf{n}_0$ has been analyzed by following a procedure similar to the one used for the elastic description in the bulk [26], i.e., by decomposing the surface energy in terms of the elements of symmetry of the surface and of the NLC, and using as expansion parameter the scalar order parameter S. In the case of a flat isotropic surface, characterized by the geometrical normal $\mathbf{k}$, the surface energy is written as [11]–[13]

$$W = A_0 + A_1 k_i Q_{ij} k_j + A_2 Q_{ij} Q_{ji} + A_3 (k_i Q_{ij} k_j)^2 + A_4 k_i Q_{il} Q_{lj} k_j + \mathcal{O}(S^3), \tag{4.1}$$

at the second order in the surface order parameter. Strictly speaking, this expansion is valid only for small S. However, since the nematic-isotropic phase transition is of first order, with a finite jump of S at the critical temperature ($\Delta S \sim 0.3$ [27]), it is not clear when Eq. (4.1) works well. In Eq. (4.1) the quantities A_i are temperature independent phenomenological parameters. As discussed in Refs. [11, 13] the coefficients A_2, A_3 and A_4 of the quadratic terms in S in Eq. (4.1) arise from modification of the mean field potential between two nematic molecules near the substrate. On the contrary, the coefficient A_1 has contributions both from that effect and from the direct wall-molecule interaction [28]. Using for Q_{ij} the expression (1.9), where $\mathbf{n} = \mathbf{n}_s$, Eq. (4.1) can be rewritten in terms of the angle $\theta = \arccos(\mathbf{n}_s \cdot \mathbf{k})$, formed by $\mathbf{n}_s$ with $\mathbf{k}$, as

$$W = W_0 + W_2 P_2(\cos\theta) + W_4 P_4(\cos\theta), \tag{4.2}$$

where the coefficients $W_i = W_i(S)$, $i = 0, 2, 4$, are given by

$$W_0 = A_0 + \frac{15A_2 + 2A_3 + 5A_4}{10} S^2,$$
$$W_2 = A_1 S + \frac{8A_3 + 7A_4}{28} S^2, \quad \text{and} \quad W_4 = \frac{18}{35} A_3 S^2. \tag{4.3}$$

By assuming for the surface value of S the temperature dependence obtained in the bulk, by means of the phenomenological expansion given by Eq. (4.2) it is possible to study the surface transitions induced by the temperature [11, 12]. This model can also be refined, by taking into account that the surface scalar order parameter is different from the bulk one by means of a Landau-Ginzburg approach [13],[29]–[31].

Sen and Sullivan [13] discuss how one can derive Eq. (4.1) from a molecular mean-field theory. They assume that the free energy, F, of the system under consideration is a functional of the single-particle probability density, and that spatial variations occurs only in the z-direction [32]. F is written as the sum of three bulk contributions, the first describing the interaction between the nematic and the substrate, characterized by a potential $U[\mathbf{u}(z), z]$, and the second describing the nematic-nematic interaction, and characterized by a potential $V[\mathbf{u}(z_1), \mathbf{u}(z_2); |z_1 - z_2|]$, where the subscripts 1 and 2 are relevant to the two nematic interacting molecules. The third term is a functional of the orientational distribution. To reduce the non-local two-body term to a local one, they make a formal gradient expansion of the single-particle probability density. After that, the effective surface energy generating Eq. (4.1) is defined by comparing the gradient expansion of the free energy with the full starting expression for F, according to a standard recipe [33]. In this way, the surface energy receives contributions only from the direct interactions of the nematic molecules with the substrate, via $U(\mathbf{u}, z)$ evaluated at the surface, and from the incomplete nematic-nematic interaction. Following this scheme it is possible to obtain the correct number of terms of the surface energy. However, as underlined above, the main limit of Eq. (4.1) is that it is an expansion in power series of the scalar order parameter S, and this quantity is never very small.

4.1.2 Mean field approach

We consider now the temperature dependence of the anchoring energy using an approach based on the mean field theory. We neglect all the inhomogeneities and assume, furthermore, that the surface potential is short range.

A surface molecule of the NLC is submitted to the interaction with the other nematic molecules and to the interaction with the substrate. The relevant interaction energies will be indicated with V_N and V_s, respectively. The separation of the total energy of a surface molecule in a bulk part and in a surface part is not well defined, because the nematic symmetry is broken near the surface.

In the bulk the mean field energy due to the interaction of a nematic molecule with the other nematic molecules is of the kind $V_N = V_N(\mathbf{n} \cdot \mathbf{u})$, i.e., it depends only on the relative orientation of $\mathbf{u}$ with respect to $\mathbf{n}$. The symmetry of the interaction is SO3. Near the surface this symmetry is broken, and V_N is, in general, of the type

$$V_N = V_N(\mathbf{r}, \mathbf{u}, \mathbf{n}) = V_N(z, \mathbf{u} \cdot \mathbf{r}, \mathbf{u} \cdot \mathbf{n}), \tag{4.4}$$

where $\mathbf{r}$ is the position of a given nematic molecule, and z its distance from the surface. As discussed in Ref. [34], the functional dependence of V_N on $\mathbf{u} \cdot \mathbf{r}$ can be responsible for subsurface deformations. Near the surface it is possible to rewrite Eq. (4.4) as

$$V_N = V_N(\mathbf{n} \cdot \mathbf{u}) + \delta V_N(z, \mathbf{u} \cdot \mathbf{r}, \mathbf{u} \cdot \mathbf{n}), \tag{4.5}$$

where

$$\delta V_N(z, \mathbf{u} \cdot \mathbf{r}, \mathbf{u} \cdot \mathbf{n}) = V_N(z, \mathbf{u} \cdot \mathbf{r}, \mathbf{u} \cdot \mathbf{n}) - V_N(\mathbf{n} \cdot \mathbf{u}), \tag{4.6}$$

represents the deviation of the actual mean field energy V_N from the SO3 symmetry. The function $\delta V_N(z, \mathbf{u} \cdot \mathbf{r}, \mathbf{u} \cdot \mathbf{n}) \neq 0$ in a surface layer whose thickness is of the order of the range of the molecular forces responsible for the nematic phase. It follows that $\delta V_N(z, \mathbf{u} \cdot \mathbf{r}, \mathbf{u} \cdot \mathbf{n})$ can be considered as an "intrinsic" surface energy. The effective surface energy is then obtained by adding to the surface energy, due to the direct interaction between the nematic molecules and the substrate, the intrinsic surface energy. Here, V_s

has the meaning of effective surface energy, and V_N is a shorthand notation for $V_N(\mathbf{n}\cdot\mathbf{u})$.

We assume that V_N, due to the nematic-nematic interaction, can be evaluated in a mean field theory. It is assumed to be position independent. All variations of V_N due to the broken symmetry connected with the presence of the bounding surface are considered as surface contributions, and enter in the effective surface energy V_s.

The total energy of a surface molecule is $V = V_N + V_s$. The small parameter used to expand the surface energy in power series is $V_s/V_N \ll 1$, which corresponds to the weak anchoring situation. On the contrary, the surface scalar order parameter S is not assumed to be a small quantity. In the total energy, V_N describes the tendency of the nematic molecules, characterized by the molecular orientation $\mathbf{u}$, to be oriented along the nematic director $\mathbf{n}$. In the following it will be approximated by means of the Maier-Saupe mean field [35], V_N^M, according to which, as discussed in Section 1.2,

$$V_N^M \propto n_i q_{ij} n_j = P_2(\mathbf{n}\cdot\mathbf{u}), \tag{4.7}$$

where

$$q_{ij} = \frac{3}{2}\left[u_i u_j - \frac{1}{3}\delta_{ij}\right], \tag{4.8}$$

such that $S = \langle n_i q_{ij} n_j\rangle$. In this framework

$$V_N^M = -vP_2(\mathbf{n}\cdot\mathbf{u})S, \tag{4.9}$$

where v is a molecular parameter (that we call hereafter the Maier-Saupe parameter). A possible generalization of (4.9) is

$$V_N^H = -\sum_l v_{2l} P_{2l}(\mathbf{n}\cdot\mathbf{u}) S_{2l}, \tag{4.10}$$

where $S_{2l} = \langle P_{2l}(\mathbf{n}\cdot\mathbf{u})\rangle$ are the nematic order parameters [36].

Let us consider now the other term of the total energy, V_s. It is clear that this interaction has to describe the tendency of the surface to orient the surface nematic molecules along a given direction, $\mathbf{n}_0$. This direction depends on the symmetry of the surface and on the molecular properties of the mesophase. If the analysis is limited to non-polar media, then V_s has to be an even function of $\mathbf{u}$. It follows that V_s is, actually, a function of the tensor q and can be written, in general, as

$$V_s(\mathbf{u}) = V_s(\mathbf{q}) = \sum_k w_k(0) L_k(\mathbf{q}), \tag{4.11}$$

where $L_k(\mathbf{q})$ indicates the scalar quantities that can be built with the molecular tensor of elements q_{ij} and the elements of symmetry characterizing the surface. Each term of the expansion of $V_s(\mathbf{q})$ represents a given interaction, like induced dipole-induced dipole or quadrupole-quadrupole and so on [37]; the "intrinsic" anchoring coefficients $w_k(0)$ are physical parameters connected with the type of interaction described by $L_k(\mathbf{q})$. It is assumed that the physical properties of the substrate can be considered constants in the temperature range of the nematic phase. In this framework, since $w_k(0)$ refers to specific fundamental interactions, it is temperature independent. So, thermal effects arise only from the temperature dependence of the degree of alignment of the nematic molecules. For the previous hypothesis on the temperature independence of the physical properties of the substrate, the theory describes correctly NLCs in contact with solid substrates. On the contrary, deviations from the predictions are expected for nematic samples oriented by means of surfactants, because their thermal behavior is similar to that of the liquid crystal materials.

It is useful to describe the molecular direction and the nematic director in terms of the polar angles with respect to a Cartesian reference frame having the z-axis parallel to the geometrical normal to the flat surface and the x-axis along the possible surface anisotropy. Let Θ, Φ and θ, ϕ be the polar angles defining $\mathbf{u}$ and $\mathbf{n}$, respectively. In this framework, $V_s(\mathbf{q})$ can be rewritten as

$$V_s(\Theta, \Phi) = \sum_k w_k(0) L_k(\Theta, \Phi). \tag{4.12}$$

By decomposing the functions $L_k(\Theta, \Phi)$ in series of spherical harmonic functions $Y_k^m(\Theta, \Phi)$ one obtains

$$L_k(\Theta, \Phi) = \sum_m a_k^m Y_k^m(\Theta, \Phi). \tag{4.13}$$

Since $L_k = L_k(\mathbf{q})$, and hence $L_k(\Theta, \Phi) = L_k(\pi - \Theta, \pi + \Phi)$ for all k, one deduces that $k = 2l$, i.e., L_k for odd k is absent in the expansion of $V_s(\Theta, \Phi)$. It follows that for non-polar NLC

$$L_{2l}(\Theta, \Phi) = \sum_m a_{2l}^m Y_{2l}^m(\Theta, \Phi), \tag{4.14}$$

and the microscopic surface energy can be written as

$$V_s(\Theta, \Phi) = \sum_l w_{2l}(0) L_{2l}(\Theta, \Phi), \tag{4.15}$$

or, as follows from the discussion reported above, in the form

$$V_s(\Theta, \Phi) = \sum_l w_{2l}(0) \sum_m a_{2l}^m Y_{2l}^m(\Theta, \Phi). \tag{4.16}$$

The macroscopic anchoring energy $W(\mathbf{n}) = W(\theta, \phi)$ is obtained by averaging V_s over the molecular orientations $\mathbf{u}$, or over Θ and Φ. In the problem under consideration $V_s \ll V_N$; consequently, V_s can be treated as a perturbation.

According to the thermodynamic perturbation theory [38] one has $W(\theta, \phi) = \langle V_s(\Theta, \Phi)\rangle$. Therefore,

$$W(\theta, \phi) = \sum_l w_{2l}(0) \sum_m a_{2l}^m \langle Y_{2l}^m(\Theta, \Phi)\rangle, \tag{4.17}$$

where $\langle A\rangle = Tr(\rho A)/Tr(\rho)$, and $\rho = \exp(-\beta V_N)$ is the density matrix.

In order to derive the macroscopic surface energy $W(\theta, \phi)$ one has first to express $V_s(\Theta, \Phi)$ in terms of a polar coordinates system based on the director $\mathbf{n}$ as polar axis. The Cartesian reference frame has to be rotated in such a way that $\mathbf{z}' = \mathbf{n} = \langle \mathbf{u}\rangle$. Let ϑ ad φ be the polar angles of $\mathbf{u}$ with respect to the rotated coordinate system. In this case [39]

$$Y_l^m(\Theta, \Phi) = \sum_{m'} D_{m,m'}^l(\theta, \phi) Y_l^{m'}(\vartheta, \varphi), \tag{4.18}$$

where $D_{m,m'}^l(\theta, \phi)$ are the elements of the Wigner's matrix. Since there is axial symmetry about the direction of $\mathbf{n}$ in the unperturbed system, only the member $m = 0$ of the Y_l^m is different from zero. Consequently

$$\langle Y_l^{m'}(\vartheta, \varphi)\rangle = \langle Y_l^0(\vartheta)\rangle \delta_{m',0} \tag{4.19}$$

and, from Eq. (4.18),

$$\langle Y_l^m(\Theta, \Phi)\rangle = D_{m,0}^l(\theta, \phi)\langle Y_l^0(\vartheta)\rangle. \tag{4.20}$$

By taking into account that $D_{m,0}^l(\theta, \phi) = Y_l^m(\theta, \phi)$ [39], one has finally, as follows from Eq. (4.15) and (4.17),

$$W(\theta, \phi) = \sum_l w_{2l}(0)\langle P_{2l}(\mathbf{n} \cdot \mathbf{u})\rangle L_{2l}(\theta, \phi), \tag{4.21}$$

because $Y_{2l}^0(\vartheta, \varphi) = P_{2l}(\cos\vartheta)$. Equation (4.21) is a consequence of the fact that all anisotropic effects are regarded as perturbation, so that they do not need to be included in the computation of the average values. By comparing Eq. (4.21) with Eq. (4.15) one deduces that the temperature dependence of the parameters describing the anisotropic part of the surface energy is given by

$$w_{2l}(T) = w_{2l}(0)\langle P_{2l}(\mathbf{n} \cdot \mathbf{u})\rangle. \tag{4.22}$$

To calculate $\langle P_{2l}(\mathbf{n} \cdot \mathbf{u})\rangle$ in the mean field approximation, one assumes that $V_N = V_N^M$. Therefore, Eq. (4.22) gives

$$\frac{w_{2l}(T)}{w_{2l}(0)} = \frac{\int_0^1 P_{2l}(\mathbf{n} \cdot \mathbf{u}) \exp[\beta v P_2(\mathbf{n} \cdot \mathbf{u})S] d(\mathbf{n} \cdot \mathbf{u})}{\int_0^1 \exp[\beta v P_2(\mathbf{n} \cdot \mathbf{u})S] d(\mathbf{n} \cdot \mathbf{u})}. \tag{4.23}$$

The integrals can be easily calculated and give the required temperature dependence. If $V_N = V_N^H$, Eq. (4.22) gives simply

$$\frac{w_{2l}(T)}{w_{2l}(0)} = S_{2l}. \tag{4.24}$$

This means that in the framework of a generalized mean field theory of the kind represented by V_N^H the temperature dependence of $w_{2l}(T)/w_{2l}(0)$ coincides with the temperature dependence of the $2l$-th scalar order parameter.

According to the analysis presented above, where the macroscopic anchoring energy is given by the series expansion in spherical harmonic functions, shown in Eq. (4.21), the thermal renormalization of the anchoring coefficients is given by Eq. (4.23) or by Eq. (4.24). From these results it follows that the anchoring coefficients of the same order in the expansion have the same temperature dependence. Consequently, in the frame of the model, temperature surface transitions are possible only in a nematic sample whose anchoring energy contains contributions from different orders in the spherical harmonic functions expansion.

The ratios $\langle P_{2l}(\mathbf{n}\cdot\mathbf{u})\rangle/S$ versus S for $l = 2, 3$ and 4 in the Maier-Saupe approximation can be easily evaluated in the nematic phase, where $0.4 \leq S \leq 0.7$. A direct calculation shows that $\langle P_{2l}(\mathbf{n}\cdot\mathbf{u})\rangle/S \leq 0.1$, for $l = 3, 4$. This explains why, usually, the anisotropic part of the surface anchoring energy, given by Eq. (4.21), is well approximated by few terms [40].

As an example, consider an NLC limited by an isotropic substrate, using the simple Maier-Saupe mean field theory. In this case, only the polar angle θ enters in the description, and $L_{2l}(\theta) = P_{2l}(\cos\theta)$. From Eq. (4.23) it follows that $w_2(T)/w_2(0) = S$. This means that to the lowest order in S, the temperature dependencies of the anchoring energy deduced by means of symmetry considerations as Eq. (4.2) and the one obtained by means of the mean field agree. Effectively, according to the recipe based on the symmetry one has $W_2(T) = A_1 S$ and, according to the mean field $w_2(T) = w_2(0)S$. However, for $l = 2$ there is a discrepancy between the two approaches. In fact, according to the mean field we have

$$w_2(T)/w_2(0) = S \quad \text{and} \quad w_4(T)/w_4(0) = \langle P_4(\mathbf{n}\cdot\mathbf{u})\rangle \neq S^2, \tag{4.25}$$

whereas the approach based on the symmetry predicts the temperature dependencies given by Eq. (4.3). More precisely, it predicts a renormalization of the coefficient of $P_2(\cos\theta)$, by means of an S^2 contribution, and a temperature dependence of the coefficient of $P_4(\cos\theta)$ like S^2. Of course, in the limit of small S the two predictions agree. In fact, if $S \ll 1$ the renormalization of $P_2(\cos\theta)$ in S^2 can be neglected with respect to the linear term in S. Furthermore, in this approximation, $\langle P_4(\mathbf{n}\cdot\mathbf{u})\rangle \propto S^2$. However, in the case of large S the discrepancy between the two approaches can be significant. In Figure 4.1, the trends of $\langle P_4(\mathbf{n}\cdot\mathbf{u})\rangle$ and S^2 are shown versus S. From this figure it follows that in the nematic range $(0.4 \leq S \leq 1)$ the difference between S^2 and $\langle P_4(\mathbf{n}\cdot\mathbf{u})\rangle$ is always rather large.

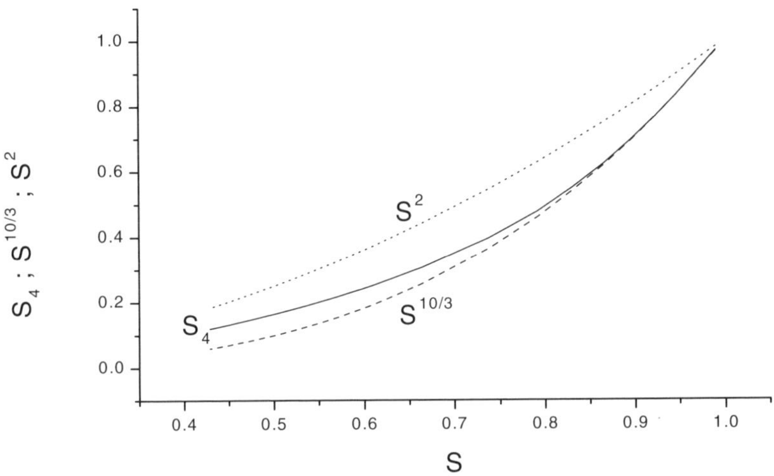

FIGURE 4.1
Dependence of $\langle P_4(\mathbf{n}\cdot\mathbf{u})\rangle = S_4$ versus S. The S-dependence of S^2, which is the next term in the surface energy expansion predicted by Landau-like analyses, and of $S^{10/3}$, predicted by the mean field approach in the large range of S, are also shown. As is evident from the figure, the Akulov-Zener law works well when the fluctuations of $\mathbf{u}$ with respect to $\mathbf{n}$ are small (large S). Reprinted with permission from Ref. [21]. Copyright (2000) by the American Physical Society.

Consider now a nematic sample bounded by a surface having monoclinic symmetry, as the one obtained by means of the SiO-oblique evaporation technique [9]. The (y, z)-plane is the evaporation plane, the normal to the substrates is parallel to the z-axis, and the direction of the grooves coincides with the x-direction. In this framework, to the second order in the spherical harmonic expansion, it is possible to write

$$L_2(\Theta, \Phi) = \sum_{m=-2}^{2} a_2^m Y_2^m(\Theta, \Phi). \tag{4.26}$$

Since [41]

$$\begin{aligned} Y_2^{\pm 2}(\Theta, \Phi) &= \sqrt{15/(32\pi)} \sin^2 \Theta \exp(\pm 2i\Phi), \\ Y_2^{\pm 1}(\Theta, \Phi) &= \mp\sqrt{15/(8\pi)} \sin \Theta \cos \Theta \exp(\pm i\Phi), \\ Y_2^{0}(\Theta, \Phi) &= \sqrt{5/(16\pi)}(3\cos^2 \Theta - 1), \end{aligned} \tag{4.27}$$

from Eq. (4.21) one obtains

$$W(\theta, \phi) = w_2(T)[C_1 \cos(2\theta) + C_2 \sin^2 \theta \cos(2\phi) + C_3 \sin(2\theta) \cos \phi], \tag{4.28}$$

where C_i represents numerical factors. This equation shows clearly that, at the second order in the spherical harmonic functions expansion of $W(\theta, \phi)$, the nematic orientation is expected to be independent of the temperature. The temperature surface transitions, in which the nematic orientation variations are driven by the surface, can be interpreted, in the frame of the present model, only by taking into account also terms of fourth order in $Y_4^m(\Theta, \Phi)$ in the microscopic surface energy $V_s(\Theta, \Phi)$.

It is possible to obtain Eq. (4.28) also in the following manner. At the second order in the Cartesian components of the nematic director $\mathbf{n}$, the surface energy is given by $W(\mathbf{n}) = (1/2)\sum_{i,j} H_{ij} n_i n_j$, where the elements H_{ij} describe the surface interaction. Explicitly, $W(\mathbf{n})$ can be written as

$$W(n_x, n_y, n_z) = H_{xx} n_x^2 + H_{yy} n_y^2 + H_{zz} n_z^2 + H_{xy} n_x n_y + H_{xz} n_x n_z + H_{yz} n_y n_z.$$

In the monoclinic symmetry under consideration, it is easy to verify that $W(n_x, n_y, n_z) = W(-n_x, n_y, n_z)$, and hence $H_{xy} = H_{xz} = 0$. Furthermore, since $n_x^2 + n_y^2 + n_z^2 = 1$, one can rewrite $W(n_x, n_y, n_z)$ in the form

$$W(n_x, n_y, n_z) = H_{xx} + (H_{yy} - H_{xx}) n_y^2 + (H_{zz} - H_{xx}) n_z^2 + H_{yz} n_y n_z.$$

Using polar coordinates one obtains Eq. (4.28).

4.1.3 Comparison with Akulov-Zener law

At low temperature, when the fluctuations are small, i.e., $-\beta V_N \gg 1$, it is possible to rewrite Eq. (4.23) in a different manner, well known in other fields of solid state physics as the Akulov-Zener law [42, 43]. In order to obtain this expression, one has to take into account that in the low temperature region, where $S \sim 1$ and hence the fluctuations of $\mathbf{u}$ with respect to $\mathbf{n}$ are very small,

$$\mathbf{n} \cdot \mathbf{u} = \cos\vartheta \sim 1 - (1/2)\vartheta^2 + \mathcal{O}(4), \tag{4.29}$$

i.e., $\vartheta \ll 1$. In this framework

$$P_2(\mathbf{n} \cdot \mathbf{u}) \sim 1 - (3/2)\vartheta^2 + \mathcal{O}(4), \tag{4.30}$$

and

$$P_4(\mathbf{n} \cdot \mathbf{u}) \sim 1 - 5\vartheta^2 + \mathcal{O}(4). \tag{4.31}$$

Consequently, by taking into account that

$$w_2(T)/w_2(0) = \langle P_2(\mathbf{n} \cdot \mathbf{u})\rangle \quad \text{and} \quad w_4(T)/w_4(0) = \langle P_4(\mathbf{n} \cdot \mathbf{u})\rangle, \tag{4.32}$$

one obtains, by assuming $V_N = V_N^M$,

$$\begin{aligned}\frac{w_2(T)}{w_2(0)} &= \frac{\int_0^\pi \left(1 - \frac{3}{2}\vartheta^2\right) \exp\left[\beta v S\left(1 - \frac{3}{2}\vartheta^2\right)\right] \vartheta d\vartheta}{\int_0^\pi \exp\left[\beta v S\left(1 - \frac{3}{2}\vartheta^2\right)\right] \vartheta d\vartheta} \\ &\sim 1 - \frac{1}{\beta v S} \sim \exp\left(-\frac{1}{\beta v S}\right),\end{aligned} \tag{4.33}$$

and

$$\begin{aligned}\frac{w_4(T)}{w_4(0)} &= \frac{\int_0^\pi (1 - 5\vartheta^2) \exp\left[\beta v S\left(1 - \frac{3}{2}\vartheta^2\right)\right] \vartheta d\vartheta}{\int_0^\pi \exp\left[\beta v S\left(1 - \frac{3}{2}\vartheta^2\right)\right] \vartheta d\vartheta} \\ &\sim 1 - \frac{10}{3\beta v S} \sim \exp\left(-\frac{10}{3\beta v S}\right).\end{aligned} \tag{4.34}$$

Since $\beta v S \gg 1$, to the integrals appearing in Eq. (4.33) and (4.34) contribute only small values of ϑ. It follows that the upper limit of the integrals, π, can be substituted with ∞. In this manner all the integrals reduce to Gaussian integrals of the type $\int_0^\infty x^3 \exp(-cx^2)dx = 1/2c^2$ and $\int_0^\infty x \exp(-cx^2)dx = 1/2c$. On the other hand, according to the Maier-Saupe theory $S = \langle P_2(\mathbf{n}\cdot\mathbf{u})\rangle$, from Eq. (4.33) and (4.34) we have $w_2(T)/w_2(0) = S$, and $w_4(T)/w_4(0) = S^{10/3}$. In Figure 4.1, $\langle P_4(\mathbf{n} \cdot \mathbf{u})\rangle$ and $S^{10/3}$ are shown versus S in the nematic phase ($0.4 \leq S \leq 1$). As follows from this figure, for large values of S,

$\langle P_4(\mathbf{n}\cdot\mathbf{u})\rangle \sim S^{10/3}$. In the same figure, S^2, which is the term predicted by Landau-like models at the second order in S, is shown.

The calculations reported above for $l = 1$ and $l = 2$ can be generalized for all l. The final result is that in the low temperature region the thermal renormalization of the anchoring coefficient is

$$\frac{w_{2l}(T)}{w_{2l}(0)} = \langle P_{2l}(\mathbf{n}\cdot\mathbf{u})\rangle = S^{\frac{l(2l+1)}{3}}. \tag{4.35}$$

Equation (4.35) has been obtained by assuming $V_N = V_N^M$. However, it holds true also in the case where $V_N = V_N^H$. To prove this, consider that $V_N = V_N^H = -\sum v_{2l}S_{2l}P_{2l}(\cos\vartheta)$. In the low temperature region, where the fluctuations of $\mathbf{u}$ with respect to $\mathbf{n}$ are small, $\vartheta \ll 1$. In this case

$$P_{2l}(\cos\vartheta) = 1 - \frac{1}{2}l(2l+1)\vartheta^2 + \mathcal{O}(4). \tag{4.36}$$

Consequently

$$V_N^H(\cos\vartheta) = -[N - M\vartheta^2] + \mathcal{O}(4), \tag{4.37}$$

where

$$N = \sum_l v_{2l}S_{2l} \quad \text{and} \quad M = \frac{1}{2}\sum_l l(2l+1)v_{2l}S_{2l}. \tag{4.38}$$

It follows that

$$S_{2l} = \langle P_{2l}(\mathbf{n}\cdot\mathbf{u})\rangle = 1 - \frac{1}{2}l(2l+1)\frac{\int_0^\infty \vartheta^3 e^{-M\vartheta^2}d\vartheta}{\int_0^\infty \vartheta e^{-M\vartheta^2}d\vartheta}, \tag{4.39}$$

from which one obtains

$$S_{2l} = 1 - \frac{l(2l+1)}{2M} \sim \exp\left\{-\frac{l(2l+1)}{2M}\right\}. \tag{4.40}$$

In particular the main nematic order parameter is found to be

$$S = \langle P_2(\mathbf{n}\cdot\mathbf{u}) = \exp\left(-\frac{3}{2M}\right). \tag{4.41}$$

By substituting Eq. (4.41) into Eq. (4.40) one has, finally,

$$S_{2l} = S^{\frac{l(2l+1)}{3}}, \tag{4.42}$$

which coincides with Eq. (4.35).

Expression (4.35) can be compared with the Akulov-Zener law for ferromagnetic materials well-known in solid state physics [44, 45]

$$\frac{\mathcal{L}_n(T)}{\mathcal{L}_n(0)} = \left(\frac{M_s(T)}{M_s(0)}\right)^{\frac{n(n+1)}{2}}, \tag{4.43}$$

where $M_s(T)$ is a magnetization, $\mathcal{L}_n(T)$ is the n-order coefficient of the magnetic crystallographic anisotropy, magnetostriction, etc.

It is not difficult to show that expression (4.35) follows immediately from the Akulov-Zener law. Indeed, in the low temperature range where $S \sim 1$ and hence $\vartheta \ll 1$, $P_1(\mathbf{n}\cdot\mathbf{u}) = \mathbf{n}\cdot\mathbf{u} \sim 1 - (1/2)\vartheta^2$, $P_2(\mathbf{n}\cdot\mathbf{u}) \sim 1 - (3/2)\vartheta^2 \sim P_1^3(\mathbf{n}\cdot\mathbf{u})$. Consequently, $S(T) = \langle P_2(\mathbf{n}\cdot\mathbf{u})\rangle \sim \langle P_1(\mathbf{n}\cdot\mathbf{u})\rangle^3 = q^3$. Since $M_s(T)/M_s(0) = \langle P_1(\mathbf{n}\cdot\mathbf{u})\rangle = q$, Eq. (4.35) yields

$$\frac{w_{2l}(T)}{w_{2l}(0)} = q^{l(2l+1)}, \tag{4.44}$$

which is just the Akulov-Zener formula for $w_{2l}(T)$.

Although the proof of Eq. (4.33) and (4.34) is rigorously valid only for small perturbations and low temperatures, an exact calculation performed for magnetic materials [46] shows that the power law is fairly accurate even for relatively large perturbations, and is roughly followed almost up to the critical temperature.

At arbitrary temperatures one has

$$\frac{w_{2l}(T)}{w_{2l}(0)} = \langle P_{2l}(\mathbf{n}\cdot\mathbf{u})\rangle = \mathcal{L}_{2l}(\mathcal{Z}^{-1}(S(T)), \tag{4.45}$$

where

$$\mathcal{L}_n = \frac{\int_0^1 P_n(\mathbf{n}\cdot\mathbf{u})\exp[P_2(\mathbf{n}\cdot\mathbf{u})/\tau]d(\mathbf{n}\cdot\mathbf{u})}{\int_0^1 \exp[P_2(\mathbf{n}\cdot\mathbf{u})/\tau]d(\mathbf{n}\cdot\mathbf{u})}, \tag{4.46}$$

$$\mathcal{Z} = \frac{\int_0^1 P_2(\mathbf{n}\cdot\mathbf{u})\exp[P_2(\mathbf{n}\cdot\mathbf{u})/\tau]d(\mathbf{n}\cdot\mathbf{u})}{\int_0^1 \exp[P_2(\mathbf{n}\cdot\mathbf{u})/\tau]d(\mathbf{n}\cdot\mathbf{u})} = S \tag{4.47}$$

and

$$\tau = \frac{1}{\beta v S(T)} = \frac{k_B T}{v S(T)}, \tag{4.48}$$

is a reduced temperature. This parametric representation allows one to deduce $\langle P_{2l}(\mathbf{n}\cdot\mathbf{u})\rangle$ as a function of S, without knowing the interaction parameter v.

4.1.4 Averaging with continuous Hamiltonian

The previous results can be obtained also by using the procedure of averaging with continuous Hamiltonian (elastic energy density). According to Eq. (4.21) the macroscopic anchoring energy can be expanded in series of $L_{2l}(\theta,\phi)$. The expansion coefficients are the anchoring coefficients $w_{2l}(T) = w_{2l}(0)\langle P_{2l}(\mathbf{n}\cdot\mathbf{u})\rangle$ thermally renormalized by the presence of $\langle P_{2l}(\mathbf{n}\cdot\mathbf{u})\rangle$. Using the local reference frame, in which $\mathbf{n}\cdot\mathbf{u} = \cos\vartheta$, where $\vartheta = \vartheta(\mathbf{r}_s)$ and $\mathbf{r}_s$ is the position vector of a surface nematic molecule, $P_{2l}(\mathbf{n}\cdot\mathbf{u})$ can be decomposed as

$$P_{2l}(\mathbf{n}\cdot\mathbf{u}) = P_{2l}(\cos\vartheta) = \sum_{m=0}^{l} b_{2m}^{2l}\cos(2m\vartheta), \tag{4.49}$$

where b_{2m}^{2l} are numerical factors [41]. From Eq. (4.49) it is possible to obtain the thermal average of $P_{2l}(\mathbf{n}\cdot\mathbf{u})$ using the procedure of averaging with a continuous Hamiltonian. The subscript *loc* is used to indicate the thermal average obtained in this framework. One has

$$\langle P_{2l}(\mathbf{n}\cdot\mathbf{u})\rangle = \sum_{m=0}^{l} b_{2m}^{2l}\langle\cos(2m\vartheta)\rangle_{loc}. \tag{4.50}$$

In the local reference frame, where $\mathbf{n}\parallel\mathbf{z}'$, the elastic energy density is $f_h = (1/2)K(\nabla\vartheta)^2$, where K is the elastic constant. In fact, as shown in Chapter 1, the elastic energy density of an NLC is given by the Frank expression (1.56). The total elastic energy of an NLC is obtained by integrating f_{Frank} over the entire volume of the sample, considering only bulk terms, namely

$$H_0 = F = \frac{1}{2}\int_V [K_{11}(\nabla\cdot\mathbf{n})^2 + K_{22}(\mathbf{n}\cdot\nabla\times\mathbf{n})^2 + K_{33}(\mathbf{n}\times\nabla\times\mathbf{n})^2]\,d\mathbf{r}. \tag{4.51}$$

In the one-constant approximation, Eq. (4.51) becomes

$$H_0 = \frac{1}{2}K\int_V [(\nabla\cdot\mathbf{n})^2 + (\nabla\times\mathbf{n})^2]\,d\mathbf{r}. \tag{4.52}$$

In the local reference frame in which $\mathbf{n}$ coincides with the polar axis (z-axis), the fluctuations of $\mathbf{u}(\mathbf{r})$ at any point $\mathbf{r}$ are described by small non-zero components $u_x(\mathbf{r})$ and $u_y(\mathbf{r})$. At the second order in $u_x(\mathbf{r})$ and $u_y(\mathbf{r})$ Eq. (4.52) reads

$$\begin{aligned} H_0 = \frac{1}{2}K\int_V \{&[u_{x,x}(\mathbf{r}) + u_{y,y}(\mathbf{r})]^2 + [u_{x,y}(\mathbf{r}) - u_{y,x}(\mathbf{r})]^2 + [u_{x,z}(\mathbf{r})]^2 \\ &+ [u_{y,z}(\mathbf{r})]^2\}\,d\mathbf{r}, \end{aligned} \tag{4.53}$$

where $u_{i,j}(\mathbf{r}) = \partial u_i(\mathbf{r})/\partial x_j$. One can expand $u_i(\mathbf{r})$ in exponential Fourier's series as follows

$$u_i(\mathbf{r}) = \sum_{\mathbf{q}} u_i(\mathbf{q})\exp(i\mathbf{q}\cdot\mathbf{r}), \tag{4.54}$$

where $u^*(\mathbf{q}) = u_i(-\mathbf{q})$ because $u_i(\mathbf{r})$ are real quantities. By substituting Eq. (4.54) into Eq. (4.53) one obtains

$$H_0 = \frac{KV}{2} \sum_{\mathbf{q}} [|u_x(\mathbf{q})|^2 + |u_y(\mathbf{q})|^2]. \tag{4.55}$$

In the limit of small fluctuations $u_z(\mathbf{r}) = 1 - (1/2)[u_x^2(\mathbf{r}) + u_y^2(\mathbf{r})]$ and also $u_z(\mathbf{r}) = \cos\vartheta(\mathbf{r}) = 1 - (1/2)\vartheta^2(\mathbf{r})$. Consequently, $\vartheta^2(\mathbf{r}) = u_x^2(\mathbf{r}) + u_y^2(\mathbf{r})$. A simple calculation shows that $|\vartheta(\mathbf{q})|^2 = |u_x(\mathbf{q})|^2 + |u_y(\mathbf{q})|^2$, where $\vartheta(\mathbf{q})$ are the coefficients of the exponential Fourier's expansion of $\vartheta(\mathbf{r})$. Therefore, Eq. (4.55) is equivalent to

$$H_0 = \frac{KV}{2} \sum_{\mathbf{q}} q^2 |\vartheta(\mathbf{q})|^2. \tag{4.56}$$

It can be rewritten as

$$\frac{KV}{2} \sum_{\mathbf{q}} q^2 [A^2(\mathbf{q}) + B^2(\mathbf{q})], \tag{4.57}$$

where $A(\mathbf{q})$ and $B(\mathbf{q})$ are the Fourier's coefficients of the expansion of $\vartheta(\mathbf{r})$ in terms of $\cos(\mathbf{q}\cdot\mathbf{r})$ and $\sin(\mathbf{q}\cdot\mathbf{r})$, as shown below. A direct calculation shows that Eq. (4.56) can be obtained by considering f_h and decomposing $\vartheta(\mathbf{r})$ in Fourier's series. The expression for f_h is valid in the harmonic approximation. Hence it holds true only for $\vartheta \ll 1$. By decomposing ϑ in Fourier's series one obtains

$$\vartheta = \sum_{\mathbf{q}} [A(\mathbf{q}) \cos(\mathbf{q}\cdot\mathbf{r}) + B(\mathbf{q}) \sin(\mathbf{q}\cdot\mathbf{r})]. \tag{4.58}$$

The elastic energy $H_0 = \int_V f_h d^3\mathbf{r}$, where V is the volume of the nematic sample using Fourier's expansion of ϑ, is given by (4.57). The thermal average of $\cos(n\vartheta)$ is

$$\langle\cos(n\vartheta)\rangle_{loc} = \frac{\int \mathcal{D}(\vartheta)\cos(n\vartheta)\exp(-\beta H_0)}{\int \mathcal{D}(\vartheta)\exp(-\beta H_0)}. \tag{4.59}$$

By writing $\cos(n\vartheta)$ in the exponential form, and taking into account the expression for H_0 written above, one has

$$\langle\cos(n\vartheta)\rangle_{loc} = \mathrm{Re}\left[\frac{\prod_{\mathbf{q}} \int dA(\mathbf{q})dB(\mathbf{q})e^{-\gamma_q}}{\prod_{\mathbf{q}} \int dA(\mathbf{q})dB(\mathbf{q})e^{-\nu_q}}\right]. \tag{4.60}$$

In Eq. (4.60) $\gamma_q = \gamma_q^A + \gamma_q^B$, where

$$\begin{aligned}\gamma_q^A &= \frac{1}{2}\beta V K q^2 \left[A(\mathbf{q}) - i\frac{n}{\beta V K}\frac{\cos(\mathbf{q}\cdot\mathbf{r})}{q^2}\right]^2 \\ &+ \frac{n^2}{2\beta V K}\frac{\cos^2(\mathbf{q}\cdot\mathbf{r})}{q^2},\end{aligned} \tag{4.61}$$

$$\gamma_q^B = \frac{1}{2}\beta V K q^2 \left[B(\mathbf{q}) - i\frac{n}{\beta V K}\frac{\sin(\mathbf{q}\cdot\mathbf{r})}{q^2}\right]^2 + \frac{n^2}{2\beta V K}\frac{\sin^2(\mathbf{q}\cdot\mathbf{r})}{q^2}, \tag{4.62}$$

and

$$\nu_q = \frac{1}{2}\beta V K q^2 [A^2(\mathbf{q}) + B^2(\mathbf{q})]. \tag{4.63}$$

Taking into account that $\int_0^\infty dx \exp(-\lambda x^2) = \sqrt{\pi/\lambda}$, the thermal averages become

$$\langle \cos(n\vartheta)\rangle_{loc} = \exp\left(-\frac{n^2}{2\beta V K}\sum_{\mathbf{q}}\frac{1}{q^2}\right). \tag{4.64}$$

Since

$$\frac{1}{V}\sum_{q=q_{min}\sim 0}^{q_{max}=2\pi/a}\frac{1}{q^2} = \frac{1}{8\pi a}, \tag{4.65}$$

it follows that $\langle \cos(n\vartheta)\rangle_{loc} = \exp[-(1/2)n^2 t]$, where $t = 1/(4\pi\beta K a)$ and a is a mesoscopic length of the order of the nematic-isotropic coherence length [40]. Therefore,

$$\langle \cos(2m\vartheta)\rangle_{loc} = \exp[-2m^2 t],$$

and, consequently, using Eq. (4.50),

$$\langle P_{2l}(\mathbf{n}\cdot\mathbf{u})\rangle = \sum_{m=0}^{l} b_{2m}^{2l}\exp[-2m^2 t], \tag{4.66}$$

which represents a generalization of the result reported in [40].

The expression used above for f_h is valid only in the quadratic approximation for the elastic energy. It follows that the thermal fluctuations of $\mathbf{u}$ with respect to $\mathbf{n}$ have to be small. This implies that $\vartheta \ll 1$, and hence, as follows from the expression for $\langle \cos(n\vartheta)\rangle_{loc}$ written above, $t \ll 1$. In this approximation, $\exp[-2m^2 t] \sim \cos(2m\sqrt{t})$. By substituting this result into Eq. (4.66) one obtains

$$\langle P_{2l}(\mathbf{n}\cdot\mathbf{u})\rangle = \sum_{m=0}^{l} b_{2m}^{2l}\cos(2m\sqrt{t}). \tag{4.67}$$

It is useful, for further considerations, to put $\sqrt{t} = \langle\vartheta\rangle$. In this manner Eq. (4.67) reads

$$\langle P_{2l}(\mathbf{n}\cdot\mathbf{u})\rangle = \sum_{m=0}^{l} b_{2m}^{2l}\cos(2m\langle\vartheta\rangle) = P_{2l}(\cos\langle\vartheta\rangle), \tag{4.68}$$

if Eq. (4.49) is taken into account. The macroscopic anchoring energy can be rewritten, as follows from Eq. (4.21) and Eq. (4.68), as

$$w(\theta,\phi) = \sum_{l=0}^{\infty} w_{2l}(0)P_{2l}(\cos\langle\vartheta\rangle)L_{2l}(\theta,\phi). \tag{4.69}$$

By taking into account that $\langle\vartheta\rangle \ll 1$

$$\langle P_{2l}(\mathbf{n}\cdot\mathbf{u})\rangle \sim \exp\left\{-\frac{l(2l+1)}{2}\langle\vartheta\rangle^2\right\}, \tag{4.70}$$

and

$$\langle P_2(\mathbf{n}\cdot\mathbf{u})\rangle = S \sim \exp\left\{-\frac{3}{2}\langle\vartheta\rangle^2\right\}. \tag{4.71}$$

From Eq. (4.71) the thermal average of ϑ is connected to S by the relation $\langle\vartheta\rangle^2 = -(2/3)\ln S$. Substituting this result into Eq. (4.70) one finally obtains

$$\langle P_{2l}(\mathbf{n}\cdot\mathbf{u})\rangle = S^{l(2l+1)/3}, \tag{4.72}$$

which coincides with the Akulov-Zener law reported above in the low temperature region.

This model has been compared with the experimental data in Refs. [21], [47], and [48] and demonstrated a good agreement.

4.2 Temperature dependence of the surface tension near the nematic-isotropic transition

In this section, we consider the influence of a surface on the nematic-isotropic phase transitions by means of a local self-consistent approach [49]. In the bulk, the system undergoes first-order phase transition. However, near the substrate, a boundary layer can appear in which the ordering is very different from the bulk one. Theoretically, the existence of this boundary layer in a nematic sample was predicted by Sheng [50], using a Landau-de Gennes theory. In the work of Sheng [50] the nematic order at the surface was assumed perfect and temperature independent. It has been shown that there is a critical thickness of the sample below which the transition from the nematic phase to the isotropic phase becomes continuous. Subsequently, experimental works have verified the dependence of the nematic-isotropic phase transition on the thickness of the sample [51, 52]. To take into account the influence of the surface treatment on the nematic-isotropic phase transition [53], Sheng extended his model to consider also the surface energy [54]. It was assumed to describe the direct interaction between the liquid crystal with the substrate. According to this more general model, the surface order parameter was found temperature dependent. A detailed study of the nematic wall interface was carried out through a molecular model of the nematic free energy [32].

The Maier-Saupe mean field theory has been used to investigate the behavior of a nematic liquid crystal in the presence of an external field [55] and near the wall in the presence of a surface ordering field [24, 25]. The basic mechanisms involved in these phenomena were discussed in these works. The analyses were performed by assuming that the substrate was treated in such a manner to induce homeotropic orientation to the nematic liquid crystal. In this case, the biaxiality due to the presence of the surface can be neglected. Furthermore, the surface potential describing the interaction between the substrate and the liquid crystal was assumed as short range in nature and represented in the form $-G\delta(z)S$, where z is the coordinate normal to the surface, S is the scalar order parameter and G is a constant denoting the strength of the potential.

In this section, we reanalyze the problem of a semi-infinite sample in the framework of the Maier-Saupe mean field theory by considering a more realistic form for the potential, which takes into account the van der Waals nature of the surface field, and the incomplete interaction among the nematic molecules due to the presence of a surface. This fact leads to significant modifications in the critical surface behavior of a liquid-crystalline system: while it does not change the qualitative aspects of the phase diagram, the quantitative description is more rigorous. In particular, the determination of the temperature dependence of the surface tension near the nematic-isotropic phase transition can be easily done, and the general trends agree with experimental data [56]. It is possible to show that, according to the strength of the surface field, the surface tension presents a minimum just before the transition, followed by a strong variation at the transition.

Let us consider a sample of nematic liquid crystal bounded on one side by a substrate in such a way that the solid-NLC interface is defined in $z = 0$, and the sample is assumed to be uniform in the x and y directions. As in [24, 25, 55] it is supposed that the easy axis of the interface is normal to it, and that the nematic liquid crystal can be considered a uniaxial medium also at the surface. If θ denotes the angle between the long axis of a molecule with the z axis, the van der Waals interaction with the surface can be expressed by

$$V_s(z) = -H(z)P_2(\theta) = -\frac{h}{z^3}P_2(\theta), \tag{4.73}$$

where $P_2(\theta)$ denotes the Legendre polynomial and h is a constant coupling. The interaction of a nematic molecule with the substrate is minimum for

$\theta = 0$, i.e., for a homeotropic alignment at the surface, as assumed above. $V_s(z)$ has the functional form similar to the interaction of an NLC with an external field of amplitude $H(z)$ and anisotropy equal to one.

The bulk potential, taking into account the incomplete interaction between the nematic molecules, can be put in the form

$$V_N(\theta, z) = -v \left[1 - \left(\frac{r_0}{2z} \right)^3 \right] P_2(\theta) S(z), \tag{4.74}$$

where $S(z)$ is now the z-dependent scalar order parameter, whereas v is the Maier-Saupe molecular parameter already introduced [see also Eqs. (1.33) and (1.34)]. r_0 is a typical molecular dimension which is defined as the lowest molecular cut-off of the van der Waals interaction. Equations (4.73) and (4.74) hold in the r^{-6} van der Waals approximation for the molecular interactions. In this manner, the effective total potential is given by

$$V(\theta, \zeta) = V_s + V_N = -v \left[\left(1 - \frac{1}{\zeta^3} \right) S(\zeta) + \frac{\mu}{\zeta^3} \right] P_2(\theta), \tag{4.75}$$

where we have introduced the reduced quantities

$$\zeta = 2\frac{z}{r_0}, \qquad 2 \le \zeta < \infty, \tag{4.76}$$

and

$$\mu = \frac{8h}{v r_0^3}, \tag{4.77}$$

which denotes the strength of the surface field measured in units of the strength of the bulk field. The total macroscopic potential, obtained by statistical average, is then given by

$$V = \langle V(\theta, \zeta) \rangle = -v \left[\left(1 - \frac{1}{\zeta^3} \right) S^2(\zeta) + \frac{\mu}{\zeta^3} S(\zeta) \right]. \tag{4.78}$$

This kind of ordering potential favors perpendicular alignment at the wall, and we do not consider competing potentials.

Notice that in this approach the z-dependence of the macroscopic potential comes from two sources: (1) the van der Waals surface field and (2) the incomplete interaction among the NLC molecules. As it follows from Eq. (4.78), it is impossible to compensate the incomplete nematic-nematic interaction by means of a surface interaction. In fact, the first contribution depends on S^2, whereas the second one is linear in this quantity. In particular, to $\mu = 0$ does not correspond the usual bulk nematic molecular potential.

To explore the thermodynamic effects of this kind of potential, in the framework of the Maier-Saupe model, we determine the temperature dependence of the scalar order parameter S through the local self-consistency relation

$$S(\zeta) = \frac{1}{Z} \int_{-1}^{1} P_2(\cos\theta) e^{-\beta V(\theta,\zeta)} d(\cos\theta). \tag{4.79}$$

Furthermore, in (4.79) Z is the single molecule orientational partition function, defined as

$$Z(\zeta) = \int_{-1}^{1} e^{-\beta V(\theta,\zeta)} d(\cos\theta). \tag{4.80}$$

The equilibrium value of the scalar order parameter is determined by choosing the solution which minimizes the free energy. The free energy for unit area is written as

$$\begin{aligned} f = \frac{F}{A} &= \int_{r_0}^{\infty} f_V(z) dz \\ &= \frac{r_0}{2} \int_2^{\infty} \rho \left[-\frac{1}{\beta} \ln Z(\zeta) + \frac{1}{2} v \left(1 - \frac{1}{\zeta^2} \right) S(\zeta)^3 \right] d\zeta, \end{aligned} \tag{4.81}$$

where ρ is the particle density of the fluid. It will be assumed, in a first approximation, as position and temperature independent. The surface tension connected to the nematic order can be easily obtained through the relation

$$\gamma(T) = \int_{r_0}^{\infty} [f_V(z) - f_b]\, dz, \tag{4.82}$$

where $f_b = f_V(z \to \infty)$ refers to the bulk value of $f_V(z)$. The total surface tension of an NLC is $\gamma_t(T) = \gamma_i(T) + \gamma(T)$, where $\gamma_i(T)$ is the isotropic part. As in normal liquid, it is a decreasing function of T, and does not present critical behavior close to the nematic-isotropic phase transition temperature (T_{NI}). On the contrary, $\gamma(T)$ is connected with the spatial variation of the scalar order parameter $S = S(z)$. This contribution to $\gamma_t(T)$ is expected to exhibit a critical behavior at T_{NI}. To complete the analysis we also introduce an adsorption parameter as is done in [24], defined as

$$\Gamma(T) = \int_{r_0}^{\infty} [S(z) - S_b]\, dz, \tag{4.83}$$

where S_b is the bulk value of the scalar order parameter. The function Γ is a measure of the extra order near the wall.

The trends of $S(z)$ versus z are exhibited in Figure 4.2 for a temperature slightly below the bulk transition temperature for illustrative values of the parameter μ. $S(z)$ changes drastically in a surface layer whose thickness is of the order of several molecular dimensions, and tends to the bulk value for large z. Notice however that the order parameter has a positive slope for

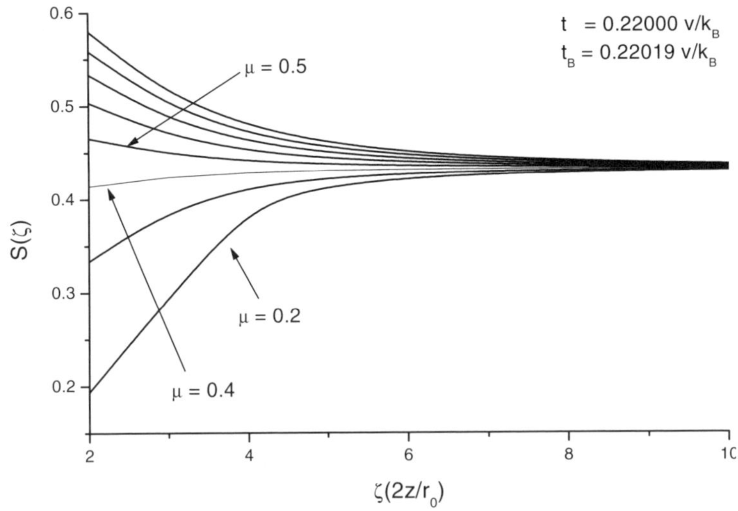

FIGURE 4.2
The spatial dependence of the order parameter for a temperature close to the bulk nematic-isotropic transition. Reprinted with permission from Ref. [49]. Copyright (2002) by the American Physical Society.

$\mu < \mu_0$ and a negative one for $\mu > \mu_0$. The existence of μ_0 is easily understood because $dS(z)/dz = 0$ at least in two circumstances: when $z \longrightarrow \infty$, i.e., in the bulk, where the order parameter has a constant value, or in the case in which its z-dependence is absent. This is possible when $\mu_0 = S_0(T)$. Therefore, for the case depicted in Figure 4.2, the value of μ corresponding to the inversion of the slope of order parameter is $\mu = 0.43274$, which coincides with the value of the order parameter in the entire sample because the z-dependence is completely lost at this value. Notice, however, that once the value of μ is fixed, there is only one temperature for which the effect of the incomplete interactions is compensated by the effect of the van der Waals field. This value of the surface field strength indicates the regions for which the interface is more or less ordered than the bulk, for a given temperature below the critical temperature.

The surface order parameter $S_0(T) = S(z = 0, T)$ as a function of the reduced temperature ($t = k_B T/v$) is reported in Figure 4.3, for a few values of the surface field μ defined in (4.77). An evident feature in the behavior of $S_0(T)$ is that it is discontinuous at the surface transition temperature until $\mu = \mu^* \approx 0.07$. In this region the surface transition temperature is always lower than the bulk transition temperature. As pointed out earlier, the effect of the incomplete interaction is to favor the destabilization of the homeotropic pattern near the wall, working in the model as an extra repulsive interaction

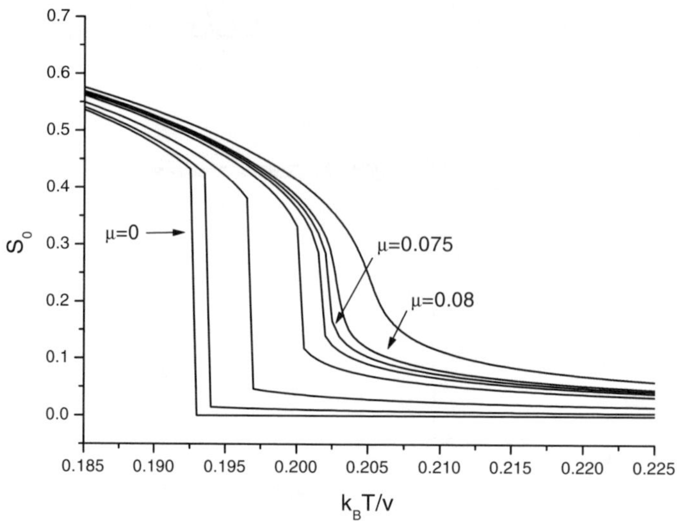

FIGURE 4.3
Surface order parameter $S_0 = S(z = 0, T)$ versus reduced temperature $t = k_B T/v$ for different values of the surface field strength μ. Reprinted with permission from Ref. [49]. Copyright (2002) by the American Physical Society.

which tends to destroy the ordering in the vicinity of the interface. For $\mu > \mu^*$, the discontinuous transition disappears and $S_0(T)$ becomes a continuous function of the temperature.

In Figure 4.4, the surface tension, in $\gamma_0 = \rho v r_0/2$ units, versus the reduced temperature for different values of μ is reported. The case of $\mu = 0$ corresponds to an NLC in contact with the vacuum. This case is not important from the experimental point of view, for obvious reasons. By assuming $\rho \approx 2.5\,10^{21}$ cm^{-3} [57], $v = k_B T_{NI}/0.22019 \approx 2.6\,10^{-13}$ erg [27], and $r_0 \approx 20\mathring{A}$, we obtain $\gamma_0 \approx 65$ erg/cm^2, for PAA. In this case, γ is estimated one order of magnitude smaller than the surface tension detectable by experimental techniques [57].

In Figure 4.5, we show the surface tension $\gamma(T)$ predicted by the model and we compare it with the experimental data concerning the PAA liquid crystal in contact with its vapor, as reported in [56]. For the fit we have assumed that close to T_{NI} ($t = 0.22019 v/k_B$), the isotropic part of the surface tension ($\gamma_i(T)$) is practically temperature independent, which represents a rough approximation. As is evident in this framework the agreement is rather good for $T/T_{NI} < 1.04$. For large temperatures, the disagreement between the theory and the experimental data is probably connected to the hypotheses of a constant density of the liquid crystal, and of a γ_i temperature independent, that do not work well in this temperature range. It is possible to compare the theory with the data of $\gamma(T)$ for different liquid crystals as reported by [57], showing a different behavior in the low temperature region. As shown in

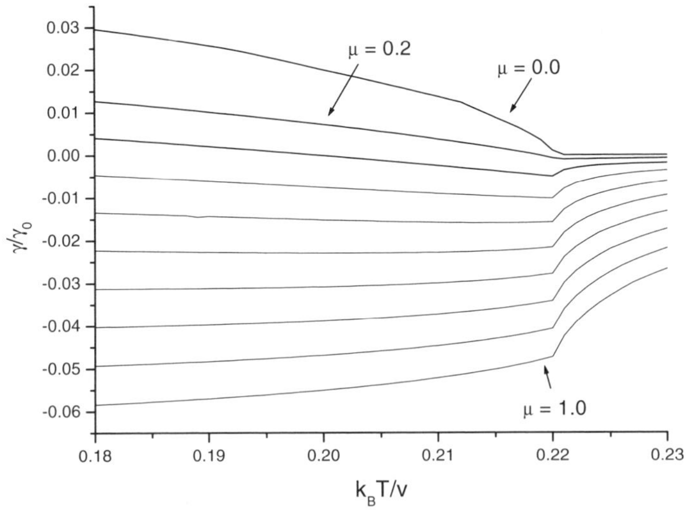

FIGURE 4.4
Surface tension γ/γ_0 versus the reduced temperature $t = k_{\rm B}T/v$ for different values of the surface field strength μ, where $\gamma_0 = \rho v r_0/2 \approx 65$ erg/cm^2. Reprinted with permission from Ref. [49]. Copyright (2002) by the American Physical Society.

Figure 4.4, it is connected to the different values of μ. Note, in particular, that for $\mu > 0.3$, $\gamma(T)$ presents a minimum in the nematic phase and it starts to increase just close to T_{NI}. We underline that, as pointed out in page 84 of Ref. [57], the model represents one of the first successful attempts to interpret the experimental data of $\gamma(T)$, obtained long ago.

Finally, in Figure 4.6 we present the trend of the adsorption function $\Gamma(T)$ versus the temperature $\mu = 0.3$. Γ presents a pronounced variation in the vicinity of the transition point, as experimentally observed by [53].

4.3 Surface director gliding in lyotropic liquid crystals

The elementary constituents of a liquid-crystalline system are molecules in the case of *thermotropics*, where the mesogenic behavior is controlled by the temperature, and *micelles* in the case of *lyotropics* [58], in which the mesogenic behavior is determined by the concentration. In thermotropic liquid crystals all the known surface orientational effects are reversible: removing the distorting field, the sample recovers the original orientational state [59]. On the contrary, in lyotropic liquid crystals, it has been recently observed

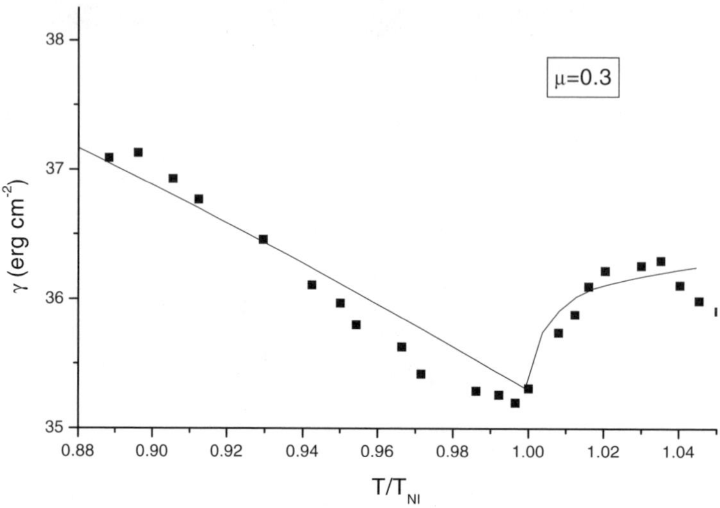

FIGURE 4.5
Surface tension γ versus T/T_{NI}, where T_{NI} is the bulk nematic-isotropic phase transition. Squares are the experimental data of Ref. [56] for PAA; solid lines are the results of the presented model. Reprinted with permission from Ref. [49]. Copyright (2002) by the American Physical Society.

that a slow gliding of a planar degenerate surface orientation can be induced by an external orienting magnetic field of the order of only a few kG [60]–[62]. This irreversible process has been interpreted by means of a phenomenological model, by supposing that at the surface a bilayer of amphiphilic molecules is present, with defects or channels as in a micellar structure [60]. This model explains the observed dependence of the relaxation time on the amplitude of the applied magnetic field, but, in order to fit the order of magnitude of these relaxation times, must introduce an effective surface viscosity connected to the bulk one by the macroscopic dimension of the objects that are supposed to rigidly participate in the surface orienting phenomenon.

In this section we discuss a microscopic approach to the surface gliding effect in terms of a thermally activated process along the lines of Ref. [63]. According to this model, the micelles are rotationally pinned to the surface by a potential displaying successive local minima in correspondence with metastable orientational states. For simplicity, it is assumed that the potential is periodic, with equispaced minima separated by the same potential barrier U_0 (see Figure 4.7a). In the absence of external orienting fields, the surface micelles rotate against the potential barriers with an attempt frequency τ_0^{-1} and they succeed in overcoming the barrier with a rate $\tau^{-1} = \tau_0^{-1}\, \exp[-U_0/(k_B T)] \ll \tau_0^{-1}$. During the jump from one metastable state to the adjacent one, a bunch of phonons is radiated, leading to an irreversible process. These jumps give rise

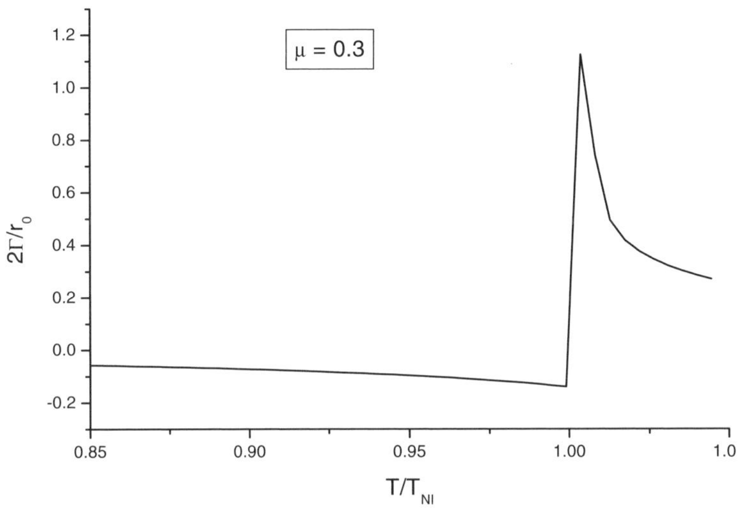

FIGURE 4.6
Adsorption function Γ versus T/T_{NI}. This function is a measure of the extra order near the wall [24]. For $T/T_{NI} > 1$ from the analysis one can also obtain the induced birefringence in the isotropic phase, defined as $\Delta\phi = \int_0^\infty S(z)dz$ [53]. This quantity presents a sharp variation close to T_{NI}. By assuming $r_0 \approx 20\mathring{A}$ and $\mu \approx 1$ (which corresponds to the experimental situation analyzed in [53], where the NLC was strongly homeotropically oriented by a surfactant) one obtains $\Delta\phi$ of the order of several angstroms, as experimentally detected. Reprinted with permission from Ref. [49]. Copyright (2000) by the American Physical Society.

to a uniform *diffusion*, which is contrasted, as we shall see in the following, by the elasticity of the nematic. An equilibrium surface distributions of angles at the surface results, which is rotationally invariant: the mean surface twist angle only depends on the initial conditions. In the presence of an external magnetic field the rotational symmetry of the surface potential is broken (see Figure 4.7b): the transition probability for rotation in the direction of the field is higher than in the opposite direction. Hence, a *drift* of the initial probability distribution in the direction of the field occurs, giving rise to the gliding process. The relaxation times are expected to be much longer than the characteristic thermal equilibration time τ, as they result from the small imbalance of the potential barriers introduced by the presence of the field. This model of the surface gliding recalls the solid friction model involving the pinning between the asperities of two solid surfaces in contact [64].

4.3.1 Surface molecular energy model

A semi-infinite nematic lyotropic sample extending in the region $z > 0$ of a Cartesian coordinate frame, with the $z = 0$ plane coincident with an isotropic solid substrate is considered. A magnetic field $\mathbf{H}$ is applied parallel to the (x, y) substrate plane at an angle $\theta_{\rm H}$ with respect to the x-axis. We suppose that the nematic lyotropic can be modeled as an assembly of layers, parallel to the solid substrate, having a thickness ℓ comparable with the molecular short diameter. Each layer is labeled with an index $i = 0, 1, 2, \ldots$, where the zero-th layer is the one in contact with the substrate.

Consider an Ising-like mean-field model [65] for the interaction between the layers. Therefore, for a given bulk layer $i > 0$, the total energy per unit surface g_i can be written as

$$\begin{aligned} g_i = &- J\left[\left(\langle \mathbf{n}_i\rangle \cdot \langle \mathbf{n}_{i+1}\rangle\right)^2 + \left(\langle \mathbf{n}_i\rangle \cdot \langle \mathbf{n}_{i-1}\rangle\right)^2\right] - J'\sum_{k=1}^{N}\left(\mathbf{n}_i^{(k)} \cdot \langle \mathbf{n}_i\rangle\right)^2 \\ &- \frac{1}{2}\frac{\chi_{\rm am}}{NA}\sum_{k=1}^{N}\left(\mathbf{n}_i^{(k)} \cdot \mathbf{H}\right)^2 . \end{aligned} \tag{4.84}$$

Here $\mathbf{n}_i^{(k)}$ is the nematic director for the k-th micelle in the i-th layer, $\langle \mathbf{n}_i\rangle$ is its average in the layer, J is the coupling constant between adjacent layers, J' is the self-coupling in each layer–favoring an homogeneous orientation–, $\chi_{\rm am}$ is the micellar diamagnetic anisotropy, that is assumed as positive; finally N is the number of micelles in each layer and A the area occupied by each micelle.

Similarly, the total energy per unit surface g_0 of the layer in contact with the solid substrate can be written as

$$g_0 = -J\left(\langle \mathbf{n}_0\rangle \cdot \langle \mathbf{n}_1\rangle\right)^2 - J'\sum_{k=1}^{N}\left(\mathbf{n}_0^{(k)} \cdot \langle \mathbf{n}_0\rangle\right)^2 - \frac{1}{2}\frac{\chi_{\rm am}}{NA}\sum_{k=1}^{N}\left(\mathbf{n}_0^{(k)} \cdot \mathbf{H}\right)^2 + U_0 , \tag{4.85}$$

where U_0 accounts for the interaction between the surface layer and the substrate.

In the following it is assumed that all the micelles lie everywhere parallel to the (x, y)-plane, such that

$$\mathbf{n}_i^{(k)} = \cos\theta_i^{(k)}\,\mathbf{x} + \sin\theta_i^{(k)}\,\mathbf{y} , \tag{4.86}$$

$$\langle \mathbf{n}_i\rangle = \cos\bar{\theta}_i\,\mathbf{x} + \sin\bar{\theta}_i\,\mathbf{y} , \tag{4.87}$$

where $\mathbf{x}$ and $\mathbf{y}$ are the x and y-axis unit versors, respectively. Therefore,

$$\sum_{k=1}^{N}\left(\mathbf{n}_i^{(k)} \cdot \langle \mathbf{n}_i\rangle\right)^2 = N\langle\cos^2(\theta_i - \bar{\theta}_i)\rangle , \tag{4.88}$$

$$\sum_{k=1}^{N}\left(\mathbf{n}_i^{(k)} \cdot \mathbf{H}\right)^2 = NH^2\langle\cos^2(\theta_i - \theta_{\rm H})\rangle . \tag{4.89}$$

By performing the expansion

$$\bar{\theta}_{i\pm 1} \cong \bar{\theta}_i \pm \frac{d\bar{\theta}_i}{dz}\,\ell , \tag{4.90}$$

it follows that

$$\left(\langle \mathbf{n}_i\rangle \cdot \langle \mathbf{n}_{i\pm 1}\rangle\right)^2 = \cos^2(\bar{\theta}_i - \bar{\theta}_{i\pm 1}) \cong 1 - \frac{1}{2}\left(\frac{d\bar{\theta}_i}{dz}\right)^2 \ell^2 . \tag{4.91}$$

Consequently, apart from inessential constants,

$$g_i = J\ell^2\left(\frac{d\bar{\theta}_i}{dz}\right)^2 - J'N\langle\cos^2(\theta_i - \bar{\theta}_i)\rangle - \frac{1}{2}\frac{\chi_{\rm am}}{A}H^2\langle\cos^2(\theta_i - \theta_{\rm H})\rangle , \tag{4.92}$$

and

$$g_0 = \frac{1}{2}J\ell^2\left(\frac{d\bar{\theta}_0}{dz}\right)^2 - J'N\langle\cos^2(\theta_0 - \bar{\theta}_0)\rangle - \frac{1}{2}\frac{\chi_{\rm am}}{A}H^2\langle\cos^2(\theta_0 - \theta_{\rm H})\rangle + U_0. \tag{4.93}$$

To connect the microscopic parameters entering in Eqs. (4.92) and (4.93) with the usual macroscopic constants appearing in the framework of the continuum elastic theory, one writes

$$\theta_i^{(k)} = \bar{\theta}_i + \Delta\theta_i^{(k)} \,, \tag{4.94}$$

where by definition $\langle \Delta\theta_i \rangle = 0$, such that

$$\langle \cos^2(\theta_i - \bar{\theta}_i) \rangle \cong 1 - \frac{1}{2} \langle \Delta\theta_i{}^2 \rangle \tag{4.95}$$

and

$$\begin{aligned} \langle \cos^2(\theta_i - \theta_{\rm H}) \rangle &= \langle \cos^2(\bar{\theta}_i - \theta_{\rm H} + \Delta\theta_i) \rangle \\ &\cong \langle \cos^2(\bar{\theta}_i - \theta_{\rm H}) - \sin[2(\bar{\theta}_i - \theta_{\rm H})]\, \Delta\theta_i \rangle \\ &= \cos^2(\bar{\theta}_i - \theta_{\rm H}) \,. \end{aligned} \tag{4.96}$$

Consequently, the total energy per unit surface of the i-th bulk layer (4.92) becomes, apart from inessential constant terms,

$$g_i = J\ell^2 \left(\frac{d\bar{\theta}_i}{dz} \right)^2 + \frac{1}{2} J' N \langle \Delta\theta_i{}^2 \rangle - \frac{1}{2} \frac{\chi_{\rm am}}{A} H^2 \cos^2(\bar{\theta}_i - \theta_{\rm H}) \,. \tag{4.97}$$

According to the continuum elastic theory, in the one-constant approximation the bulk energy density is given by (1.76) (for $\phi' = 0$), by taking into account the interaction energy with the magnetic field (1.72), i.e.,

$$f = \frac{1}{2} K \, (\nabla\theta)^2 - \frac{1}{2} \chi_{\rm a} H^2 \cos^2(\theta - \theta_{\rm H}). \tag{4.98}$$

In this geometry, where θ mainly depends on the distance z from the surface, one can make the approximation

$$(\nabla\theta)^2 = \left(\frac{\partial\theta}{\partial x} \right)^2 + \left(\frac{\partial\theta}{\partial y} \right)^2 + \left(\frac{\partial\theta}{\partial z} \right)^2 \cong \frac{\langle \Delta\theta^2 \rangle}{A} + \left(\frac{\partial\theta}{\partial z} \right)^2 . \tag{4.99}$$

By comparing Eq. (4.97) with ℓf given by (4.98) and (4.99), one arrives at

$$J = \frac{K}{2\ell} \,, \quad J' = \frac{K\ell}{2NA} \,, \quad \chi_{\rm am} = \chi_{\rm a} \, A \, \ell \,. \tag{4.100}$$

By neglecting in Eq. (4.98) the small variations of the twist angle θ in the (x, y)-plane, standard calculations give for the actual stable profile [59]

$$\left(\frac{d\theta}{dz} \right)^2 = \frac{\chi_{\rm a} H^2}{K} \sin^2(\theta - \theta_{\rm H}) \,. \tag{4.101}$$

Then, from (4.93) one has

$$g_0 \, A = \langle \mathcal{U} \rangle \,, \tag{4.102}$$

where $\mathcal{U}$ can be identified as the total energy of a surface micelle

$$\mathcal{U} = \frac{1}{4}\chi_{\rm am} H^2 \sin^2(\bar{\theta}_0 - \theta_{\rm H}) + \frac{1}{2}\chi_{\rm am} H^2 \sin^2(\theta_0 - \theta_{\rm H}) - \frac{1}{2} K\ell \cos^2(\theta_0 - \bar{\theta}_0) + \mathcal{U}_0, \tag{4.103}$$

with $\mathcal{U}_0 = U_0\, A$.

4.3.2 Master equation and Fokker-Planck approximation

Suppose that the possible angular positions of the surface micelles are quantized according to

$$\theta_n = n\,\delta\,, \quad (n = 0, \pm 1, \pm 2, \ldots)\,. \tag{4.104}$$

Let P_n be the probability to find a micelle with the orientation θ_n. Evidently, the time evolutions of the probabilities obey the *master equation*

$$\frac{dP_n}{dt} = W_{n-1,n}P_{n-1} + W_{n+1,n}P_{n+1} - (W_{n,n-1} + W_{n,n+1})P_n\,, \tag{4.105}$$

where $W_{n,m}$ is the transition probability between two adjacent angles θ_n and θ_m, with $n - m = \pm 1$ (see Figures 4.7). According to statistical mechanics [65]

$$W_{m,m\pm 1} = \frac{1}{\tau_0} \exp\left[\frac{\mathcal{U}(m\delta) - \mathcal{U}((m \pm 1/2)\delta)}{k_{\rm B}T}\right]\,, \tag{4.106}$$

where τ_0^{-1} the trial frequency. In (4.106) $\mathcal{U}((m \pm 1/2)\delta) - \mathcal{U}(m\delta)$ represents the height of the barrier that a micelle has to overcome to jump between adjacent angular positions. For $\delta \ll 1$, according to Eq. (4.103), one can write

$$W_{m,m\pm 1} = \frac{1}{\tau_0} \exp\left\{-u_0 \mp \frac{1}{2}\left[h^2 \sin[2(\theta_m - \theta_{\rm H})] + \kappa \sin[2(\theta_m - \bar{\theta})]\right]\delta\right\}, \tag{4.107}$$

where

$$u_0 = \frac{\mathcal{U}_0((m \pm 1/2)\delta) - \mathcal{U}_0(m\delta)}{k_{\rm B}T} \tag{4.108}$$

is the height of the barrier, that we suppose constant, between two adjacent angular positions due to the surface anchoring energy; h and κ are the normalized magnetic field and elastic constant, respectively,

$$h^2 = \frac{\chi_{\rm am} H^2}{2k_{\rm B}T} \quad \text{and} \quad \kappa = \frac{K\ell}{2k_{\rm B}T}\,. \tag{4.109}$$

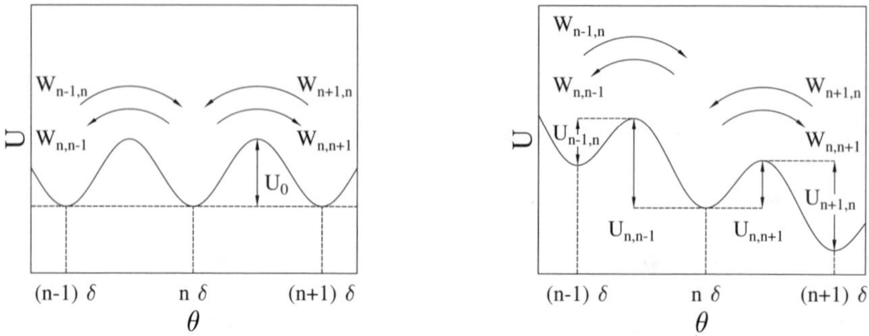

FIGURE 4.7
Schematic representation of the surface potential in the absence a) and in the presence b) of an orienting magnetic field. An extra elastic contribution is actually present, which contrasts the diffusion in the absence of field. Reprinted with permission from Ref. [63]. Copyright (1997) by the American Physical Society.

In Eq. (4.107) the subscript 0 on the angles has been omitted, for the sake of simplicity. Note that in Eq. (4.107), the term proportional to h^2 is connected with the drift of the surface micelles toward $\theta_{\rm H}$, whereas the term proportional to κ contrasts the diffusion. The height of the barrier u_0 renormalizes the trial frequency, allowing us to define the characteristic diffusion time

$$\tau = \tau_0 \exp\left(u_0\right) \ . \tag{4.110}$$

The average surface twist angle $\bar{\theta}$ in (4.107) defines the direction of the macroscopic surface director, which is the eigenvector associated with the positive eigenvalue of the traceless surface order parameter

$$q = 2\left\langle \mathbf{n} \otimes \mathbf{n} - \frac{1}{2} I \right\rangle \ , \tag{4.111}$$

where I is the identity tensor. In the diagonal frame

$$q = \begin{pmatrix} s & 0 \\ 0 & -s \end{pmatrix} \ , \tag{4.112}$$

where $0 \leq s \leq 1$ is the 2D surface order parameter

$$s = \sqrt{\left(2\langle\cos^2\theta\rangle - 1\right)^2 + \langle\sin(2\theta)\rangle^2} \ . \tag{4.113}$$

According to the previous definition

$$\bar{\theta} = \tan^{-1}\left[\frac{\langle \sin(2\theta)\rangle}{2\langle \cos^2\theta\rangle + s - 1}\right] . \tag{4.114}$$

In the limit of small jumps $\delta \to 0$, one can perform the expansions

$$P_{n\pm1} = P(\theta) \pm \frac{\partial P}{\partial \theta}\,\delta + \frac{1}{2}\frac{\partial^2 P}{\partial \theta^2}\,\delta^2 , \tag{4.115}$$

$$W_{n\pm1,n} = W(\theta,\theta'=\theta) \pm \frac{\partial W}{\partial \theta}\,\delta + \frac{1}{2}\frac{\partial^2 W}{\partial \theta^2}\,\delta^2 , \tag{4.116}$$

$$W_{n,n\pm1} = W(\theta,\theta'=\theta) \pm \frac{\partial W}{\partial \theta'}\,\delta + \frac{1}{2}\frac{\partial^2 W}{\partial \theta'^2}\,\delta^2 , \tag{4.117}$$

with

$$W(\theta,\theta'=\theta) = \frac{1}{\tau} , \tag{4.118}$$

$$\frac{\partial W}{\partial \theta} = \frac{1}{2\tau}\frac{du}{d\theta} , \tag{4.119}$$

$$\frac{\partial W}{\partial \theta'} = -\frac{1}{2\tau}\frac{du}{d\theta} , \tag{4.120}$$

$$\frac{\partial^2 W}{\partial \theta^2} = \frac{1}{4\tau}\left(\frac{du}{d\theta}\right)^2 + \frac{1}{2\tau}\frac{d^2u}{d\theta^2} , \tag{4.121}$$

$$\frac{\partial^2 W}{\partial \theta'^2} = \frac{1}{4\tau}\left(\frac{du}{d\theta}\right)^2 - \frac{1}{2\tau}\frac{d^2u}{d\theta^2} , \tag{4.122}$$

where, according to (4.103),

$$u(\theta) = \frac{\mathcal{U} - \mathcal{U}_0}{K_{\mathrm{B}}T} = \frac{1}{2}h^2\sin^2(\bar{\theta} - \theta_{\mathrm{H}}) + h^2\sin^2(\theta - \theta_{\mathrm{H}}) + \kappa\sin^2(\theta - \bar{\theta}) . \tag{4.123}$$

In this continuum-θ approximation, the master equation (4.105) therefore reduces to the following *Fokker-Planck* equation for the probability density $\mathcal{P}(\theta,t)$

$$\frac{\partial \mathcal{P}}{\partial t'} = \frac{\partial}{\partial \theta}\left(\frac{du}{d\theta}\mathcal{P}\right) + \frac{\partial^2 \mathcal{P}}{\partial \theta^2} , \tag{4.124}$$

being $t' = D\,t$ the normalized time, where D is the rotational surface diffusion coefficient

$$D = \lim_{\delta \to 0}\frac{\delta^2}{\tau} . \tag{4.125}$$

4.3.3 Analytical and numerical results

By considering the Fokker-Planck approximation (4.124), it readily follows the steady state solution [66]

$$\mathcal{P}_s(\theta) = \frac{\exp[-u(\theta)]}{\displaystyle\int_0^\pi \exp[-u(\theta)]\, d\theta} \ . \tag{4.126}$$

As expected, it is a Boltzmann distribution. For a sufficiently well-peaked distribution, one can set

$$\bar{\theta} \cong \langle \theta \rangle = \int_0^\pi \theta\, \mathcal{P}(\theta)\, d\theta \ , \tag{4.127}$$

such that, from (4.124)

$$\frac{d\bar{\theta}}{dt'} = -\left\langle \frac{du}{d\theta} \right\rangle \ . \tag{4.128}$$

By neglecting the fluctuations [66], one then arrives at the following approximate evolution equation of the average angle

$$\frac{d\bar{\theta}}{dt'} = h^2 \sin[2(\theta_{\rm H} - \bar{\theta})] \ , \tag{4.129}$$

whose solution, for $\bar{\theta}(t' = 0) = 0$, is

$$\bar{\theta}(t') = \theta_{\rm H} - \tan^{-1}\left[\exp(-2h^2 t') \tan\theta_{\rm H}\right] \ . \tag{4.130}$$

According to Eq. (4.130), the (unnormalized) relaxation time $\tau_{\rm H}$ is inversely proportional to the square of the applied magnetic field

$$\tau_{\rm H} = \frac{1}{D\, h^2} \ . \tag{4.131}$$

This H^{-2} power-law is in agreement with the experimental findings [60]–[62]. In the next section we shall perform a quantitative comparison of our theoretical predictions with the experimental data obtained by the São Paulo group [60]–[62].

In Figure 4.8, the steady state probability P_s obtained by numerical integration of Eq. (4.105), in the absence of magnetic field, is shown for two values of the normalized elastic constant (solid line: $\kappa = 10$, dashed line: $\kappa = 200$). The average twist angle $\bar{\theta}$ is evidently arbitrary and solely depends on the initial conditions. For δ sufficiently small (here $\delta = \pi/100$) this numerical solution practically coincides with the analytical solution (4.126) of the Fokker-Planck continuum approximation. According to Eqs. (4.123) and (4.126), the width of the steady state distribution is inversely proportional to the square root of the elastic constant.

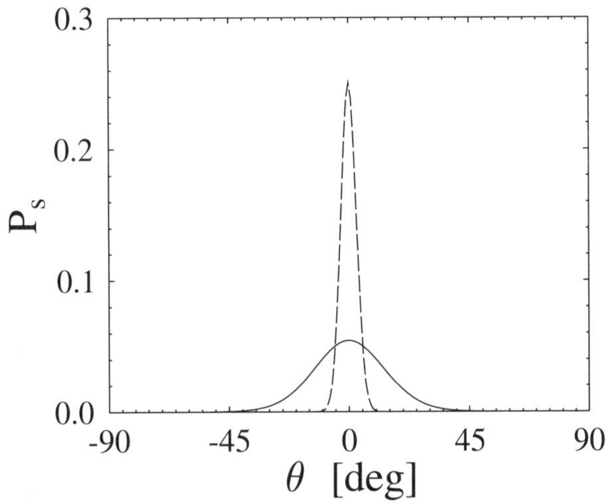

FIGURE 4.8
Steady state probability distribution P_s for $h = 0$, $\delta = \pi/100$ and two different values of κ (solid line: $\kappa = 10$, dashed line: $\kappa = 200$). The numerical solution of the master equation and the analytical result of the Fokker-Planck approximation coincide. Reprinted with permission from Ref. [63]. Copyright (1997) by the American Physical Society.

The time-evolution of the average twist angle $\bar{\theta}$ is shown in Figures 4.9, for different values of the normalized magnetic field h and normalized elastic constant κ, for the same value $\delta = \pi/100$ as before, with $\theta_{\rm H} = \pi/4$ and $\bar{\theta}(t' = 0) = 0$, where $t' = Dt$ is the previously defined normalized time. The initial probability distribution $P(t' = 0, \theta)$ is taken as the steady state solution $P_{\rm s}(\theta)$ in the absence of magnetic field. In Figure 4.9a, the amplitude of the normalized magnetic field is $h = 1$. The long (resp. short) dashed line is the numerical solution corresponding to $\kappa = 10$ (resp. $\kappa = 200$). The solid line is the analytical approximate solution (4.130) of the Fokker-Planck equation: it does not depend on κ, i.e. on the width of the steady state distribution, since the fluctuations in deriving (4.130) have been neglected. As is evident, the relaxation time somehow depends on κ. However, this effect can be taken into account by simply rescaling the magnetic field h: this is shown by the circles in the same Figure 4.9a, which correspond to the analytical solution (4.130) with $h' = 0.95\,h$. This rescaling is h-independent, as shown in Figure 4.9b, in which $h = 0.1$. The insets in Figures 4.9 show the probability distribution for three successive values of time. For larger values of the magnetic field h, the drift of the distribution is accompanied by a squeezing, while for smaller h the evolution is practically a rigid translation. This is evident from Eqs. (4.123) and (4.126), which show that under stationary conditions, where $\bar{\theta} = \theta_{\rm H}$, the width of the distribution is inversely proportional to $\sqrt{h^2 + \kappa}$. However, as one can see in the following, in practical cases $h^2 \ll \kappa$ and therefore the field-induced squeezing of the probability distribution is completely negligible.

4.3.4 Temperature behavior and comparison with the experimental data

The experimental data of the time evolution of the surface director gliding show a rather strong dependence of the relaxation time $\tau_{\rm H}$ on temperature, close to the first-order clearing point $T_{\rm c}$ (see Table II of Ref. [61]). This dependence cannot be explained by the temperature variation of the diamagnetic anisotropy $\chi_{\rm a}$: in fact, according to Eq. (4.131), $\Delta\tau_{\rm H}/\tau_{\rm H} = -\Delta\chi_{\rm a}/\chi_{\rm a}$. Now $\chi_{\rm a}$ decreases as the temperature increases, in a way essentially proportional to the bulk nematic order parameter S [59]; this would therefore give an increase in the relaxation time $\tau_{\rm H}$ with temperature, contrary to what is observed experimentally [61]. Moreover, for typical variations of the nematic order parameter [59], the effect due to the temperature variations of $\chi_{\rm a}$ is about 20 times smaller than the observed temperature variations of $\tau_{\rm H}$.

In order to explain this temperature dependence, one assumes that the depth of the potential well u_0 [see Eq. (4.108)] depends on the nematic order parameter S. In the isotropic phase, $S = 0$, one expects $u_0 = 0$. For increasing values of S we expect u_0 to increase. Therefore, by expanding u_0 in power series of S, the leading term expansion will be simply

$$u_0 = \alpha S \,. \tag{4.132}$$

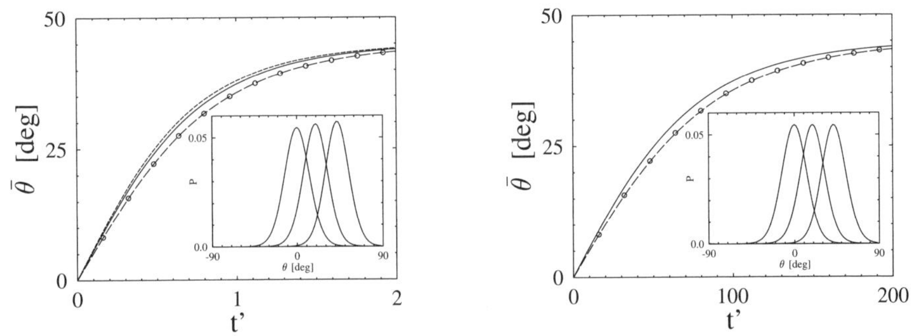

FIGURE 4.9
Time evolution of the average twist angle $\bar{\theta}$ as a function of the normalized time t' for $\delta = \pi/100$ and $\theta_{\rm H} = \pi/4$. a) $h = 1$ with $\kappa = 10$ (long-dashed line) and $\kappa = 200$ (short-dashed line). The solid line (resp. circles) is the analytical solution (4.130) with $h = 1$ (resp. $h = 0.95$). The inset shows the probability distribution $P(\theta, t')$ corresponding to the long-dashed curve as a function of θ for $t' = 0$, 0.4, and 1.6 (from left to right). b) Same as a) but for $\kappa = 10$ and $h = 0.1$ (long-dashed line); the analytical solution is plotted for $h = 0.1$ (solid line) and $h = 0.095$ (circles). The probability distributions in the inset are for $t' = 0$, 41, and 163 (from left to right). Reprinted with permission from Ref. [63]. Copyright (1997) by the American Physical Society.

This mean-field approximation is similar to the behavior for the angular anchoring energy proposed in Ref. [54].

In the nematic phase, for $T < T_c$, S is approximatively given by [27]

$$S = \Delta \sqrt{1 - \frac{T}{T_0}} \,, \tag{4.133}$$

where $T_0 - T_c \cong 1\,°\mathrm{C}$ [59]. From Eqs. (4.132) and (4.110), neglecting the small temperature dependence of χ_a, at first order in $T - T_c$ the characteristic diffusion time τ varies with temperature according to the linear dependence

$$\tau(T) = \tau(T_c) + \beta(T - T_c) \,, \tag{4.134}$$

where

$$\beta = \frac{\tau(T_c)}{2(T_0 - T_c)} \ln\left[\frac{\tau_0}{\tau(T_c)}\right] . \tag{4.135}$$

Consequently, the temperature dependence of the magnetic relaxation time (4.131) describing the surface director gliding writes

$$\tau_H(T) = \tau_H(T_c) + \beta_H(T - T_c) \,, \tag{4.136}$$

where

$$\beta_H = \frac{\beta}{\delta^2 h^2} \,. \tag{4.137}$$

The experimental results reported in [61] give $\beta_H \cong -56\,\mathrm{s}/°\mathrm{C}^{-1}$ for $H = 8\,\mathrm{kG}$. According to [67], $\chi_a = 10^{-8}$ (cgs units), while the molecular density $n = 1/(A\ell)$ is of the order of $n \cong 3 \times 10^{19}\,\mathrm{cm}^{-3}$ [68]. Therefore, for $H = 8\,\mathrm{kG}$ we have $h^2 \cong 2.6 \times 10^{-7}$. We note also that $\ell \cong 60\,\mathring{A}$ [69]; hence, with the elastic constant $K \cong 10^{-6}\,\mathrm{dyn}$, $\kappa \cong 8$. We can estimate the elementary angular jump δ as the ratio between the distance p between two pinning centers and the length L of a micelle; taking $p \cong 2\,\mathring{A}$ and $L \cong 100\,\mathring{A}$ [69], one obtains $\delta \cong p/L \cong 2 \times 10^{-2}$ rad. The trial frequency τ_0^{-1} is of the order of the molecular vibration frequencies, $\tau_0^{-1} \cong 10^{14}\,\mathrm{Hz}$ [70]. With these values, from Eq. (4.137) we get $\beta \cong -5.8 \times 10^{-9}\,\mathrm{s}/°\mathrm{C}^{-1}$; consequently, from Eq. (4.135), we have $\tau(T_c) \cong 10^{-9}\,\mathrm{s}$. Hence, at $H = 8\,\mathrm{kG}$, $\tau_H(T_c) = \tau(T_c)/(\delta^2 h^2) \cong 9.7\,\mathrm{s}$. For $T_c - T = 4\,°\mathrm{C}$ then $\tau_H(T) \cong 234\,\mathrm{s}$, in good agreement with the experimental value $\tau_H(T) \cong 250\,\mathrm{s}$ reported in Ref. [61]. The comparison with the full data presented in Ref. [61] is shown in Figure 4.10. We note also that with these data at $T = T_c$ we get a depth of the potential well per micelle of the order of $u_0\, k_B T \cong 0.3\,\mathrm{eV}$, which is a reasonable value for the interaction energy between one micelle and the substrate.

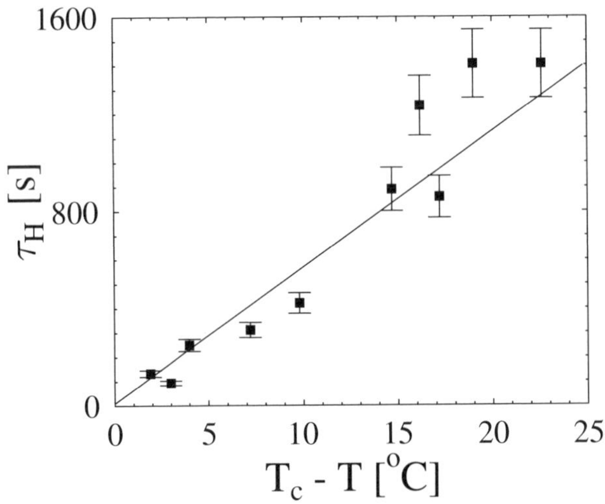

FIGURE 4.10
Comparison between the experimental data (squares) and the theoretical prediction (full line) for the relaxation time τ_H as a function of temperature. The error bars represent a 10% error on the experimental data. Reprinted with permission from Ref. [63]. Copyright (1997) by the American Physical Society.

Conclusions

Some considerations about the presented model are now in order. This model is based on the theory of thermally activated processes. The micelles are supposed to be pinned at the surface by a periodic potential having equally spaced minima separated by barriers having the same height. The application of an external magnetic field induces a critical state which does not correspond to an equilibrium situation and therefore tends to relax. This phenomenon is well known in the case of magnetism as "after-effect" or "magnetic-viscosity", and in the case of superconductivity as "creep of flow" [71].

A possible extension of the model is to consider a distribution of barrier heights centered around some mean value: such a distribution is expected to modify the details of the time evolution of the average orientation, but not to considerably change the dependence of the relaxation time on the applied magnetic field. The model we have discussed could also be extended to Berreman-like geometric anchorings [72], where the surface anchoring has a geometrical contribution connected to the topography of the surface. In this case the pinning potential should be substituted by the total elastic energy connected with different surface geometries.

[1] Bouchiat MA and Langevin-Cruchon D. Molecular order at free surface of a nematic liquid crystals from light reflectivity measurements. *Physics Letters A*, **34**, 331 (1971).

[2] Chiarelli P, Faetti S, and Fronzoni L. Structural transition at the free surface of the nematic liquid crystals MBBA and EBBA. *Journal de Physique*, **44**, 1061 (1983).

[3] Chiarelli P, Faetti S, and Fronzoni L. Critical behavior of the anchoring energy of the director at the free surface of a nematic liquid crystal. *Physics Letters A*, **101**, 31 (1984).

[4] di Lisi GA, Rosenblatt C, Griffin AC, and Hari U. Behavior of the anchoring strength coefficient near a structural transition at a nematic substrate interface. *Liquid Crystals*, **7**, 353 (1990).

[5] Flatischler K, Komitov L, Lagerwall ST, Stebler B, and Strigazzi A. Surface induced alignment transition in a nematic layer with symmetrical boundary conditions. *Molecular Crystals and Liquid Crystals*, **198**, 119 (1991).

[6] Komitov L, Lagerwall ST, Sparavigna A, Stebler B, and Strigazzi A. Surface transition in a nematic layer with reverse pretilt. *Molecular Crystals and Liquid Crystals*, **223**, 197 (1992).

[7] Patel JS and Yokoyama H. Continuous anchoring transition in liquid crystals. *Nature*, **362**,525 (1993).

[8] Moldovan R, Frunza S, Beica T, and Tintaru M. Temperature induced surface transitions in liquid crystal tilted homogeneous cells. *Liquid Crystals*, **20**, 331 (1996).

[9] Jagemalm P, Barbero G, Komitov L, and Strigazzi A. Symmetry rules and temperature-induced anchoring transitions. *Physics Letters A*, **235**, 621 (1997).

[10] di Lisi GA, Rosenblatt C, Akins RB, Griffin AC, and Hari U. Anchoring strength coefficient of a monomer and its dimer at a polymer-coated interface. *Liquid Crystals*, **11**, 63 (1992).

[11] Sluckin TJ and Poniewierski A. In *Fluid and Interfacial Phenomena.* Edited by Croxton CA. John Wiley & Sons, Chichester, 1986.

[12] Barbero G, Gabbasova Z, and Osipov MA. Surface order transition in nematic liquid crystals. *Journal de Physique II*, **1**, 691 (1991).

[13] Sen AK and Sullivan DE. Landau-de Gennes theory of wetting and orientational transitions at a nematic-liquid substrate interface. *Physical Review A*, **35**, 1391 (1987).

[14] McMullen WE. Interfacial polar ordering and anomalous tilt angles at isotropic-nematic interfaces. *Physical Review A*, **40**, 2649 (1989).

[15] McMullen WE and Moore BG. Theoretical studies of the isotropic-nematic interface. *Molecular Crystals and Liquid Crystals*, **198**, 107 (1991).

[16] Holyst R and Poniewierski A. Orientation of liquid crystal molecules at the nematic-isotropic interface and the nematic free surface. *Molecular Crystals and Liquid Crystals*, **192**, 65 (1990).

[17] Sharlow MF and Gelbart WM. On the parallel perpendicular transition for a nematic phase at a wall. *Liquid Crystals*, **11**, 25 (1992).

[18] Holyst R and Poniewierski A. Director orientation at the nematic phase-isotropic phase interface for the model of hard spherocylinders. *Physical Review A*, **38**, 1527 (1988).

[19] Poniewierski A and Holyst R. Nematic alignment at a solid substrate–the model of hard spherocylinders near a hard wall. *Physical Review A*, **38**, 3721 (1988).

[20] McMullen WE. Molecular theory of the isotropic-nematic interface–hard spherocylinders. *Physical Review A*, **38**, 6384 (1988).

[21] Barbero G and Zvezdin AK. Thermal renormalization of the anchoring energy of nematic liquid crystals. *Physical Review E*, **62**, 6711 (2000).

[22] Barbero G, Madhusudana NV, and Durand G. Anchoring energy for nematic liquid crystals: An analysis of the proposed forms. *Zeitschrift fur Naturforschung A*, **39**, 1066 (1984).

[23] Parsons JD. Structural critical point at free surface of a nematic liquid crystal. *Physical Review Letters*, **41**, 877 (1978).

[24] Poniewierski A and Sluckin TJ. Statistical mechanics of a simple model of the nematic liquid crystal-wall interface. *Molecular Crystals and Liquid Crystals*, **111**, 373 (1984).

[25] Poniewierski A and Sluckin TJ. Statistical mechanics of a simple model of the nematic liquid crystal-wall interface II. *Molecular Crystals and Liquid Crystals*, **126**, 143 (1985).

[26] Berreman DW and Meiboom S. Tensor representation of Oseen-Frank strain energy in uniaxial cholesterics. *Physical Review A*, **30**, 1955 (1984).

[27] Priestley EB, Wojtowicz PJ, and Sheng P (Editors). *Introduction to Liquid Crystals*. Plenum Press, New York, 1975.

[28] Sheng P. Boundary-layer phase transition in nematic liquid crystals. *Physical Review A*, **26**, 1610 (1982).

[29] Faetti S, Gatti M, Palleschi V, and Sluckin TJ. Almost critical behavior of the anchoring energy at the interface between a nematic liquid crystal and a SiO substrate. *Physical Review Letters*, 55, 1681 (1985).

[30] Faetti S and Palleschi V. The twist elastic constant and anchoring energy of the nematic liquid crystal 4-normal-octyl-4'-cyanobiphenyl. *Liquid Crystals*, **2**, 261 (1987).

[31] Alexe-Ionescu AL, Barberi R, Barbero G, Beica T, and Moldovan R. Surface energy for nematic liquid crystals–a new point of view. *Zeitschrift fur Naturforschung A*, **47**, 1235 (1992).

[32] da Gama MMT. The interfacial properties of a model of a nematic liquid crystal I–The nematic-isotropic and the nematic-vapor interfaces. *Molecular Physics*, **52**, 585 (1984); The interfacial properties of a model of a nematic liquid crystal II–Induced orientational order and wetting transitions at a solid fluid interface. *Molecular Physics*, **52**, 611 (1984).

[33] Teletzke GF, Scriven LE, and Davis HT. Wetting transitions II–1st order or 2nd order. *Journal of Chemical Physics*, **78**, 1431 (1983).

[34] Rajteri M, Barbero G, Galatola P, Oldano C, and Faetti S. Van der Waals induced distortions in nematic liquid crystals close to a surface. *Physical Review E*, **53**, 6093 (1996).

[35] Maier W and Saupe A. Eine einfache molekulare theorie des nematischen kristallinflussigen zustandes. *Zeitschrift fur Naturforschung A*, **13** 564 (1958); Eine einfache molekular-statistiche theorie der nematischen kristallinflussigen phase I. *Zeitschrift fur Naturforschung A*, **14**, 882 (1959); Eine einfache molekular-statistiche theorie der nematischen

kristallinflussigen phase II . *Zeitschrift fur Naturforschung A*, **15**, 287 (1960).

[36] R. Humphries, P. G. James, and G. R. Luckurst, Molecular field treatment of nematic liquid crystals. *Journal of Chemical Society - Faraday Transactions II*, **68**, 1031(1972).

[37] Israelachvili J. *Intermolecular and Surface Forces.* Academic Press, London, 1998.

[38] Landau LD and Lifchitz EM. *Statistical Physics.* MIR, Moscow, 1986.

[39] Rose ME. *Elementary Theory of Angular Momentum.* John Wiley & Sons, New York, 1957.

[40] Fournier JB and Galatola P. Coarse-grained surface energies and temperature-induced anchoring transitions in nematic liquid crystal. *Physical Review Letters*, **82**, 4859 (1999).

[41] Gradshtein IS and Ryzhik IM. *Table of Integrals, Series and Products.* Academic Press, New York, 1965.

[42] Akulov N. The quantum theory of the temperature dependence of the magnetization curve. *Zeitschrift fuer Physik*, **100**, 197 (1936).

[43] Zener C. Classical theory of the temperature dependence of magnetic anisotropy energy. *Physical Review*, **96**, 1335 (1954).

[44] Zvezdin AK, Mateev VM, and Popov AI. *Rare Earth Ions in Magnets.* Nauka, Moscow, 1985.

[45] Pincus P. Temperature dependence of anisotropy energy in antiferromagnets. *Physical Review*, **113**, 769 (1959); Kittel C and van Vleck JH. Theory of the temperature dependence of the magnetoelastic constants of cubic crystals. *Physical Review*, **118**, 1231 (1960).

[46] Callen ER. Temperature dependence of ferromagnetic uniaxial anisotropy coefficients. *Journal of Applied Physics*, **33**, 832 (1962).

[47] Barbero G, Jägemalm P, and A. K. Zvezdin. Temperature-induced surface transitions in nematic liquid crystals by evaporated SiO_x. *Physical Review E*, **64**, 021703 (2001).

[48] Vilfan M and Copic M. Temperature dependence of azimuthal anchoring strength measured by dynamic light scattering. *Molecular Crystals and Liquid Crystals*, **375**, 155 (2001).

[49] Barbero G and Evangelista LR. Local self-consistent approach to the phase transition at the nematic liquid crystal-wall interface. *Physical Review E*, **65**, 031708 (2002).

[50] Sheng P. Phase transition in surface-aligned nematic films. *Physical Review Letters*, **37**, 1059 (1976).

[51] Miyano K. Surface-induced ordering of a liquid crystal in the isotropic phase. *Journal of Chemical Physics*, **71**, 4108 (1979).

[52] Mada H and Kobayashi S. Surface and bulk order parameters of nematic liquid crystals. *Applied Physics Letters*, **35**, 4 (1979).

[53] K. Miyano, Wall-induced pretransitional birefringence–new tool to study boundary aligning forces in liquid crystals. *Physical Review Letters*, **43**, 51 (1979).

[54] Sheng P. Boundary-layer phase transition in nematic liquid crystals. *Physical Review A*, **26**, 1610 (1982).

[55] Wojtowickz PJ and Sheng P. Critical point in magnetic field temperature phase diagram of nematic liquid crystals. *Physics Letters A*, **48**, 235 (1974).

[56] Krishnaswamy S and Shashidar R. In *Proceedings of the International Liquid Crystals Conference, Bangalore, 1973*; Pramana Supplement I, p. 247.

[57] Chandrasekhar S. *Liquid Crystals.* Cambridge University Press, Cambridge, 1977.

[58] Charvolin J. Aggregates of amphiphilic molecules in lyotropic liquid crystals. *Nuovo Cimento D*, **3**, 3 (1984).

[59] de Gennes PG and Prost J. *The Physics of Liquid Crystals.* Clarendon Press, Oxford, 1994.

[60] Oliveira EA, Neto AMF, and Durand G. Gliding anchoring of lyotropic nematic liquid crystals on amorphous glass surfaces. *Physical Review E*, **44**, R825 (1991).

[61] Oliveira EA and Neto AMF. Anchoring properties of lyotropic liquid crystals near the nematic-isotropic transition. *Physical Review E*, **49**, 629 (1994).

[62] Turchiello RD and Oliveira EA. Behavior of the director reorientation time in glass surfaces of lyotropic liquid crystals in the nematic to biaxial transition. *Physical Review E*, **54**, 1618 (1996).

[63] Galatola P, Barbero G, and Zvezdin AK. Thermal relaxation model of surface director gliding in lyotropic liquid crystals. *Physical Review E*, **55**, 4314 (1997).

[64] Tanguy A and Nozières P. First order bifurcation landscape in a 2D geometry: The example of solid friction. *Journal de Physique I*, **6**, 1251 (1996).

[65] Gopal ESR. *Statistical Mechanics and Properties of Matter: Theory and Applications.* Chichester, Ellis Horwood, 1974.

[66] Gardiner CW. *Handbook of Stochastic Methods for Physics, Chemistry and the Natural Sciences.* Springer, Berlin, 1983.

[67] Kroin T and Neto AMF. Bend periodic distortion of the texture in nematic lyotropic liquid crystals with and without ferrofluid. *Physical Review A*, **36**, 2987 (1987).

[68] Zhou E, Stepanov M, and Saupe A. Curvature elasticity and rotational viscosity of the nematic decylammonium chloride system. *Journal of Chemical Physics*, **88**, 5137 (1988).

[69] Galerne Y, Neto AMF, and Liebert L. Microscopic structure of the uniaxial and biaxial lyotropic nematics. *Journal of Chemical Physics*, **87**, 1851 (1987).

[70] Herzberg G. *Spectra of Diatomic Molecules.* Van Nostrand, Princeton, 1950.

[71] de Gennes PG. *Superconductivity of Metals and Alloys.* Benjamin, New York, 1976.

[72] Berreman DW. Solid surface shape and alignment of an adjacent nematic liquid crystal. *Physical Review Letters*, **28**, 1683 (1972).

5

SURFACE ADSORPTION OF PARTICLES: GENERAL CHARACTERISTICS

In this chapter the phenomenon of surface adsorption of particles is discussed. The general problem of adsorption of neutral particles is stated. The rates of adsorption and desorption are introduced permitting the presentation of some known isotherms (Langmuir, BET and Temkin). Simple kinetic equations at the interfaces are constructed in order to investigate dynamical aspects of the adsorption phenomena (as will be done in Chapter 10). A statistical approach for the adsorption problem–based on the two-level approximation–is discussed in details, underlining the main limitations of the Langmuir approximation. An application of the two-level formalism is proposed in connection with the adsorption of magnetic grains in magnetic fluids, in order to illustrate the relevance of the adsorption phenomena on the nematic ordering, in general, followed by a careful analysis of the meaning of the two-level system approximation. The subsequent part of the chapter is dedicated to the discussion of a model for the adsorption of molecules at the surface, responsible for the anchoring transition in liquid crystalline systems containing azobenzene materials. In these systems, reversible transformations in the molecular conformation from *cis* to *trans* forms occur upon illumination. According to the concentration of *cis*-isomers, a transition from planar to homeotropic alignment can take place at the surface. This phenomenon–connected to the adsorption phenomenon–can be the basis for the photo-manipulation of the anchoring energy and this point is discussed in details, in connection with experimental data.

5.1 Introduction

The surfaces of real fluid systems, even if cleaned and polished, are usually covered with adsorption sites, to which particles of the fluid may attach themselves [1]. This phenomenon is due to interatomic forces acting between the particles of the system and the surface particles, and can be present in a broad variety of physical and chemical systems [2]. The phenomenon is also relevant in liquid crystalline systems with variable molecular shape, where reversible transformations between different molecular structures can be induced by temperature variation [3, 4] as well as by absorption of light [5, 6].

The relevance of the phenomenon reinforces the necessity to take into account the adsorption phenomenon in order to correctly interpret the experimental measurements performed on a variety of different samples.

For what concerns the physics of liquid crystalline systems, three kinds of adsorption phenomena can play a relevant role and will be considered in this book:

(1) The adsorption of magnetic grains by the surface in magnetic fluids, even in the absence of external field [7, 8];

(2) The surface adsorption of the molecules themselves, giving rise to polar (or quadrupolar) effects near the boundaries [9]; or giving rise to anchoring transitions, as is the case of adsorption of *cis*-molecules in liquid crystalline systems formed by azobenzene materials [10];

(3) Selective adsorption of ions due to electrochemical forces at the surface in liquid crystalline systems containing ionic impurities (i.e., a weak electrolyte [11]).

In what follows we present some general concepts and definitions regarding the phenomenon of adsorption to be used through this book. The importance of the adsorption phenomenon on the surface properties of NLC was firstly recognized by Pieranski and Jerome [12, 13] and by Teixeira and Sluckin [14, 15].

5.2 Adsorption at interfaces

Adsorption phenomena have attracted the attention since the first decades of the last century. The name of Irving Langmuir is associated with the research in this area, because he made extensive pioneering studies of the adsorption of gases onto metal surfaces. It is important to distinguish between *absorption*, in which the molecules go to the interior of a substance, and *adsorption*, in which the molecules are stuck to the surface. *Desorption* is the reverse of the adsorption and requires that the adsorbed particle gains energy enough to break the binding with the surface. A molecule may bind at the surface as a chemisorbed or as a physisorbed species. In the first case we refer to the *chemisorption* process, in which a chemical bond is formed between the particle being adsorbed (from the gas or the liquid state that forms the adsorbate) and the substrate or adsorbent. In the second case we refer to the physical adsorption or *physisorption* process, in which the particles are bound to the substrate by physical forces, retaining their identity. The energy involved in the physisorption process will be called adsorption energy. Adsorption requires a particle to lose energy during its collision with the surface. Some of the particles bombarding the surface will bounce back off the surface, but at any particular density a certain fraction will remain, giving rise to a coverage of the surface. The covering ratio or *coverage* is defined as

$$\sigma_{\mathrm{R}} = \frac{\sigma}{\sigma_0} = \frac{\text{Number of surface sites occupied}}{\text{Total number of surface sites}}. \tag{5.1}$$

The rate at which the first adsorbed layer is built up decreases as the layer reaches a saturation in the filling process. The entire process to arrive at the equilibrium situation is characterized by some adsorption time, as we discuss in Chapter 10.

In this book we are concerned with the adsorption of particles coming from a liquid phase, i.e., from solution. In this case, the situation is different because the surface is already covered with the solvent molecules which may interact more or less strongly with it. In the case of liquid crystalline system doped with dye molecules or in the presence of ionic charges, an adsorbed molecule has to compete with a liquid crystal molecule for the occupation of a site at the surface and will be adsorbed only if this process becomes energetically favorable. The rate at which the equilibrium can be attained can be slower in this case.

The coverage of the surface when dynamic equilibrium is reached depends on the density of the phase. The variation of σ_{R} with density at a given temperature is called adsorption isotherm. The simplest isotherm is the Langmuir isotherm, that gives the relation between the coverage of the first layer and the density at a particular temperature. This isotherm is constructed on three basic assumptions:

1. The adsorption occurs only in a first layer (monolayer coverage);
2. All adsorbing sites are equivalent, and the surface is uniform (it is perfectly flat on a microscopic scale);
3. The adsorption energy of one site is independent of occupancy of neighboring sites.

The adsorbed molecules are assumed to be in dynamical equilibrium with the molecules in the surrounding [16] and the adsorption process can be described as a chemical reaction:

$$\mathrm{A(bulk)} + \mathrm{M(surface)} \rightleftharpoons \mathrm{AM(surface)}$$

The rate of adsorption is proportional to the density of A, and the number of adsorbing sites at the surface, i.e,

$$\frac{d\sigma}{dt} = \kappa_a \, \rho \, (\sigma_0 - \sigma), \tag{5.2}$$

where κ_a is the rate constant for adsorption, ρ is the bulk density of adsorbate, just in front of the adsorbing surface, and $\sigma_0 - \sigma$ is the total number of free sites. The rate of desorption is proportional to the number of adsorbed species, i.e.,

$$\frac{d\sigma}{dt} = -\kappa_d \, \sigma \tag{5.3}$$

where κ_d is the rate constant for desorption. Let us introduce the reduced quantities $\sigma_{\rm R} = \sigma/\sigma_0$ and $\rho_{\rm R} = \rho/\rho_0$, where ρ_0 is the bulk density of the adsorbate, in the absence of adsorption. At equilibrium the net rate of adsorption is zero, implying the equality between the absolute values of (5.2) and (5.3); solving for $\sigma_{\rm R}$, one obtains the Langmuir isotherm:

$$\sigma_{\rm R} = \frac{\alpha \, \rho_{\rm R}}{1 + \alpha \, \rho_{\rm R}}, \tag{5.4}$$

where α is a parameter governing the steady state and expressed as

$$\alpha = \frac{\kappa \tau \rho_0}{\sigma_0},$$

if we define $\tau = 1/\kappa_d$ and $\kappa = \kappa_a \, \sigma_0$. The first parameter represents a characteristic time associated with the desorption process. The second parameter is connected with the adsorption phenomenon. Notice that $\kappa \, \tau$ has the dimensions of a length. From (5.4) one obtains

$$\rho_{\rm R} = \frac{1}{\alpha} \frac{\sigma_{\rm R}}{1 - \sigma_{\rm R}}. \tag{5.5}$$

For adsorption with dissociation, the rate of adsorption is proportional to the density and to the probability that both atoms will find sites, which is proportional to the square of the number of vacant sites

$$\frac{d\sigma}{dt} = \kappa_a \, \rho \left[(\sigma_0 - \sigma)\right]^2 . \tag{5.6}$$

For the same case, the rate of desorption is proportional to the frequency of encounters of atoms on the surface, and has the form

$$\frac{d\sigma}{dt} = -\kappa_d \, \sigma^2. \tag{5.7}$$

The equilibrium condition gives, in this case, another isotherm, i.e.,

$$\sigma_{\rm R} = \frac{(\alpha \rho_{\rm R})^{1/2}}{1 + (\alpha \rho_{\rm R})^{1/2}}, \tag{5.8}$$

which shows a weaker density dependence with respect to the Langmuir isotherm. Likewise, from (5.8) we have

$$\rho_{\mathrm{R}} = \frac{1}{\alpha}\left(\frac{\sigma_{\mathrm{R}}}{1-\sigma_{\mathrm{R}}}\right)^2. \tag{5.9}$$

There are other isotherms that arise depending on the change in the assumptions made above. If, for instance, the first layer acts as a substrate for further physisorption, one can obtain the so-called BET isotherm (from the names of Stephen Brunauer, Paul Emmett, and Edward Teller), which states

$$\frac{V}{V_{\mathrm{mol}}} = \frac{c\,z}{(1-z)[1-(1-c)z]}, \tag{5.10}$$

where $z = \rho/\rho^*$, with ρ^* being the equilibrium density of the adsorbate and c a constant related to the adsorption and desorption rate constants. A remarkable feature of these isotherms is that they can rise indefinitely as the density increases, because there is no limit to the amount of material that may condense when multilayer coverage may occur. Another isotherm can be obtained when one considers that the energetically most favorable sites are occupied first, giving the result

$$\sigma_{\mathrm{R}} = c_1 \ln(c_2 \rho_{\mathrm{R}}),$$

where c_1 and c_2 are constants, and the expression, known as Temkin isotherm, was obtained under the assumption that the enthalpy changes linearly with the density; if the change is logaritmic, one obtains the Freundlich isotherm, in the form

$$\sigma_{\mathrm{R}} = c_1\, {\rho_{\mathrm{R}}}^{1/c_2}.$$

A phenomenological way to introduce an isotherm can be summarized as follows [17]. Let ΔG be the molar Gibbs energy of the adsorbate, written as $\Delta G = \Delta G^0 + \gamma\sigma_{\mathrm{R}}$, i.e., as proportional to the coverage. The constant γ is positive if the adsorbed particles repel, and negative if they attract each other. It is possible to introduce an isotherm in the form

$$\frac{\sigma_{\mathrm{R}}}{1-\sigma_{\mathrm{R}}} = c_{\mathrm{S}} \exp\left(\frac{-\Delta G^0}{RT}\right) e^{-g\sigma_{\mathrm{R}}}, \tag{5.11}$$

where c_{S} is a constant, R is the molar gas constant, and $g = \gamma/RT$. The resulting isotherm represented by (5.11) is known as Frumkin isotherm [17]. The case $g = 0$ reduces to the Langmuir isotherm we are considering, because the molar Gibbs energy does not depend on the coverage.

5.2.1 Thickness dependence of the coverage

Let us consider a sample of slab shape limited by two flat (uniform) surfaces placed at a distance d apart, in an essentially one-dimensional medium whose coordinate z is normal to the surfaces. We indicate by $\rho(z,t)$ the density of particles of the medium, in the position z, at a given time t, and $\sigma(t)$ is the surface density of particle. The equilibrium value of these quantities is defined by

$$\lim_{t\to\infty} \rho(z,t) = \rho(z) = \rho$$

and

$$\lim_{t\to\infty} \sigma(t) = \sigma.$$

In terms of these quantities, from (5.2) and (5.3) a kinetic equation can be written as

$$\frac{d\sigma}{dt} = \kappa\rho\left(1 - \frac{\sigma}{\sigma_0}\right) - \frac{1}{\tau}\,\sigma, \tag{5.12}$$

which, by using the reduced quantities introduced before, becomes

$$\frac{d\sigma_{\rm R}}{dt} = \kappa\frac{\rho_0}{\sigma_0}\,\rho_{\rm R}\,(1-\sigma_{\rm R}) - \frac{1}{\tau}\,\sigma_{\rm R}. \tag{5.13}$$

At the equilibrium ($d\sigma_{\rm R}/dt = 0$) one obtains the Langmuir isotherm, which can be shown to have a thickness dependence. In fact, since $\rho_0 d$ is the initial number of particles per unit area, the conservation of the number of particles requires, at any t, that

$$2\sigma(t) + \int_{-d/2}^{d/2} \rho(z,t)dz = \rho_0 d \tag{5.14}$$

which at equilibrium ($t \to \infty$) corresponds to

$$2\sigma + \rho\, d = \rho_0 d. \tag{5.15}$$

To obtain (5.14) and (5.15) we assumed, for simplicity, a sample whose (identical) limiting surfaces were placed at $z = \pm d/2$. From (5.15) we obtain

$$\sigma_{\rm R} + \alpha\,\delta\,(\rho_{\rm R} - 1) = 0, \tag{5.16}$$

where

$$\delta = \frac{d}{2\kappa\tau}$$

is a dimensionless thickness. By substituting (5.16) into (5.4) one obtains

$$\sigma_{\mathrm{R}}{}^2 - [1 + (1 + \alpha)\delta]\,\sigma_{\mathrm{R}} + \alpha\delta = 0, \tag{5.17}$$

whose solution, such that for $\rho_0 = 0$ vanishes, is

$$\sigma_{\mathrm{R}} = \frac{1}{2}\left\{1 + (1 + \alpha)\delta - \sqrt{[1 + (1 + \alpha)\delta]^2 - 4\alpha\delta}\right\}, \tag{5.18}$$

giving $\sigma_{\mathrm{R}} = \sigma_{\mathrm{R}}(\delta)$. Furthermore, by using (5.5) we obtain

$$\rho_{\mathrm{R}} = \frac{-1 + (-1 + \alpha)\delta + \sqrt{[1 + (1 + \alpha)\delta]^2 - 4\alpha\delta}}{2\alpha\delta}. \tag{5.19}$$

For $\delta \to 0$ ($d \ll \kappa\tau$), from (5.18) we get

$$\sigma_{\mathrm{R}} \simeq \alpha\delta, \quad \text{i.e.,} \quad \sigma_{\mathrm{eq}} = \frac{1}{2}\rho_0 d, \tag{5.20}$$

as expected. In the opposite limit of large δ (i.e., $d \gg \kappa\tau$) we obtain

$$\sigma_{\mathrm{R}} \approx \frac{\alpha}{1+\alpha} - \frac{\alpha}{(1+\alpha)^3}\frac{1}{\delta} + \mathcal{O}(\delta^{-2}) \quad \text{and} \quad \rho_{\mathrm{R}} = 1 - \frac{1}{1+\alpha}\frac{1}{\delta} + \mathcal{O}(\delta^{-2}). \tag{5.21}$$

In Figure 5.1, the behavior of σ_{R} is shown as a function of the thickness δ for an illustrative value of the parameter α.

In the limit in which σ_0 is very large, and hence $\sigma \ll \sigma_0$, for all experimental situation, $\sigma/\sigma_0 \approx 0$. In this framework Eq. (5.12) can be written as

$$\frac{d\sigma}{dt} = \kappa\rho - \frac{1}{\tau}\sigma, \tag{5.22}$$

that, in dimensionless form, may be written as

$$\frac{d\sigma_{\mathrm{R}}}{dt} = \frac{\kappa\rho_0}{\sigma_0}\rho_{\mathrm{R}} - \frac{1}{\tau}\sigma_{\mathrm{R}}. \tag{5.23}$$

The steady state values of σ_{R} and ρ_{R} are such that

$$\rho_{\mathrm{R}} = \frac{1}{\alpha}\sigma_{\mathrm{R}}. \tag{5.24}$$

This relation shows that in the considered limit ρ_{R} is proportional to σ_{R}. It can be obtained directly from Eq. (5.4) by neglecting σ_{R} with respect to 1. By means of (5.24) and (5.14), one obtains

$$\sigma_{\mathrm{R}} = \alpha\,\frac{\delta}{1+\delta} \quad \text{and} \quad \rho_{\mathrm{R}} = \frac{\delta}{1+\delta} = \frac{\sigma_{\mathrm{R}}}{\alpha}, \tag{5.25}$$

instead of (5.18) and (5.19)

In Figures 5.2 and 5.3, we compare (5.18) and (5.19) with (5.25) for different values of α. As follows from these curves, the agreement on the predictions of the two kinetic equations is good for small α. Since we will be interested

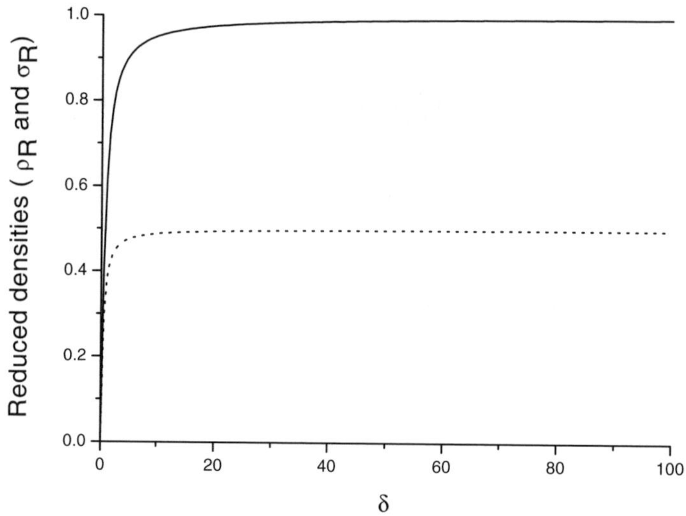

FIGURE 5.1
Reduced densities ρ_{R} (volume) and σ_{R} (surfaces) versus the thickness δ for $\alpha = 1$. Both quantities are linear in δ, for small δ and tend to a limiting value independent of δ for large δ. From Eqs. (5.5) and (5.21) one deduces that $\rho_{\mathrm{R}} \to 1$ in this limit.

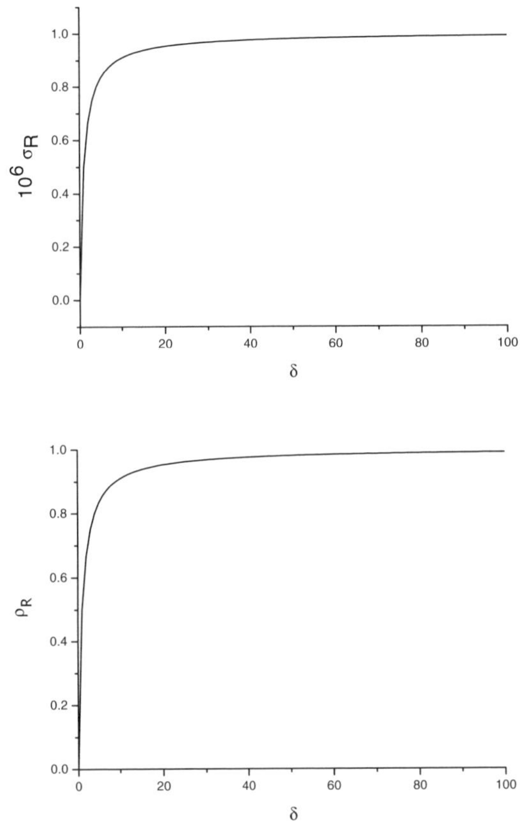

FIGURE 5.2
Comparison of the behavior of the reduced densities versus δ. (a) refers to σ_{R}, given by Eq. (5.18) and (5.25), and (b) refers to ρ_{R}, given by (5.19) and (5.25) in the case $\alpha = 10^{-6}$.

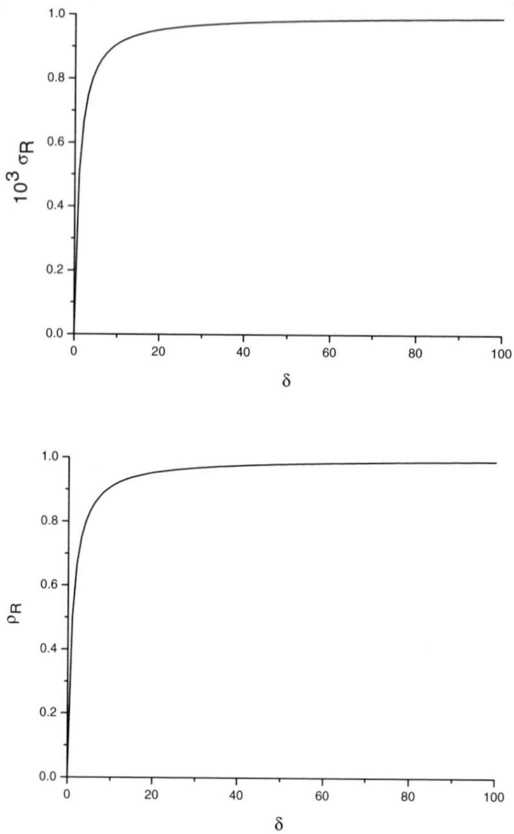

FIGURE 5.3
The same as Figure 5.2 for the case $\alpha = 10^{-3}$.

mainly in the adsorption of impurities dissolved in the liquid crystal, $\rho_0 \propto 1/L^3$ and $\sigma_0 \propto 1/b^2$, where b is typically of the order of a molecular length, whereas L is a macroscopic length. Consequently,

$$\alpha = \frac{\kappa\tau\rho_0}{\sigma_0} = \frac{\kappa\tau b^2}{L^3} \ll 1.$$

As an example we can consider the case of ions dissolved in a liquid crystal sample. Typically, one has $\kappa \sim 6 \times 10^{-6}\,\mathrm{m/s}$, $\tau \sim 10^{-3}\,\mathrm{s}$, $\rho_0 \sim 10^{19}\,\mathrm{m}^{-3}$ [18]. By assuming $b \sim 10\,\mathrm{nm}$, we get $\alpha \sim 10^{-5}$.

5.2.2 Adsorption phenomena for two adsorbates

In the previous section we have considered the adsorption phenomenon when only one type of adsorbate molecules is present. However, as stated above, in the analysis we are interested in the adsorption phenomenon in liquid containing impurities. In this case, it is necessary to take into account that the adsorbed molecules can be of the solvent as well as of the solute. Consequently, previous analysis has to be generalized at least to consider the case in which two types of molecules can be adsorbed.

We indicate by A and B the two types of molecules. In this framework, the kinetic equations at the surface, in the Langmuir approximation, are

$$\frac{d\sigma_{\rm i}}{dt} = \kappa_{\rm i}\rho_{\rm i}\left(1 - \frac{\sigma_{\rm A} + \sigma_{\rm B}}{\sigma_0}\right) - \frac{1}{\tau_{\rm i}}\sigma_{\rm i}, \tag{5.26}$$

for $i = \mathrm{A}$ and B. As we have done before, we introduce dimensionless quantities

$$\sigma_{\rm Ri} = \frac{\sigma_{\rm i}}{\sigma_0} \quad \text{and} \quad \rho_{\rm Ri} = \frac{\rho_{\rm i}}{\rho_0} \tag{5.27}$$

and obtain the kinetic equations

$$\frac{d\sigma_{\rm Ri}}{dt} = \frac{\alpha_i}{\tau_{\rm i}}\rho_{\rm Ri}\left[1 - \sigma_{\rm RA} + \sigma_{\rm RB}\right] - \frac{1}{\tau_{\rm i}}\sigma_{\rm R,i}, \tag{5.28}$$

where

$$\alpha_i = \frac{\kappa_{\rm i}\,\tau_{\rm i}}{\sigma_0}\,\rho_{i\,0}. \tag{5.29}$$

At the equilibrium $d\sigma_{\rm Ri}/dt = 0$. Consequently,

$$\begin{aligned} (1 + \alpha_{\rm A}\,\rho_{\rm RA})\,\sigma_{\rm RA} + \alpha_{\rm A}\rho_{\rm RA}\sigma_{\rm RA} &= 0 \\ \alpha_{\rm B}\rho_{\rm RB}\sigma_{\rm RA} + (1 + \alpha_{\rm B}\,\rho_{\rm RB})\,\sigma_{\rm RB} &= 0, \end{aligned} \tag{5.30}$$

whose solution is

$$\begin{aligned} \sigma_{\rm RA} &= \frac{\alpha_{\rm A}\,\rho_{\rm RA}}{1 + \alpha_{\rm A}\,\rho_{\rm RA} + \alpha_{\rm B}\,\rho_{\rm RB}} \\ \sigma_{\rm RB} &= \frac{\alpha_{\rm B}\,\rho_{\rm RB}}{1 + \alpha_{\rm A}\,\rho_{\rm RA} + \alpha_{\rm B}\,\rho_{\rm RB}}. \end{aligned} \tag{5.31}$$

It follows that

$$\frac{\sigma_{\rm RA}}{\sigma_{\rm RB}} = \frac{\alpha_{\rm A}}{\alpha_{\rm B}}\,\frac{\rho_{\rm RA}}{\rho_{\rm RB}} = \frac{\kappa_{\rm A}\,\tau_{\rm A}\,\rho_{\rm A\,0}}{\kappa_{\rm B}\,\tau_{\rm B}\,\rho_{\rm B\,0}}\,\frac{\rho_{\rm A}}{\rho_{\rm B}}. \tag{5.32}$$

The conservation of the number of particles, in the steady state, implies that

$$\sigma_{\rm Ri} + \alpha_i\delta_i\,(\rho_{\rm Ri} - 1) = 0, \tag{5.33}$$

where, as before, $\delta_i = d/(2\kappa_{\rm i}\tau_{\rm i})$. Using (5.31), from (5.33) we get

$$\frac{\rho_{\rm RA}}{1+\alpha_{\rm A}\,\rho_{\rm RA}+\alpha_{\rm B}\,\rho_{\rm RB}} = \delta_{\rm A}\,(1-\rho_{\rm RA})$$
$$\frac{\rho_{\rm RB}}{1+\alpha_{\rm A}\,\rho_{\rm RA}+\alpha_{\rm B}\,\rho_{\rm RB}} = \delta_{\rm B}\,(1-\rho_{\rm RB})\,. \tag{5.34}$$

By means of (5.34) it is possible to obtain $\rho_{\rm Ri} = \rho_{\rm Ri}(d)$. It follows that

$$\frac{\rho_{\rm RA}}{\rho_{\rm RB}} = \frac{\delta_{\rm A}\,(1-\rho_{\rm RA})}{\delta_{\rm B}\,(1-\rho_{\rm RB})}, \tag{5.35}$$

that, for the definition of δ_i, is equivalent to

$$\frac{1}{\rho_{\rm RA}} - 1 = \frac{\kappa_{\rm A}\,\tau_{\rm A}}{\kappa_{\rm B}\,\tau_{\rm B}}\left(\frac{1}{\rho_{\rm RB}} - 1\right). \tag{5.36}$$

If $\kappa_{\rm A}\,\tau_{\rm A}/\kappa_{\rm B}\,\tau_{\rm B} > 1$, from (5.36) we obtain $\rho_{\rm RA} < \rho_{\rm RB}$, and vice versa if $\kappa_{\rm A}\,\tau_{\rm A}/\kappa_{\rm B}\,\tau_{\rm B} < 1$.

Let us consider now the particular case in which the two species are completely equivalent with respect to the adsorption phenomenon. This means that $\kappa_{\rm A}\,\tau_{\rm A} = \kappa_{\rm B}\,\tau_{\rm B}$. In this situation, Eq. (5.32) becomes

$$\frac{\sigma_{\rm A}}{\sigma_{\rm B}} = \frac{\rho_{\rm A}}{\rho_{\rm B}}. \tag{5.37}$$

This relation states that if the adsorption properties of the two species are identical, at the equilibrium, the surface density of adsorbed molecules is proportional to the bulk density of these molecules.

5.2.3 Statistical approach to the problem: Adsorption energy

The simplest statistical approach to the adsorption phenomena can be stated in the following manner. One considers an adsorbent surface having N sites at which the neutral particles of the gas may attach (see Figure 5.4). The gas is composed of neutral, non-interacting particles. In a very crude approach, the gas acts as a reservoir of particles (whose number is, therefore, not fixed). The gas has a chemical potential μ and each adsorbed molecule has energy A (compared with the ones in the reservoir, i.e., in the ideal gas).

The fundamental problem is then to determine the number of adsorbed particles in the equilibrium, for a given temperature T and pressure p.

The problem can be faced in terms of elementary statistical mechanics as a two-level system [1, 19]. One level corresponds to a state of energy $E = 0$ (in the reservoir) and another one to a state of energy $E = A(<0)$ (adsorbed molecule). The number of states is equal to the number of possible ways of distributing n molecules among the N adsorbing sites. This number is

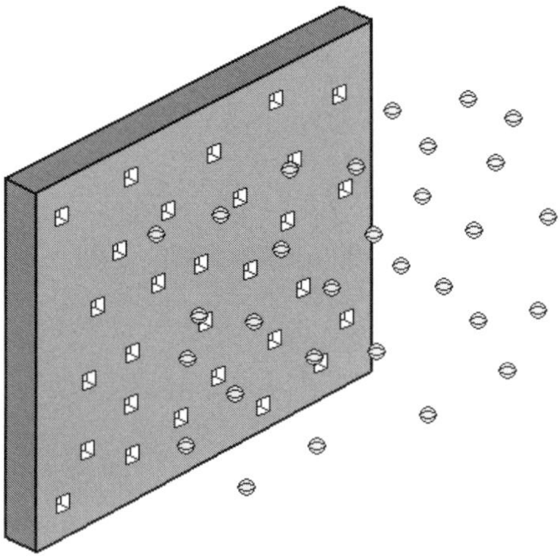

FIGURE 5.4
Adsorbent surface containing a defined number of sites, at which particles of the gas may attach.

$$g_n = \frac{N!}{n!(N-n)!}. \tag{5.38}$$

The canonical partition function is written as

$$Z_n = \frac{N!}{n!(N-n)!} e^{-\beta A n}. \tag{5.39}$$

However, the number of adsorbed molecules is not fixed. Therefore, it is better to consider the grand-canonical partition function, which is

$$\mathcal{Z} = \sum_{n=0}^{N} e^{\beta \mu n} Z_n = \sum_{n=0}^{N} \frac{N!}{n!(N-n)!} e^{\beta n(-A+\mu)}, \tag{5.40}$$

where μ is the chemical potential of the system of particles. Equation (5.40) can be connected with the binomial distribution. For this reason, it can be written in the simple form below:

$$\mathcal{Z} = \left[1 + e^{\beta(-A+\mu)}\right]^N. \tag{5.41}$$

The mean value of the number of adsorbed molecules is given by

$$\langle n \rangle = \frac{1}{\mathcal{Z}} \sum_{n=0}^{N} n Z_n e^{\beta n \mu}$$

$$= \frac{1}{\beta}\frac{\partial}{\partial \mu} \ln \mathcal{Z}. \tag{5.42}$$

The covering ratio, σ_R, is given by

$$\sigma_R = \frac{\langle n \rangle}{N} = \frac{1}{e^{-\beta(-A+\mu)} + 1}. \tag{5.43}$$

The chemical potential of the classical ideal gas (the particles in the reservoir) can be written as

$$\mu = k_B T \ln \left[\frac{p}{k_B T} \left(\frac{h^2}{2\pi m k_B T} \right)^{3/2} \right], \tag{5.44}$$

where h is the Planck constant, p is the pressure and m is the mass of the particles forming the gas. In this manner, the covering ratio σ_R may be put in the form

$$\sigma_R = \frac{p}{p + p_0(T)}, \tag{5.45}$$

where

$$p_0(T) = k_B T \left(\frac{2\pi m k_B T}{h^2} \right)^{3/2} e^{A/k_B T}. \tag{5.46}$$

Equation (5.45) is the Langmuir isotherm introduced in (5.4) written in terms of the pressure and the absolute temperature.

Notice that in the present problem, for a fixed pressure, when $T \to 0$, $\sigma_R \to 1$, i.e., all the potentially adsorbed particles are, in fact, actually adsorbed; on the contrary, when $T \to \infty$, $\sigma_R \to 0$. If $|A| \gg 1$, $\sigma_R \to 1$. Therefore, for a fixed temperature the coverage ratio may attain a saturation value (in this case, all the adsorbing sites can be occupied by a molecule of the gas).

There is final limitation to be considered in the approach we are discussing: The hypothesis that the bulk of the system acts as a reservoir works for several systems, under convenient conditions, but is not operating in the case of real liquid crystalline samples. For instance, in the case of selective ion adsorption, which will be discussed in the next chapter, we have to consider that the number of potentially adsorbed particles has to be conserved. Furthermore, the electro-neutrality condition in these systems has to be considered. In this case, the correct approach is to work with a defined number of particles in the bulk. This fact leads to a more complex approach, which will be discussed in details in the next chapters.

5.2.4 Adsorption of magnetic grains in magnetic fluids

One illustrative example of the relevance of the adsorption phenomena to the nematic ordering can be found in the context of ferrofluids [8]. Ferrofluids are colloidal suspensions of magnetic grains (typical size of 10 nm) coated with surfactant agents or electrically charged, dispersed in a liquid carrier [7, 20]. These fluids are isotropic without applying external fields. When a magnetic field is applied to a ferrofluid sample, it becomes birefringent due to the orientation of the isolated magnetic grains and even due to a secondary aggregation of them. In a birefringence experiment with ferrofluids based on iron oxide grains, a small birefringence has been observed without any applied magnetic field and has been interpreted by the presence of oriented ferrofluid layers stabilized by the glass surfaces of the cell.

To give a theoretical support to this surface-stabilized order in the magnetic fluid, grains were assumed to be adsorbed by the surfaces, which act like a surface field, and this process (similarly to the small scale aggregation phenomenon observed when the ferrofluid is subjected to a magnetic field) leads to the formation of small aggregates. Molecular dynamic simulations [21] have shown that strongly interacting dipolar spheres can form a nematic phase. It means that dipolar forces alone can generate an orientationally ordered liquid state if the grain density is larger than a critical value [21, 22]. Due to the adsorption phenomenon, the density of small aggregates near the substrate becomes larger than the critical one, which induces the formation of the ferronematic phase. Therefore, in the problem we are analyzing, the ordered phase seems to be built up through a small scale chain formation mechanism.

One can show that, by taking a reasonable Hamaker constant value [11], adsorption phenomena could give a complete coverage of the surfaces and induce a magnetic grain concentration larger in the vicinity of the surfaces, than that in the bulk. If the density of the aggregates, formed by the surface effect, is larger than a critical one, a nematic phase is observed. This ordered phase, yielded and stabilized by the surface field, vanishes into the bulk. Experimentally it was observed that when a ferrofluid wets a glass surface, magnetic grains stuck to the glass, even after washing the surface with distilled water. To remove the grains from the surface it is necessary to wash it with concentrated chlorhydric acid for a few minutes, which is confirmed by AFM observations [7].

Consider a slab of thickness d, filled with ferrofluid. The van der Waals interaction between magnetic grains and the substrate leads to an adsorption phenomenon. It is assumed that the adsorbing surfaces are in equilibrium with the ferrofluid at pressure p. From a phenomenological point of view it is reasonable to assume also that the fraction of adsorbed grains (the coverage) is given by

$$\frac{\sigma}{\sigma_0} = \frac{v}{1 + ve^{A}}, \tag{5.47}$$

where σ is the number of occupied sites at the surface, and v are two positive phenomenological parameters which depend weakly on the temperature, and A is the adsorption energy in $k_{\rm B}T$ units. Notice that when $A \to -\infty$, $\sigma/\sigma_0 \to v$, i.e., $\sigma \to v\sigma_0$. In the opposite limit, in which $A \to \infty$, $\sigma \to 0$, as expected. This expression recalls the classical formula obtained in the ideal gas approximation for a two-level system (as discussed in details in Section 5.2.3). In any case, it is possible to face the problem according to the following scheme.

In what follows we deal with the following quantities: $\mathcal{N}$ is the number of particles in the system; $\mathcal{N}_B$ is the number of sites in the bulk and $\mathcal{N}_\mathrm{S}$ is the number of surface adsorption sites. In thermodynamical equilibrium, at a given absolute temperature, T, η_B denotes the number of particles in the bulk, whereas η_S stands for the number of particles in the surface. According to Maxwell-Boltzmann (MB) statistics, $\eta_\mathrm{B} = \mathcal{N}_\mathrm{B} e^{\mu}$ and $\eta_\mathrm{S} = \mathcal{N}_\mathrm{S} e^{\mu - A}$, with μ being a normalization factor (in $k_\mathrm{B}T$ units) to be deduced, and A the surface adsorption energy (as above). The conservation of the number of particles imposes that $\eta_\mathrm{B} + \eta_\mathrm{S} = \mathcal{N}$. From these relations we obtain

$$e^{-\mu} = \frac{\mathcal{N}_\mathrm{B} + \mathcal{N}_\mathrm{S} e^{-A}}{\mathcal{N}}. \tag{5.48}$$

Since μ is fixed by the total number of particles $\mathcal{N}$, in the following we call it chemical potential of the system. Consequently,

$$\eta_\mathrm{B} = \mathcal{N}_\mathrm{B} \left(\frac{\mathcal{N}}{\mathcal{N}_\mathrm{B} + \mathcal{N}_\mathrm{S} e^{-A}} \right) \tag{5.49}$$

and

$$\eta_\mathrm{S} = \mathcal{N}_\mathrm{S} \left(\frac{\mathcal{N} e^{-A}}{\mathcal{N}_\mathrm{B} + \mathcal{N}_\mathrm{S} e^{-A}} \right). \tag{5.50}$$

In the simple analysis presented above, the chemical potential μ is deduced by the conservation of the number of particles $\mathcal{N}$. In this respect, it differs from the analysis performed in the previous section, where only the adsorbed particles are considered, and no mention of $\mathcal{N}$ is made. The analysis presented in Section 5.2.3 is then relevant to an adsorbing surface in contact with a reservoir. But, in real problems, the control parameter is $\mathcal{N}$. Consequently, the correct statistics has to be performed, in the Maxwell-Boltzmann approach, as reported above.

The number of sites in the volume is $\mathcal{N}_B = \rho_0 d S$, where ρ_0 is the bulk density of magnetic grains in the absence of adsorption, d is the thickness of the slab and S its surface area. Actually, ρ_0 is the bulk density of magnetic grains in the limit of infinite thickness. The number of sites on the surface is $\mathcal{N}_\mathrm{S} = 2\sigma_0 S$, where σ_0 is a surface density of sites, as before. It depends on the properties of the adsorbing surfaces. By means of (5.49) and (5.50), we obtain that the number of particles in the bulk, η_B, and at the surface, η_S, are given by

$$\eta_\mathrm{B} = \frac{\rho_0 d}{\rho_0 d + 2\sigma_0 e^{-A}} \mathcal{N} \tag{5.51}$$

$$\eta_S = 2\,\frac{\sigma_0 e^{-A}}{\rho_0 d + 2\sigma_0 e^{-A}}\mathcal{N}. \tag{5.52}$$

By introducing the bulk and surface densities in the steady state, by means of the relations $\eta_B = \rho_{eq}\,d\,S$ and $\eta_S = 2\sigma_{eq}\,d\,S$, and taking into account that $\mathcal{N} = \rho_0\,d\,S$, we obtain

$$\rho_{eq} = \frac{\rho_0}{1 + 2(\sigma_0/\rho_0 d)e^{-A}} \tag{5.53}$$

and

$$\sigma_{eq} = \frac{\sigma_0\,e^{-A}}{1 + 2(\sigma_0/\rho_0 d)e^{-A}}. \tag{5.54}$$

By comparing (5.54) with (5.25) we obtain

$$\kappa\tau = \frac{\sigma_0}{\rho_0}e^{-A} \tag{5.55}$$

connecting the adsorption-desorption coefficients with the properties of the surface and of the adsorbate. From Eqs. (5.47) and (5.54) we have

$$v = \frac{\rho_0 d}{2\sigma_0}. \tag{5.56}$$

Notice that according to this simple model, if $A \to -\infty$, $\sigma_{eq} = \rho_0 d/2$, if $\rho_0 d/2 < \sigma_0$, or $\sigma_{eq} = \sigma_0$, if $\rho_0 d/2 > \sigma_0$.

The quantity A entering in the above expressions is due to the van der Waals interaction between the magnetic grains and the glass surfaces. By assuming spherical grains with radius R, the dispersion interaction is given by [11]

$$E = -\frac{1}{6}\frac{R}{X}A_H, \tag{5.57}$$

where A_H is the Hamaker constant and X the minimum distance between the sphere and the surface. When the sphere is in contact with the substrate, X is a molecular dimension. By assuming $X \approx 0.2\,\text{nm}$, taking into account that $A_H \approx 10^{-19}\,\text{J}$ and $R \approx 5\,\text{nm}$ we obtain $A_H \approx -100$. In this limit $\eta_S \approx \mathcal{N}$. If $\mathcal{N}_S = 1/(\pi R^2)$, we can conclude that the bulk concentration in the surface layer is $1/(2\pi R^3) = 2 \times 10^{17}$ grains/cm^3. This concentration indicates that as the isolated grains are almost spherical, no nematic order due to steric interactions between them is expected. On the other hand, at this high concentration, dipolar interaction between grains [22] favors the aggregation process and the magnetic grains can give rise to a ferronematic phase [21, 22]. In the present approach, this ferronematic phase is a surface phase, because the van der Waals interaction is already comparable with the thermal agitation when $X \approx 2R$. However, the chains formed at the surface favor the formation of chains also in the bulk. It means that we have a propagation of the surface nematic order to the bulk.

5.2.5 Adsorption energy as a non-local quantity

Two-level systems are widely used in physics and are of particular interest for the description of simple systems of non-interacting particles, as spins in magnetic fields, dissociation of particles in liquids, *cis-trans* isomerization in liquid crystal molecules, and many others. In the framework of this simple approach the partition function and the statistical averages are easily evaluated and their meanings are transparent, as we have seen in Section 5.2.3. The application of this description to the adsorption phenomenon of neutral particles requires some care, even if at first sight the surface and the bulk can be considered as a two-level system. In fact, in this situation the forces responsible for the adsorption are short range, and, hence, localized at the surfaces. In this manner, in the statistical mechanics description there are two kinds of sites, which are not equivalent. It follows that a simple application of a two-level description can give rise to apparently absurd results if the problem is not well formulated.

In this section we analyze the adsorption phenomenon by taking into account the kinetic of the adsorption that occurs at the adsorbing surfaces. We will show that a two-level system can be successfully interpreted in the case of short-range adsorption energy only if an effective adsorption energy, which depends on the actual thickness of the sample, is introduced [23]. To this aim we consider a sample in the shape of a slab of thickness d, limited by two adsorbing surfaces placed at $z \pm d/2$, where z is the coordinate normal to the surfaces. The physical quantities are supposed as depending only on the $z-$coordinate. The actual adsorption energy of the surface, with respect to the particles, is assumed as being localized at the surface. In other words, its penetration range is assumed as vanishing. Our discussion works well for neutral particles, as dyes dissolved in liquid crystalline samples. Of course it does not work well when ionic particles are considered due to the long-range character of the electrostatic interaction, but this problem will be discussed in the next chapters.

Consider a liquid in which is dissolved a quantity of neutral particles that can be adsorbed by the limiting surfaces. In the absence of adsorption the particles are uniformly distributed in the sample with a density ρ_0. When the adsorption phenomenon at the surfaces takes place, there is a migration of particles from the bulk to the surfaces. The bulk distribution of particles changes with time until a steady state is reached. In this final configuration, the bulk density of particles is again homogeneous but a certain quantity of particles passed to the surface.

If one considers now the steady state distribution of particles by means of the Maxwell-Boltzmann statistics, the system can be approximated by a two-level system: the bulk and the surface. The relevant energies are indicated by $E_B = 0$, for the bulk, and $E_{\rm S} = \tilde{A} < 0$, in $k_{\rm B}T$ units, for the surface. If $\mathcal{N} = \rho_0 d\, S$ is the total number of particles at the equilibrium $\eta_{\rm B} = \mathcal{N} e^{\mu}$ and $\eta_{\rm S} = \mathcal{N}\, e^{\mu - \tilde{A}}$, where μ is determined by imposing the condition $\eta_{\rm B} + \eta_{\rm S} = N$.

Simple calculations give

$$e^{\mu} = \frac{1}{1+e^{-\tilde{A}}}. \tag{5.58}$$

It follows that

$$\eta_{\rm B} = \mathcal{N}\,\frac{1}{1+e^{-\tilde{A}}} \qquad \text{and} \qquad \eta_{\rm S} = \mathcal{N}\,\frac{e^{-\tilde{A}}}{1+e^{-\tilde{A}}}. \tag{5.59}$$

By taking into account that $\eta_{\rm B} = \rho_{\rm eq} d\, S$ and $\eta_{\rm S} = 2\sigma_{\rm eq} S$, because the system has two limiting surfaces, from Eq. (5.59) one obtains

$$\rho_{\rm eq} = \rho_0\,\frac{1}{1+e^{-\tilde{A}}} \qquad \text{and} \qquad \sigma_{\rm eq} = \frac{1}{2}\,\frac{e^{-\tilde{A}}}{1+e^{-\tilde{A}}}\,\rho_0 d. \tag{5.60}$$

To investigate the physical consequences of the phenomenon of adsorption, a kinetic equation at the limiting surfaces has to be imposed. As discussed above, if the surface density of adsorbed particles is very small with respect to the surface density of sites, the balance equation at the boundary is

$$\frac{d\sigma}{dt} = \kappa\,\rho(-d/2,t) - \frac{1}{\tau}\sigma(t). \tag{5.61}$$

We limit now our investigation to the steady state, where ρ tends to $\rho_{\rm eq}$ and $\sigma \to \sigma_{\rm eq}$. In this situation, from Eq. (5.61) one obtains $\sigma_{\rm eq} = \kappa\,\tau\,\rho_{\rm eq}$. The conservation of the number of particles, Eq. (5.14), becomes $2\sigma_{\rm eq} + \rho_{\rm eq}\, d = \rho_0 d$. It follows that $\rho_{\rm eq}$ and $\sigma_{\rm eq}$, characterizing the steady state are, respectively, given by

$$\rho_{\rm eq} = \frac{\rho_0}{1+2\,\kappa\,\tau/d} \quad \text{and} \quad \sigma_{\rm eq} = \frac{\kappa\tau/d}{1+2\kappa\tau/d}\,\rho_0 d. \tag{5.62}$$

By comparing the expressions of $\rho_{\rm eq}$ and $\sigma_{\rm eq}$, given by Eqs. (5.60), with the ones given by Eqs. (5.62), it follows that

$$e^{-\tilde{A}} = 2\,\frac{\kappa\tau}{d}. \tag{5.63}$$

Since the adsorption phenomenon is expected to be a local property of the adsorbing surface, the product $\kappa\,\tau$ is expected to be thickness independent. This implies that in a two-level approach to the adsorption phenomenon, the effective adsorption energy A depends on the thickness according to the law

$$-\tilde{A} = -U_0 - \ln\frac{\Lambda}{d}, \tag{5.64}$$

where U_0 and Λ are intrinsic properties of the adsorbing surfaces. More precisely, U_0 is the intrinsic adsorption energy in $k_{\rm B}T$ units and Λ a microscopic length indicating the penetration of the adsorption energy. By substituting

(5.64) in (5.63) we obtain $\kappa\tau/\Lambda = e^{U_0}$. This relation connects the intrinsic parameter of the adsorption present in the phenomenological equation describing the kinetics with the parameters characterizing the adsorption energy. Equation (5.64) indicates that if the adsorption phenomenon is faced by means of a two-level system, the adsorption energy involved in the description is no longer a local property of the adsorbing surfaces.

This apparent surprising result can be easily understood by taking into account the results of the previous section. According to the analysis presented there the number of particles in the bulk and on the surface are given by (5.49) and (5.50), respectively. They can be rewritten in the form

$$\eta_{\rm B} = \frac{1}{1 + (\mathcal{N}_{\rm S}/\mathcal{N}_B)e^{-A}}\mathcal{N} \tag{5.65}$$

and

$$\eta_{\rm S} = \frac{(\mathcal{N}_{\rm S}/\mathcal{N}_B)e^{-A}}{1 + (\mathcal{N}_{\rm S}/\mathcal{N}_{\rm B})e^{-A}}\mathcal{N}. \tag{5.66}$$

By comparing (5.65) and (5.66) with (5.59) we get

$$\frac{\mathcal{N}_{\rm S}}{\mathcal{N}_{\rm B}}e^{-A} = e^{-\tilde{A}}, \tag{5.67}$$

from which

$$\tilde{A} = A - \ln\frac{\mathcal{N}_{\rm S}}{\mathcal{N}_B}. \tag{5.68}$$

Since, as discussed before, $\mathcal{N}_{\rm S}/\mathcal{N}_{\rm B} = \sigma_0/(\rho_o d) = \Lambda/d$, from (5.68) we reobtain (5.64).

This simple example permits us to conclude that the two-level model can be successfully used for the description of the adsorption phenomenon with the following remark. If the Boltzmann distribution is written by means of the number of sites, in the bulk and on the surface, no special care is necessary. On the contrary, if the Boltzmann distribution is written only in terms of the number of particles, the adsorption energy has to be considered as a non-local quantity. This non-locality is then connected with the circumstance that, actually, the number of sites in the bulk is proportional to the thickness of the sample. In a pure two-level model this dependence of the bulk number of sites on the thickness is neglected. To take, in some manner, into account this dependence, we have to admit that the adsorption energy is a non-local quantitiy, according to (5.64).

5.3 Surface adsorption and anchoring transition

The bulk properties of thermotropic liquid crystals (LCs) depend on the molecular structure and on the molecular interactions. The liquid crystal surface properties, which are of vital importance for the alignment of LCs and thus for the appearance and the operation of devices, depend, in addition, on the solid surface-liquid crystal interactions. Hence, both bulk and surface physical properties of the liquid crystal can effectively be controlled by external factors that result in changes of the molecular structure.

Reversible changes in the molecular structure may take place due to photoisomerisation [24, 25]. Some organic materials, such as azobenzenes, which also may exhibit liquid crystalline properties, undergo, upon light illumination, a *trans*- to *cis*-isomerisation. The photoisomerisation process, however, may result in changes of the net molecular dipole moment and it may or may not result in changes of the molecular shape [26]. As a consequence, changes of bulk and surface liquid crystal properties may take place. The photoisomerisation process might be also reversible, if there is no chemical reaction or material degradation under light illumination. Therefore, the photoisomerisation of liquid crystal attracts the interest of many searchers since the physical (bulk and surface) properties of the liquid crystals can be controlled reversibly by light without any changes of the chemical content of the photosensitive liquid crystal materials. Photo-induced changes of the liquid crystal physical properties, however, could directly (e.g., photo-induced phase transitions and anchoring transitions) or indirectly (e.g., electric field assisted photo-induced effects) lead to significant changes of the optical appearence of the liquid crystal devices. These changes could be very fast (in order of $100\,\mu$s or even less) or very slow (a couple of minutes or hours) depending on the origin of the photo-induced effect, the photochromic material, and the conditions of light illumination. The photo-induced effects in liquid crystals are very attractive for applications in photonic devices.

In the initial state of the photosensitive material, all the molecules are in the *trans* state, and the planar orientation is achieved. With the UV exposure time the concentration of *cis*-isomers increases. The *cis*-isomers having larger polarity are more strongly attracted by the cell surfaces through the polar-polar interactions. As a consequence, the surface density of the *cis*-isomers increases with the exposure time. Since, for steric reasons, *cis*-isomers favor homeotropic orientation, a planar-homeotropic anchoring transition could be induced by the adsorption phenomenon.

To explain this fact a model in which the adsorption of the *cis*-isomers at the surfaces is described by assuming the mixture of *cis* and *trans*-isomers as a solution of ideal gases will be discussed [10].

Consider that the anchoring free energy of the liquid crystal on the solid surface depends only on the polar angle θ (again, the angle between the normal and the surface nematic director). The anisotropic part of the surface free energy, f_s, per unit area, can be written in the first approximation as

$$f_s = \frac{1}{2} B \cos^2 \theta, \tag{5.69}$$

with

$$B = N_{2s} W_2 - N_{1s} W_1, \tag{5.70}$$

where $N_{2s}(N_{1s})$ are the numbers of *trans* (*cis*) molecules adsorbed at the surface and W_1 (W_2) are the corresponding surface free energies, per surface particles. Equation (5.70) holds in the hypothesis of homogeneous absorption of UV through the sample. Minimizing the surface free energy (5.69) with respect to θ two solutions are obtained: 1) if $B > 0$ the planar orientation ($\theta = \pi/2$) (favored by *trans* molecules) is stable; 2) if $B < 0$ the homeotropic orientation ($\theta = 0$) (favored by *cis* molecules) is stable.

In this manner, the problem is to calculate the number of *trans* (N_{2s}) and *cis* (N_{1s}) molecules adsorbed at the surface as a function of time.

In the initial state (at $t = 0$) all the molecules of the system $N_2(0)$ are in the *trans* state. One considers that the number of sites N (of possible adsorbed molecules) is constant. This statement constitutes an approximation since the excluded areas of *trans* and *cis* molecules are different. With this approximation, the total number of molecules in the bulk, namely,

$$N_b = N_2(0) - N \tag{5.71}$$

is also constant. Considering the typical geometry of the experiment (a sandwich-type cell consisting of two parallel glass plates, as discussed below) one can roughly estimate the ratio K between the numbers of adsorbed and bulk molecules. The density of azobenzene molecules is of the order of 10^{26} m^{-3}, so that the total number of molecules in the bulk is given by $N_2(0) \approx 10^{26} \times 2 \times 10^{-6} \times S = 2 \times 10^{20} S\,\text{m}^{-2}$, where S is the area of the glass plate and $d = 2 \times 10^{-6}$ m is the thickness of the cell. Supposing that the adsorbed molecules (with radii of the order of 10^{-10} m) form a monomolecular layer, the number of these molecules is $N \approx 10^{18} S\,\text{m}^{-2}$, so that

$$K = \frac{N}{N_b} \approx \frac{N}{N_2(0)} \approx 0.01. \tag{5.72}$$

The time evolution of the number of particles of the two constituents can be determined by assuming that the reaction kinetic describing the *trans-cis* isomerization is first order. In this framework the kinetic equation for the *trans* molecules is

$$\frac{dN_2}{dt} = -p_{\text{tc}} N_2 + p_{\text{ct}} N_1, \tag{5.73}$$

where $p_{\rm tc}$ is connected with the probability of the *trans-cis* transition, whereas $p_{\rm ct}$ with the probability of the *cis-trans* transition. Equation (5.73) has to be solved by taking into account that

$$N_2(t) + N_1(t) = N_2(0), \tag{5.74}$$

which represents the conservation law of the number of particles. By means of (5.74) it is possible to rewrite (5.73) in the form

$$\frac{dN_2}{dt} = p_{\rm ct} N_2(0) - (p_{\rm tc} + p_{\rm ct}) N_2(t). \tag{5.75}$$

The functions $N_1(t)$ and $N_2(t)$, solutions of (5.74) and (5.75), are

$$\begin{aligned} N_2(t,x) &= N_2(0)\left[x + (1-x)e^{-t/\tau}\right], \\ N_1(t,x) &= N_2(0)(1-x)\left[1 - e^{-t/\tau}\right], \end{aligned} \tag{5.76}$$

where $\tau = 1/(p_{\rm tc} + p_{\rm ct})$ is a characteristic time and $x = p_{\rm tc}/(p_{\rm tc} + p_{\rm ct})$ is a parameter that controls the fraction of *trans*/*cis* isomers after the illumination. An estimation of x can be done by measuring the shift of the critical nematic-isotropic phase transition, as we will discuss later in this section. Increasing N_1 the nematic properties of the mixture change continuously. There exists a critical concentration of *cis*-isomers such that for concentrations larger than this critical value the mixture is in the isotropic phase. In what follows we assume to be well below this critical concentration. With this parametrization at every moment of time t the conservation of the number of particles is achieved:

$$N_2(0) = N_1(t) + N_2(t) = N_{1b}(t) + N_{2b}(t) + N_{1s}(t) + N_{2s}(t), \tag{5.77}$$

where

$$N_b(t) = N_{1b}(t) + N_{2b}(t) = \text{constant}, \tag{5.78}$$

represents the total number of molecules in the bulk, and

$$N_s(t) = N_{1s}(t) + N_{2s}(t) = \text{constant}, \tag{5.79}$$

the total number of molecules at the surface.

The concentration $c_{\rm s}$ of *cis* molecules at the surface is defined by the relation

$$c_{\rm s} = \frac{N_{1s}}{N}. \tag{5.80}$$

The concentration C_b of *cis* molecules in the bulk can be expressed with respect to the number of sites by

$$C_b = \frac{N_{1b}}{N_b}. \tag{5.81}$$

By considering relations (5.78) and (5.79), the time dependence of C_b is given by

$$C_b = \frac{N_1(t)}{N_2(0)} - Kc_s \approx \frac{N_1(t)}{N_2(0)}, \tag{5.82}$$

because $Kc_s \ll 1$ and the second term can be neglected. The problem is now to obtain the relation between the concentration of *cis* molecules in the bulk, C_b, and that of the surface, c_s.

By assuming that the transfer rate of energy into the system is quite small, it follows that in every moment the system (bulk and surface) is in a thermodynamic equilibrium state. This assumption implies that the diffusion process is fast enough, as will be supposed in the analysis. The main idea is the following: one considers an isothermal solution of two different components (*cis* and *trans*) which may exist as two distinct phases, bulk and surface, with an infinitely sharp interface between them.

The bulk free energy density has the following form:

$$g_b(C_b) = C_b g_1 + (1 - C_b) g_2 + \frac{k_B T}{v}\left[C_b \ln C_b + (1 - C_b)\ln(1 - C_b)\right], \tag{5.83}$$

where

$$v = \frac{N_{1b} v_1 + N_{2b} v_2}{N_b} \tag{5.84}$$

is the molecular volume, which is assumed to be constant (an approximation since N_{1b} and N_{2b} depend on time and $v_1 \neq v_2$).

The first two terms in Eq. (5.83) correspond to the contribution to the free energy density due to the individual free energy densities of the two components and the last term is due to the decrease in energy associated with the mixing of the two components, under our assumption that bulk phase is an ideal solution. A more complex form might include energy of mixing terms proportional to $C_b(1 - C_b)$.

We re-scale the free energy density by introducing the following dimensionless quantities:

$$f_b = g_b \frac{v}{k_B T} \quad \text{and} \quad f_i = g_i \frac{v}{k_B T}, \quad i = 1, 2. \tag{5.85}$$

The dimensionless bulk free energy density is given by

$$f_b(C_b) = C_b f_1 + (1 - C_b) f_2 + C_b \ln C_b + (1 - C_b)\ln(1 - C_b). \tag{5.86}$$

Minimizing f_b subject to the constraint that the number of *cis* molecules is conserved, we obtain

$$A = \frac{\partial f_b}{\partial C_b} = f_1 - f_2 + \ln \frac{C_b}{1 - C_b}, \tag{5.87}$$

where A is the corresponding Lagrange multiplier (proportional to the chemical potential).

We take the free energy density corresponding to the surface phase as

$$gs(c_s) = c_s g_1 + (1 - c_s) g_2 + \frac{k_B T}{v} \left[c_s \ln c_s + (1 - c_s) \ln(1 - c_s) + c_s \varepsilon_1 + (1 - c_s) \varepsilon_2 \right], \tag{5.88}$$

where $\varepsilon_1 (\varepsilon_2) < 0$ is the adsorption energy of a *cis* (*trans*) molecule. Since the *cis* molecule is more attracted by the surface, $|\varepsilon_1| > |\varepsilon_2|$.

Using again the same scaling of Eq. (5.85) and including $\epsilon_i = \varepsilon_i v / k_B T$, for $i = 1, 2$, one can minimize Eq. (5.88) with respect to c_s to obtain

$$A = \frac{\partial f_s}{\partial c_s} = f_1 - f_2 + \ln \frac{c_s}{1 - c_s} - \Delta\epsilon, \tag{5.89}$$

where $\Delta\epsilon = \epsilon_2 - \epsilon_1 > 0$.

In the approximation $f_1 = f_2$, Eq. (5.87) and (5.89) lead to

$$\frac{C_b}{1 - C_b} = \frac{c_s}{1 - c_s} e^{-\Delta\epsilon}. \tag{5.90}$$

Using Eq. (5.82), the density of *cis* molecules at the surface is given by

$$c_s(t, x) = \left[1 + \frac{N_2(t, x)}{N_1(t, x)} e^{-\Delta\epsilon} \right]^{-1} \tag{5.91}$$

with $N_1(t, x)$ and $N_2(t, x)$ given by Eq. (5.76). This is the central result of the model. Note that if $\epsilon_1 = \epsilon_2$, and hence $\Delta\epsilon = 0$, from Eq. (5.90) and (5.91) one obtains $c_s(t, x) \propto N_1(t, x)$, as expected. In fact, in the case of equal adsorption energies the concentrations of adsorbed molecules are proportional to the bulk concentrations.

When $t = 0$ one has $c_s(0, x) = 0$ (the *cis* molecules do not exist). In the limit of long time exposure, i.e., $t \to \infty$ the concentration of *cis* molecules is controlled by the parameter x:

$$c_s(\infty, x) = \left[1 + \frac{x}{1 - x} e^{-\Delta\epsilon} \right]^{-1}. \tag{5.92}$$

Using Eq. (5.70), the planar-homeotropic phase transition takes place after an illumination time $t = t^*$, given by

$$\frac{N_2(t^*, x)}{N_1(t^*, x)} = \frac{W_1}{W_2} e^{-\Delta\epsilon}. \tag{5.93}$$

Substitution of Eq. (5.76) into Eq. (5.93) yields

$$t^* = \tau \frac{(1-x)[1-(W_1/W_2)e^{-\Delta\epsilon}]}{(1-x)(W_1/W_2)e^{-\Delta\epsilon} - x}. \tag{5.94}$$

Even if the experimental results for τ are available, the estimation of t^* is difficult because the result depends also on x, which is a phenomenological parameter connected with the recombination process. To have an order of magnitude of t^* one can assume $\tau \approx 10^2$ s, $x \approx 0.1$, $W_1 \approx W_2$, and $\Delta\epsilon \approx 1$. With these values in Eq. (5.94) one obtains $t^* \approx 100$ s, in good agreement with the experimental value of $t^* \approx 120$ s.

Conclusions

Some concluding remarks are now in order. Evidently, the simple approach sketched in this chapter has some limitations, because it is based on the simplifying hypothesis: (1) the adsorbed gas behaves ideally in the vapor phase (in the reservoir), (2) the adsorbed gas is confined to a monomolecular layer, (3) the surface is homogeneous, and then, the affinity of each binding site for adsorbing molecules is the same, (4) there is no interaction between adsorbate molecules, and (5) the adsorbed gas is localized [27]. Most of these assumptions are false, because the surfaces of solids are not in general uniform, the interactions between adsorbed particles are not negligible (this is evident in the case of ion adsorption), the adsorbed molecules can move about on the surface and multilayer adsorption is common in physical adsorption as we have pointed out above.

If we adopt the phenomenological point of view, however, some of the above limitations are not so decisive, and can be circumvented by treating the adsorption energy as a parameter characterizing the surface. More precisely, the fact that the adsorbed gas is localized, and confined in a monomolecular layer is not decisive in determining other macroscopic parameters characterizing the system. Furthermore, the fact that a real surface is never truly homogeneous, can be partially taken into account by considering different adsorption energies for different species of particles.

In any case, the assumption that the adsorbed particles do not interact among themselves is not realistic in general. In the case of a non-diluted system, or in the case of charged particles among which long-range interactions are present, this approximation is surely wrong, because the interactions play a crucial role. Of course, for a very diluted system this approximation can be used with some success, as is done in the statistical mechanics of ideal gases.

[1] Garrod C. *Statistical Mechanics and Thermodynamics.* Oxford University Press, Oxford, 1995.

[2] Lyklema J. *Fundamental of Interface and Colloid Sciences.* Academic Press, London, 1993.

[3] Deloche B and Cabane B. Coupling of hydrogen bonding to orientational fluctuation modes in liquid-crystal PHBA. *Molecular Crystals and Liquid Crystals*, **19**, 25 (1972).

[4] Barbero G, Evangelista LR, and Petrov MP. Influence of the dimerization process on the nematic ordering. *Physics Letters A*, **256**, 399 (1999).

[5] Ichimura K. In *Photochromism: Molecules and Systems.* Edited by Dürr H and Bouas-Laurent H. Elsevier, Amsterdam, 1990.

[6] Komitov L, Ichimura K, and Strigazzi A. Light-induced anchoring transition in a 4,4'-disubstituted azobenzene nematic liquid crystal. *Liquid Crystals*, **27**, 51 (2000).

[7] Matuo C, Bourdon A, Bee A, and Neto AMF. Surface-induced ordering in ionic and surfacted magnetic fluid. *Physical Review E*, **56**, 1310 (1997).

[8] Barbero G, Bourdon A, Bee A, and Neto AMF. Surface stabilized nematic phase in a magnetic fluid. *Physics Letters A*, **259**, 314 (1999).

[9] Jerome B. Surface effects and anchoring in liquid crystals. *Reports on Progress in Physics*, **54**, 391 (1991).

[10] Barbero G and Popa-Nita V. Model for the planar-homeotropic anchoring transition induced by *trans-cis* isomerization. *Physical Review E*, **61**, 6696 (2000).

[11] Israelachvili J. *Intermolecular and Surface Forces.* Academic Press, London, 1998.

[12] Pieranski P and Jerome B. Adsorption-induced anchoring transitions at nematic liquid crystal-crystal interfaces. *Physical Review E*, **40**, 317 (1989).

[13] Pieranski P and Jerome B. Anchoring transitions. *Molecular Crystals and Liquid Crystals*, **199**, 167 (1991).

[14] Teixeira PIC and Sluckin TJ. Microscopic theory of anchoring transitions at liquid crystal surfaces. *Journal of Chemical Physics*, **97**, 1498 (1992).

[15] Teixeira PIC and Sluckin TJ. Microscopic theory of anchoring transitions at the surfaces of pure liquid crystals and their mixtures. II. The effect of surface adsorption. *Journal of Chemical Physics*, **97**, 1510 (1992).

[16] Atkins PW. *Physical Chemistry.* Oxford University Press, Oxford, 2004.

[17] Schmickler W. *Interfacial Electrochemistry.* Oxford University Press, Oxford, 1996.

[18] Zhang H and D'Have K. Surface trapping of ions and symmetric addressing scheme for FLCDs. *Molecular Crystals and Liquid Crystals*, **351**, 27 (2000).

[19] Kubo R. *Statistical Mechanics.* North Holland, Amsterdam, 1967.

[20] Massart R. Brevet Fraçais 79-188-42 (1979); US Patent 4,329,24 (1982); Preparation of aqueous magnetic liquids in alkaline and acidic media. *IEEE Transactions on Magnetics*, **17**, 1247 (1981).

[21] Wey D and Patey GN. Ferroelectric liquid crystal and solid phases formed by strongly interacting dipolar soft spheres. *Physical Review A*, **46**, 7783 (1992).

[22] Gosh B and Dietrich S. Ferroelectric phase in Stockmayer fluids. *Physical Review E*, **50**, 3814 (1994).

[23] Barbero G and Evangelista LR. Two-level description for the adsorption phenomenon: the meaning of the adsorption energy. *Physics Letters A*, **324**, 224 (2004).

[24] Sackmann E. Photochemically induced reversible color changes in cholesteric liquid crystals. *Journal of American Chemical Society*, **93:25**, 7088 (1971).

[25] Ikeda T, Sasaki T, and Ichimura K. Photochemical switching of polarization in ferroelectric liquid crystal films. *Nature*, **361**, 428 (1993).

[26] Komitov L, Ruslim C, Matsuzawa Y, and Ichimura K. Photo-induced anchoring transitions in a nematic doped with azo dyes. *Liquid Crystals*, **27**, 1011 (2000).

[27] Daniels F and Alberty RA. *Physical Chemistry.* John Wiley & Sons, New York, 1966.

6

SELECTIVE IONIC ADSORPTION IN NEMATIC LIQUID CRYSTALS

In this chapter, the influence of adsorbed ions and the resulting surface electric field on the anchoring properties of NLC is analyzed. Some elements of the theory of a weak electrolyte are presented to establish the conceptual framework in which the subsequent investigations will be developed. The main elements of the Poisson-Boltzmann theory are presented in a concise way. The consequences of the linear approximation for this theory are investigated. In particular, we consider the influence of the adsorption phenomenon on the anchoring energy to explain the thickness dependence of this quantity. Further, in order to investigate the effect of the surface electric field on the molecular orientation, we present a model that takes into account two physical mechanisms for the coupling of the nematic director with the surface electric field: (i) the dielectric anisotropy and (ii) the coupling of the quadrupolar component of the flexoelectric coefficient with the field gradient. It is shown that for sufficiently large surface fields there can be a spontaneous curvature distortion in the cell even when the anchoring energy is infinitely strong. We also discuss the director distortion when the anchoring energy of the surface is finite.

6.1 Exponential approximation for the electric field of ionic origin

6.1.1 Introduction

The precise nature and the origin of the anchoring energy in NLC are still subjects of many fundamental and experimental studies and cannot be considered as a solved problem [1, 2]. To explain the thickness dependence [1, 3] as well as the bias voltage dependence [4] of the anchoring energy found in some nematic liquid crystal samples the phenomenon of selective ion adsorption has been invoked [5]–[8]. The influence of selective ion adsorption on the anisotropic part of the anchoring energy strength has been discussed by several authors in recent years [3, 4, 5], [9]–[14]. According to this point of view [5, 7], the adsorption phenomenon is responsible for an ionic separation inside the liquid. To this charge separation is connected an electric field distribution across the sample. The coupling of this field with the dielectric and flexoelectric properties of the liquid crystal gives rise to a dielectric energy density, localized near to the limiting surfaces, on mesoscopic thicknesses. This energy can be considered as a surface energy, which renormalizes the anisotropic part of the interfacial energy characterizing the nematic liquid crystal-substrate interface.

The origins of the different contributions to the energy density of dielectric nature can be easily understood by means of symmetry considerations. To this end, let us consider an NLC submitted to an imposed electric field **E**. The tensor characterizing the NLC is Q_{ij} defined by Eq. (1.9), which has quadrupolar symmetry. By means of Q_{ij} and the electric field **E** we can build the scalar

$$f_{\rm D} = c_1 E_i Q_{ij} E_j, \tag{6.1}$$

which is connected to the dielectric anisotropy ϵ_a of the NLC medium. By comparing (6.1) with (1.64) we deduce that $c_1 \propto \epsilon_a$. Furthermore, if Q_{ij} is position dependent, it is possible to build the vector $Q_{ij,j}$. The coupling of $Q_{ij,j}$ with **E** gives rise to the scalar

$$f_{\rm F} = c_2\, Q_{ij,j} E_i, \tag{6.2}$$

linear in the electric field. It corresponds to the Meyer flexoelectric term, connected with the polar properties of the NLC. Finally, since the liquid crystal has quadrupolar symmetry, if **E** is not homogeneous, it is possible to build the scalar

$$f_{\rm Q} = c_3\, Q_{ij} E_{i,j}, \tag{6.3}$$

which represents the electric energy density of a quadrupolar medium. All the contributions listed above are localized in the surface layer where **E** is different from zero. If the NLC is homogeneous (**n** is position independent) $f_{\rm F} = 0$ and the electric contribution to the energy density reduces to $f_{\rm D} + f_{\rm Q}$. This simple case will be considered first in this chapter. Then, we will consider the case where **n** is not homogeneous across the sample. In this situation, all the contributions discussed above exist. The bulk equilibrium equation involves an effective flexoelectric coefficient which is a combination of the dipolar and of the quadrupolar ones. For the sake of simplicity, when $\mathbf{n} = \mathbf{n}(\mathbf{r})$ we will assume that c_3 can be neglected. However, the general case can be solved without difficulties by changing c_2 in $c_2 - c_3$, and introducing a surface contribution of the type $c_3 Q_{ij} E_j \nu_i$, where ν_i is the geometrical normal to the outward directed surface.

In the next chapters we consider the problem of ion adsorption in the framework of Poisson-Boltzmann theory and with particular attention to the influence of adsorption phenomena on the molecular orientation in a liquid crystalline system, and on the effect of the adsorption on the experimental determination of the anchoring energy. The theory is developed for an isotropic fluid and, in this sense, is general. However, the agreement obtained between the predictions regarding the anchoring energy and the experimental data available is rather good, indicating that the approach presented in this book works in the correct direction to explain important characteristics of a real NLC sample.

6.1.2 The Poisson-Boltzmann equation: Exponential approximation for the electric field

We particularize the analysis to a planar surface in contact with a solution. The solution (ions coming from impurities) is modelled as point ions embedded in a dielectric continuum representing the solvent (in our applications it is the liquid crystalline system in the nematic phase). We suppose that the solution contains positive and negative ions. We choose our coordinate system such that the surface is situated in the plane at $z = 0$. The potential inside the sample, denoted by $V(z)$, is given by the Poisson's equation:

$$\frac{d^2V}{dz^2} = -\frac{\rho(z)}{\epsilon}, \tag{6.4}$$

where $\rho(z)$ is the charge density in the electrolyte, ϵ is the dielectric constant. Let $n_+(z)$ and $n_-(z)$ denote the densities of positive and negative ions, of charge $\pm q$. The density is

$$\rho(z) = q\left[n_+(z) - n_-(z)\right]. \tag{6.5}$$

The density of ions depends on the potential $V(z)$. We choose $V(\infty) = 0$ as the reference point. According to the Boltzmann statistics we have:

$$n_+(z) = n_0 e^{-qV(z)/k_BT} \quad \text{and} \quad n_-(z) = n_0 e^{qV(z)/k_BT}, \tag{6.6}$$

where n_0 is the bulk density of ionic impurities of an infinite sample (i.e., when $z \to \infty$, $V(z) \to 0$ and, therefore, $n_\pm \to n_0$). Substitution of Eq. (6.6) into Eq. (6.5), and then into (6.4), yields

$$\frac{d^2V}{dz^2} = 2\frac{q}{\epsilon} n_0 \sinh\left(\frac{qV(z)}{k_BT}\right), \tag{6.7}$$

which is a nonlinear differential equation determining the potential $V(z)$. Being nonlinear, this equation is difficult to solve. In this chapter, we follow the Debye-Hückel approximation, which consists in linearizing the right-hand side, by expanding the hyperbolic function up to first order, namely

$$\frac{d^2V(z)}{dz^2} = \frac{1}{\lambda_0}V(z), \tag{6.8}$$

where

$$\lambda_0 = \left(\frac{\epsilon k_B T}{2 n_0 q^2} \right)^{1/2} \tag{6.9}$$

is the *Debye's screening length.* An important feature of this quantity is that it depends on the *bulk* density of ions. As we will show in Chapter 7, there is another *characteristic length* connected with the *surface* density of ions. These lengths are shown to be connected by means of a relation that involves the activation energy of the chemical reaction giving rise to the presence of ions in the sample.

The solution of (6.8), satisfying boundary condition $V(z \to \infty) \to 0$, has the form

$$V(z) = V_s e^{-z/\lambda_0}, \tag{6.10}$$

where the constant V_s is fixed by the charge conservation condition, namely:

$$\int_0^\infty \rho(z) dz = -\sigma, \tag{6.11}$$

where σ is the surface charge density on the substrate. Within this approximation the problem is formally solved. The electric potential profile is given by

$$V(z) = \frac{\sigma}{\epsilon} \lambda_0 e^{-z/\lambda_0}, \tag{6.12}$$

while the charge density is obtained through (6.8) in the form:

$$\rho(z) = -\frac{\sigma}{\lambda_0} e^{-z/\lambda_0}. \tag{6.13}$$

In this manner, the excess charge on the surface is balanced by a space-charge layer, which decays exponentially in the solution. Consequently, a double layer is formed, as is shown schematically in Figure 6.1.

The configuration of charges shown in Figure 6.1 has a capacity. The interfacial capacity per unit area is known as the double-layer capacity and can be written as

$$C = \epsilon / \lambda_0. \tag{6.14}$$

It is the capacity of a parallel-plate capacitor with the plate separation given by the Debye's screening length. There is also a measurable quantity, known as differential capacity, which is defined as $C = d\sigma/dV$ and is also known as Gouy-Chapman capacity [15].

The validity of this linear approximation will be discussed in details in Chapter 7. But it is obvious that it can work well only for $qV/k_B T \ll 1$, which implies that $V \ll k_B T/q \approx 25\,\text{mV}$ for room temperature and monovalent ions.

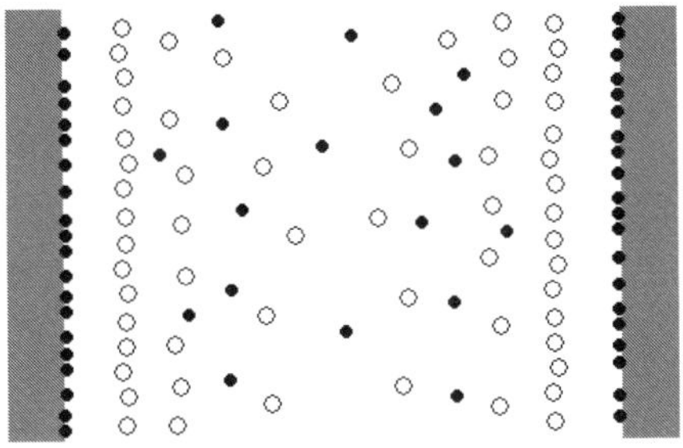

FIGURE 6.1
Schematic representation of the Debye double layer showing the space-charge separation in the sample.

Real liquid crystal samples may involve electric potentials higher than this one.

In any case, the next section will be dedicated to exploring the exponential field approximation in connection with the problem to investigate the thickness dependence of the anchoring energy in an NLC sample. The same approximation will be used in the subsequent section to explore the surface electric field-induced instabilities in a homeotropically oriented sample containing ionic impurities.

6.1.3 Thickness dependence of anchoring energy: Langmuir approximation

In this section, a complete model, in the framework of the approximation discussed in the preceding section, is presented [5]. It incorporates a new important electric term, connected with the flexoelectric properties of the medium. The model starts from the consideration of the selective ion adsorption which occurs when the solid substrate is in contact with the NLC. For instance, the positive ions are attracted by the solid surface, whereas the negative ions are repelled. In this situation, a double layer, in the sense discussed in the preceding section, of thickness of the order of λ_0, exists near the bounding surfaces.

If we denote again by σ the electric charge density adsorbed on the surface locate at $z = 0$, as in the preceding section the electric field $E(z)$ can by obtained from (6.12) in the form

$$E(z) = -\frac{dV}{dz} = \frac{\sigma}{\epsilon} e^{-z/\lambda_0}, \tag{6.15}$$

where, now, ϵ has to be interpreted as the average relative dielectric constant of the NLC. The electric field is normal to the surface, by symmetry.

The NLC is an anisotropic material having $\epsilon_{\|} \neq \epsilon_{\perp}$, such that the dielectric anisotropy $\epsilon_a \neq 0$. Therefore, the electric field given by (6.15) has an orienting effect on the NLC. The related dielectric energy density is given by (1.59) and can be written as

$$f_{\mathrm{D}} = -\frac{1}{2}\epsilon_a (\mathbf{n} \cdot \mathbf{E})^2 = -\frac{1}{2}\epsilon_a E^2(z) \cos^2\theta, \tag{6.16}$$

where, as defined in Chapter 1, the tilt angle θ is the angle formed by the director $\mathbf{n}$ with the normal to the surface. The dielectric energy per unit area is obtained by integrating (6.16) from 0 to ∞, by taking into account (6.15). Simple calculations give

$$F_{\mathrm{D}} = \int_0^\infty f_{\mathrm{D}} dz = -\frac{1}{4}\epsilon_a \left(\frac{\sigma}{\epsilon}\right)^2 \lambda_0 \cos^2\theta, \tag{6.17}$$

in the hypothesis that θ is position independent.

As we have seen in Chapter 2, a solid substrate is characterized by an easy direction $\mathbf{n_0}$ and an anchoring strength W such that the surface energy can be approximated by the Rapini-Papoular expression (2.9). In this manner, the effective surface energy is obtained by adding (2.9) and (6.17). Notice, however, that W is thickness independent, whereas f_{D} is thickness dependent. This follows from the fact that the surface density of charges depends on the volume of the NLC sample.

Another contribution to the electric energy density connected with a field given by (6.15) is

$$f_{\mathrm{Q}} = \tilde{e} Q_{ij} E_{i,j}, \tag{6.18}$$

according to (6.3), where the constant c_3 has been indicated by $\tilde{e}$. In the simplest case in which the electric field is of the form $\mathbf{E}(z) = E(z)\mathbf{k}$, where $\mathbf{k}$ is the normal to the surface, and $E(z)$ is given by (6.15), f_{q} can be written as

$$f_{\mathrm{Q}} = e\left(\cos^2\theta - \frac{1}{3}\right)\frac{dE}{dz}, \tag{6.19}$$

where $e = (3/2)S\tilde{e}$ is the total flexoelectric coefficient [16, 17]. The dielectric energy, having quadrupolar origin, per unit surface, is obtained by integrating (6.19) from 0 to ∞. By supposing, as before, the NLC orientation to be position independent, from (6.19) one obtains

$$\begin{aligned} F_{\mathrm{Q}} &= \int_0^\infty f_{\mathrm{q}} dz = -e\left(\cos^2\theta - \frac{1}{3}\right)\frac{\sigma}{\epsilon} \\ &= -e\frac{\sigma}{\epsilon}\cos^2\theta + \text{constant}, \end{aligned} \tag{6.20}$$

where the constant term is not important in the analysis, since it is independent of the NLC orientation.

The total energy per unit surface, playing the role of the *effective* anchoring energy, is given by

$$F_{\text{eff}} = F_{\text{s}} + F_{\text{D}} + F_{\text{Q}}. \tag{6.21}$$

Consider now the case in which the easy axis is parallel to $\mathbf{k}$. This means that the surface tries to impose a homeotropic alignment. In this situation (6.21) is written as

$$F_{\text{eff}} = -\frac{1}{2} W_{\text{eff}} \cos^2 \theta, \tag{6.22}$$

with

$$W_{\text{D}} = \frac{1}{2}\epsilon_a \frac{\sigma^2}{\epsilon}\lambda_0, \quad W_{\text{Q}} = 2e\frac{\sigma}{\epsilon}, \quad \text{and} \quad W_{\text{eff}} = W + W_{\text{D}} + W_{\text{Q}}. \tag{6.23}$$

According to the sign of ϵ_a and e, W_{eff} may increase or decrease when the ion adsorption takes place. In the case in which $\epsilon_a < 0$ and the quadrupolar term is neglected, the ion adsorption gives rise to a destabilizing term, independently of the sign of the adsorbed charges, since this term is quadratic in σ. On the contrary, when the quadrupolar term is considered, the sign of σ is important. Different situations are discussed in the following.

(a) In the case in which $\epsilon_a < 0$, $e < 0$, and $\sigma > 0$ the trend of W_{eff} versus σ is monotonically decreasing (see Figure 6.2). The effective anchoring strength vanishes at a critical σ_c given by

$$\sigma_c = 2\frac{\epsilon}{|\epsilon_a|}\left\{-1 + \left[1 + \frac{|\epsilon_a|}{2|e|^2}\lambda_0 W\right]^{1/2}\right\}. \tag{6.24}$$

For $\sigma < \sigma_c$ the easy axis is parallel to $\mathbf{k}$. For $\sigma > \sigma_c$ the easy axis is normal to $\mathbf{k}$ (because $W_{\text{eff}} < 0$).

(b) In the case $\epsilon_a < 0$, $e > 0$, and $\sigma > 0$, the trend of W_{eff} versus σ is not monotonic. It presents a maximum for

$$\sigma_M = 2\frac{e}{\lambda_0}\frac{\epsilon}{\epsilon_a}. \tag{6.25}$$

For $\sigma < \sigma_M$ the stabilizing effect of the quadrupolar contribution dominates the destabilizing effect of the unusual dielectric coupling. For $\sigma > \sigma_M$ the dielectric energy gives the most important term. W_{eff} vanishes at a critical density

$$\sigma_c = 2\frac{e}{\lambda_0}\frac{\epsilon}{|\epsilon_a|}\left\{1 + \left[1 + \frac{|\epsilon_a|}{2e^2}\lambda_0 W\right]\right\}. \tag{6.26}$$

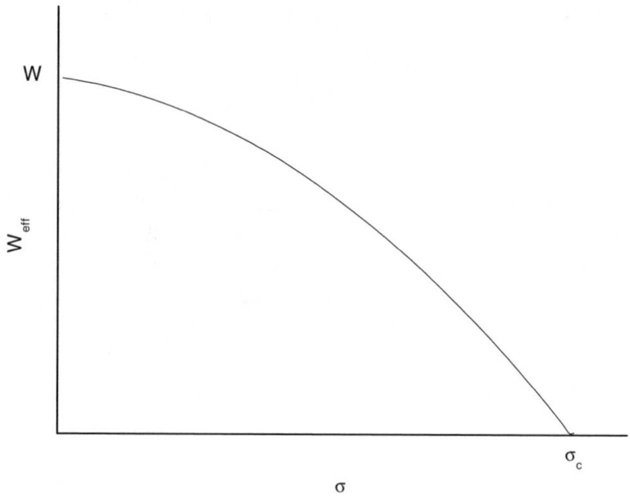

FIGURE 6.2
W_{eff} versus σ for $\epsilon_a < 0$, $e < 0$, and $\sigma > 0$. The dielectric and quadrupolar terms destabilize the initial homeotropic orientation. Hence W_{eff} decreases when σ increases. The homeotropic orientation is unstable for $\sigma > \sigma_c$. Reprinted with permission from Ref. [5]. Copyright (1993) by the American Physical Society.

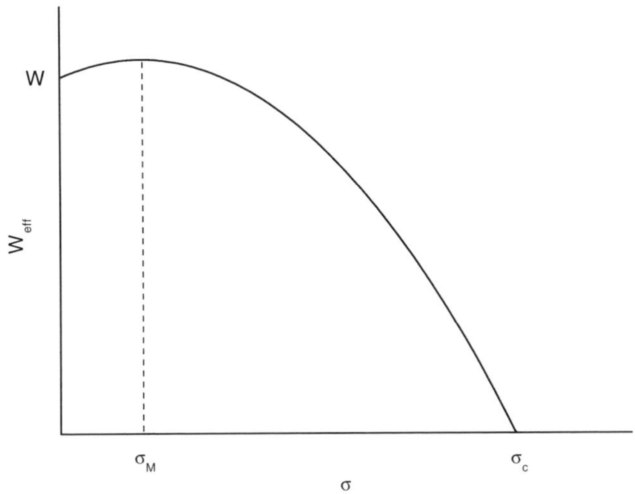

FIGURE 6.3
W_{eff} versus σ for $\epsilon_a < 0$, $e > 0$, and $\sigma > 0$. The dielectric term destabilizes the initial homeotropic orientation, whereas the quadrupolar term stabilizes it. For $\sigma < \sigma_M$, the quadrupolar term dominates the dielectric one and hence $W_{\text{eff}} > W$. The homeotropic orientation is unstable for $\sigma > \sigma_c$. Reprinted with permission from Ref. [5]. Copyright (1993) by the American Physical Society.

As before, for $\sigma < \sigma_c$ the easy axis is parallel to $\mathbf{k}$, and for $\sigma > \sigma_c$ it is perpendicular to it. The trend of W_{eff} versus σ is shown in Figure 6.3.

(c) Consider now the case in which $\epsilon_a > 0$, $e < 0$, and $\sigma > 0$. In this situation, W_{eff} versus σ presents a minimum for

$$\sigma_m = 2\frac{|e|}{\lambda_0}\frac{\epsilon}{\epsilon_a}. \tag{6.27}$$

The W_{eff} $(\sigma = \sigma_m)$ is given by

$$W_{\text{eff}}(\sigma = \sigma_m) = W - 2\frac{|e|^2}{\lambda_0 \epsilon \epsilon_a}. \tag{6.28}$$

It is interesting to observe that, if

$$W_0 = 2\frac{|e|^2}{\lambda_0 \epsilon \epsilon_a} > W, \tag{6.29}$$

there is a double transition homeotropic $\to$ planar *to* homeotropic for the critical densities

$$\sigma_{1,2} = \sigma_m \left[1 \pm \sqrt{1 - W/W_0}\right]. \tag{6.30}$$

For $\sigma < \sigma_m$ the destabilizing effect of the quadrupolar term dominates the stabilizing effect of the usual dielectric interaction. It is vice versa for $\sigma > \sigma_m$. These cases are shown in Figures 6.4(a) and 6.4(b).

(d) The case in which $\epsilon_a > 0$, $e > 0$, and $\sigma > 0$ shows a monotonic increase of W_{eff} versus σ, since both electric terms stabilize the initial orientation. This situation is shown in Figure 6.5.

(e) The case in which $\epsilon_a < 0$, $e < 0$, and $\sigma < 0$, is similar to case (b).

(f) The case in which $\epsilon_a < 0$, $e > 0$, and $\sigma < 0$, is similar to case (a).

(g) The case in which $\epsilon_a > 0$, $e > 0$, and $\sigma < 0$, is similar to case (c).

The case in which the easy direction is perpendicular to **k** can be analyzed in the same way.

Thickness dependence of W_{eff}

A complete investigation of the thickness dependence of the surface density of charges will be developed in the next chapters, where the nonlinear Poisson-Boltzmann equation is taken into account. An approximated result can be presented here if we assume that the surface density is governed by the Langmuir isotherm law, Eq. (5.8), and we identify the characteristic length $\kappa\tau = \lambda$, i.e.,

$$\sigma = \Sigma \frac{d/2\lambda_0}{1 + d/2\lambda_0}, \tag{6.31}$$

where Σ depends on the conductivity of the liquid crystals, on the adsorption energy, and on the number of free sites on the surface. We can deduce from (6.31) that the trends of W_{eff} versus σ discussed above can be interpreted as the trend of W_{eff} versus the thickness of the NLC sample.

One observes that the two terms of electric origin appearing in Eq. (6.21), i.e., F_{D} and F_{Q}, have usually the same order of magnitude. In the case of 5CB analyzed by Blinov and Kabaenkov [18], one has $\epsilon_a \approx 13$, $\epsilon \approx 12$ [19], $e \approx 10^{-10}$ SI units [20], $\sigma < \sigma_0 \approx 10^{-7}$ C/m^2, and $\lambda_0 \approx 0.6$ μm [21]. Consequently,

$$\frac{W_{\text{Q}}}{W_{\text{D}}} = 4\frac{e\,\epsilon}{\epsilon_a \sigma \lambda_0} > \frac{1}{2}, \tag{6.32}$$

showing that the quadrupolar contribution to F_{eff} is of the same order as the ordinary dielectric contributions.

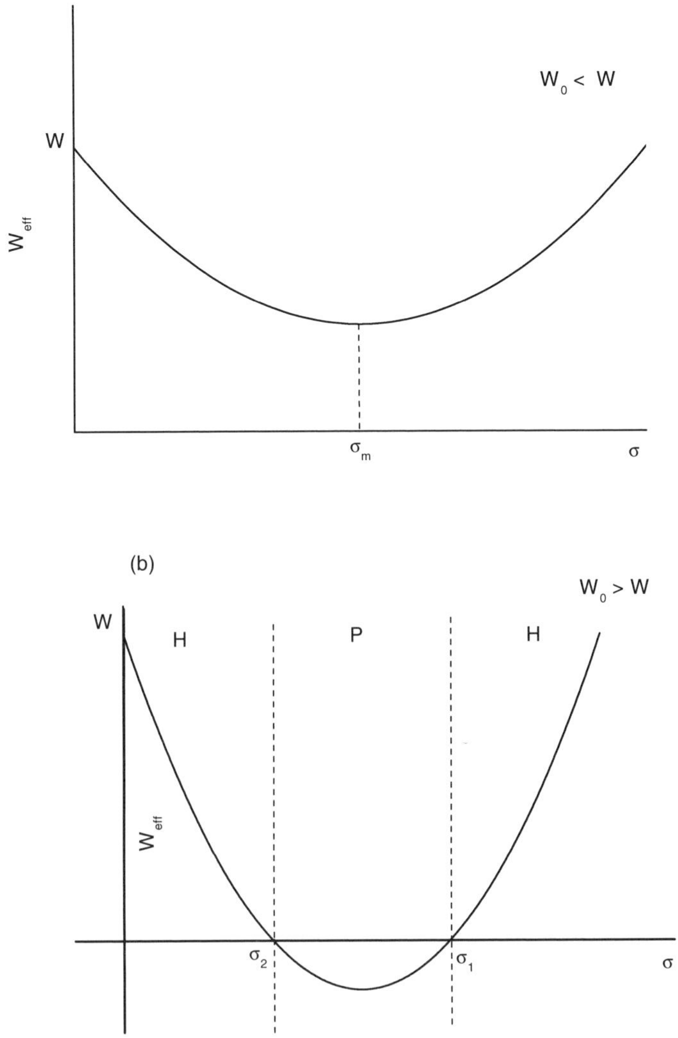

FIGURE 6.4
W_{eff} versus σ for $\epsilon_a > 0$, $e < 0$, and $\sigma > 0$. In this situation the dielectric term stabilizes the initial homeotropic orientation, whereas the quadrupolar term destabilizes it. For large σ, W_{eff} is larger than W. (a) if $W_0 = 2e^2/(\lambda_0\epsilon\epsilon_a) < W$, the effective anchoring energy has a non-monotonic trend versus σ, but the initial homeotropic easy direction remains stable. (b) If $W_0 > W$ the quadrupolar term dominates the dielectric one for $\sigma < \sigma_1$. In this situation a double surface transition H (homeotropic) $\rightarrow$ P (planar) $\rightarrow$ H (homeotropic) is predicted. Reprinted with permission from Ref. [5]. Copyright (1993) by the American Physical Society.

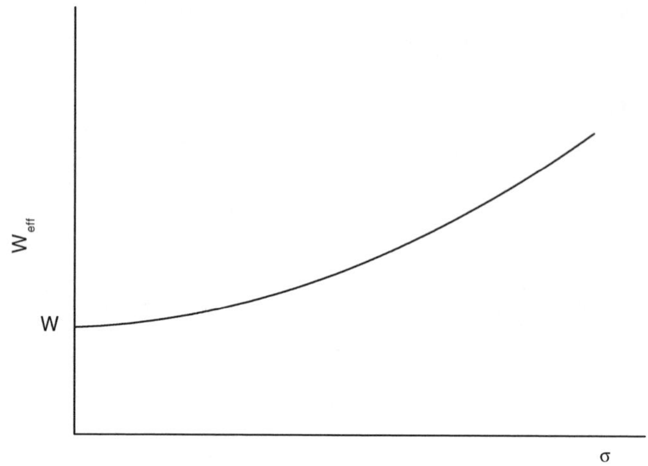

FIGURE 6.5
W_{eff} versus σ for $\epsilon_a > 0$, $e > 0$ and $\sigma > 0$. The dielectric and quadrupolar terms stabilize the initial homeotropic orientation. Hence W_{eff} increases with σ. Reprinted with permission from Ref. [5]. Copyright (1993) by the American Physical Society.

In Chapter 8, we will consider also the presence and the influence of an adsorption energy on the dielectric contribution to the anchoring energy. We will show that this quantity may have an important role in determining the precise dependence of the anchoring energy on the thickness of the sample.

6.2 Analysis of the influence of the surface electric field on the nematic orientation

The influence of the surface electric field on the nematic orientation has been recently analyzed in Refs. [22, 23] by assuming for $E(z)$ an exponential behavior. In Ref. [22] the eigenvalues of the problem, giving the threshold fields for the destabilization of the homeotropic pattern, were numerically evaluated. It was shown also that the method based on the trial function was quite good [23]. In this section we present an alternative solution of the problem, by means of which in the harmonic approximation we can evaluate the critical fields, inducing the homeotropic-distorted transition [24]. We neglect the surface polarization. First, we limit the investigation to the case of compensated

NLC for which $\epsilon_a = 0$. The case of $\epsilon_a \neq 0$ will be considered later. Different types of surface fields are considered. The analysis then will be generalized for a sample of finite thickness d, where the penetration of the surface field b is comparable to d.

6.2.1 Basic equations of the problem

We assume first $\epsilon_a = 0$. In the limit of small deformations, close to the homeotropic configuration, the total bulk energy density reads

$$f = \frac{1}{2} K \theta'^2 + eE(z)\theta\theta', \tag{6.33}$$

where, as before, we assume that the flexoelectric contribution to the energy density reduces to the dipolar one. In this case the bulk differential equilibrium equation is

$$\theta'' + \frac{e}{K} E'(z)\theta = 0. \tag{6.34}$$

The surfaces are assumed identical. Consequently, $E(z) = -E(-z)$, and hence $E'(z) = E'(-z)$. In particular $E(-d/2) = -E(d/2) = E$. Furthermore, we suppose that the surface treatments are such to induce homeotropic alignment on both surfaces, and that the anisotropic part of the surface tension is well described by the Rapini-Papoular form, introduced in (2.8). In this framework, Eq. (6.34) has to be solved with the boundary conditions

$$\theta' - \frac{W - eE}{K}\theta = 0 \quad \text{and} \quad \theta' + \frac{W - eE}{K}\theta = 0 \tag{6.35}$$

for $z = -d/2$ and $z = d/2$, respectively. The solution of Eq. (6.34) with the boundary conditions (6.35) can be written in the form

$$\theta(z) = \alpha_e \theta_e(z) + \alpha_o \theta_o(z), \tag{6.36}$$

where α_e and α_o are two constants, whereas $\theta_e(z) = \theta_e(-z)$ and $\theta_o(z) = -\theta_o(-z)$ are the even and odd solutions of the differential equation (6.34) and of the boundary condition at $z = -d/2$

$$\theta_i' - \frac{W - eE}{K}\theta_i = 0, \tag{6.37}$$

where $i = o$ (odd) or $i = e$ (even).

Note that the differential equation (6.34), with the boundary conditions (6.35), has always the trivial solution $\theta(z) = 0$, for $-d/2 \leq z \leq d/2$, which corresponds to the homeotropic alignment. However, for particular values of E, this problem can admit also solutions different from the trivial one. The procedure to obtain the eigenvalue is the usual one. If $\theta(E, z)$ is a solution of Eq. (6.34), with a well defined parity, the boundary conditions (6.35) give the relation

$$\frac{W}{K} = \frac{e}{K}E + \left\{ \frac{\theta'(E,z)}{\theta(E,z)} \right\}_{z=-d/2}, \tag{6.38}$$

which defines the critical values of the surface field E to have the instability.

To analyze the stability of the deformed structure it is necessary to evaluate the total energy, per unit surface, of the sample for a defined θ-profile, solution of Eq. (6.34). For the problem under consideration, the bulk energy density is given by (6.33). Since

$$\theta'^2 = \frac{d}{dz}(\theta\theta') - \theta\theta'', \tag{6.39}$$

by taking into account Eq. (6.34) we get

$$\theta'^2 = \frac{d}{dz}(\theta\theta') + \frac{eE'}{K}\theta^2. \tag{6.40}$$

It follows that the bulk energy density given by (6.33) can be rewritten as

$$f = \frac{1}{2}K\left(\theta'^2 + \frac{eE}{K}\theta\theta'\right) = \frac{1}{2}K\frac{d}{dz}\left\{\theta\theta' + \frac{eE}{K}\theta^2\right\}. \tag{6.41}$$

The total energy per unit surface is given by

$$F = \int_{-d/2}^{d/2} f dz + \frac{1}{2}W_1\theta_1^2 + \frac{1}{2}W_2\theta_2^2, \tag{6.42}$$

if the sample is a slab of thickness d. By substituting (6.41) into (6.42) we get

$$F = \frac{1}{2}K\left\{\theta_2\theta_2' - \theta_1\theta_1' + \frac{e}{K}\left(E_2\theta_2^2 - E_1\theta_1^2\right)\right\} + \frac{1}{2}W_1\theta_1^2 + \frac{1}{2}W_2\theta_2^2, \tag{6.43}$$

where $E_2 = E(d/2)$ and $E_1 = E(-d/2)$. Since the surfaces are assumed identical $W_1 = W_2 = W$ and $E(z) = -E(-z)$, from which we obtain $E_2 = -E_1 = -E$. In this case the solutions of (6.34) are even or odd in z. If $\theta(z)$ is even in z, $\theta(z) = \theta(-z)$, and $\theta'(z) = -\theta'(-z)$. In particular $\theta_2 = \theta_1$, $\theta_2' = -\theta_1'$. In this case, from (6.43), we obtain

$$F = \left\{W - \left(eE + K\frac{\theta_1'}{\theta_1}\right)\right\}\theta_1'^2. \tag{6.44}$$

If $\theta(z)$ is odd in z, $\theta(z) = -\theta(-z)$, and $\theta'(z) = \theta'(-z)$. In particular $\theta_2 = -\theta_1$, $\theta_2' = \theta_1'$. In this case from (6.43) we obtain again (6.44).

Expression (6.44) shows that the presence of the linear term in θ' in the elastic energy density renormalizes the anchoring energy strength. In fact, from (6.44) it is clear that the total energy of the nematic sample reduces to a surface contribution. Since the two surfaces are assumed identical, each of them contributes to the total energy with one half of (6.44). The effective anchoring energy is then

$$W_{\text{eff}} = W - \left(eE + K\frac{\theta_1'}{\theta_1} \right). \tag{6.45}$$

Expression (6.45) shows that the renormalization of the anchoring energy is not simply $-eE$. It contains also a term connected with elastic properties of the nematic liquid crystal.

The form (6.44) for F is useful to analyze the stability of the deformed state. In fact, in a linear analysis, $\theta(z) = \alpha\theta_i(z)$, where $\theta_i(z)$ is a solution of the differential Eq. (6.34) of a given symmetry, and α an integration constant. In this framework, F is a quadratic function of α of the kind $F = a\alpha^2$, where, according to (6.44),

$$a = \left\{ W - \left(eE + K\frac{\theta_{i1}'}{\theta_{i1}} \right) \right\} \theta_{i1}'^2. \tag{6.46}$$

From (6.46) one gets the critical line (6.38). If E is smaller than the value defined by (6.38), $a > 0$, and F is minimized for $\alpha = 0$. This means that the stable state is the non-deformed one. On the contrary, if E is larger than the one defined by (6.38), $a < 0$ and F has a maximum for $\alpha = 0$. This means that the stable state is the deformed one. From the point of view of the phase transitions, the critical curve $W = W(E)$ divides the (W, E) plane in two regions. The one below the curve corresponds to the deformed configurations whereas the one upper the curve to the homeotropic configuration.

If the sample is a half space the total energy per unit surface is given by

$$F = \int_0^\infty f dz + \frac{1}{2} W_1 \theta_1^2. \tag{6.47}$$

In this case, by using (6.43) and taking into account that $\lim_{z\to\infty} E(z) = 0$, and $\lim_{z\to\infty} \theta' = 0$, we reobtain for F expression (6.44). In the following, we will consider different approximations of the problem.

6.2.2 First approximation

Let us assume that the field $E(z)$, due to the limiting surfaces, can be approximated by the function

$$E(z) = \begin{cases} -(E/b)(z + z^*) \\ 0 \\ -(E/b)(z - z^*) \end{cases} \tag{6.48}$$

for $-d/2 \le z \le -z^*$, $-z^* \le z \le z^*$, and $z^* \le z \le d/2$, respectively. In (6.48), $z^* = d/2 - b$ and b is a typical length connected with the penetration of the surface field into the bulk. In this approximation $E'(z) = -(E/b)$ in the surface layers, and $E'(z) = 0$ in the bulk. We assume first that $eE > 0$. In this framework the bulk differential equations of the problem are

$$\theta'' - \nu^2\theta = 0 \quad \text{and} \quad \theta'' = 0, \tag{6.49}$$

in the surface layers $-d/2 \leq z \leq -z^*$, $z^* \leq z \leq d/2$, and in the bulk $-z^* \leq z \leq z^*$, respectively. In Eq. (6.49) we put

$$\nu = \sqrt{\frac{eE}{Kb}}. \tag{6.50}$$

Solution θ has to satisfy the boundary conditions (6.35) and be continuous, with continuous first derivative, at $z = \pm z^*$. The function $\theta_e(z)$ for the present problem is

$$\theta_e(z) = \begin{cases} \alpha_e e^{-\nu z}\left[1 + e^{2\nu(z+z^*)}\right] \\ 2\alpha_e e^{\nu z^*} \\ \alpha_e e^{\nu z}\left[1 + e^{-2\nu(z-z^*)}\right] \end{cases} \tag{6.51}$$

for $-d/2 \leq z \leq -z^*$, $-z^* \leq z \leq z^*$, and $z^* \leq z \leq d/2$, respectively. The equation determining the eigenvalue of the problem is, according to Eq. (6.38),

$$\frac{W_e}{K} = \frac{e}{K}E - \sqrt{\frac{eE}{Kb}} \tanh \sqrt{\frac{eE}{K}b}. \tag{6.52}$$

Note that from Eq. (6.52) it follows that $W_e < eE$. Consequently, the critical field to have the instability is larger than the one to have the surface instability, given by $W = eE$. In fact, if $E(z)$ is assumed position independent, the effect of the linear term is only to renormalize the surface anchoring energy, whereas in the present case it plays also a stabilizing effect on the homeotropic orientation of the bulk.

The function $\theta_o(z)$ for the same problem is

$$\theta_o(z) = \begin{cases} \alpha_o e^{-\nu z}\left\{1 - \frac{1-\nu z^*}{1+\nu z^*} e^{2\nu(z+z^*)}\right\} \\ -\nu\alpha_o e^{\nu z^*}\frac{2z}{1+\nu z^*} \\ -\alpha_o e^{\nu z}\left\{1 - \frac{1-\nu z^*}{1+\nu z^*} e^{-2\nu(z-z^*)}\right\} \end{cases} \tag{6.53}$$

for $-d/2 \leq z \leq -z^*$, $-z^* \leq z \leq z^*$, and $z^* \leq z \leq d/2$, respectively. The equation determining the eigenvalue of the problem is now

$$\frac{W_o}{K} = \frac{e}{K}E - \sqrt{\frac{eE}{Kb}}\,\frac{1 + \sqrt{eEb/K}\,[(d/2b) - 1]\tanh\sqrt{eEb/K}}{\sqrt{eEb/K}\,[(d/2b) - 1] + \tanh\sqrt{eEb/K}}. \tag{6.54}$$

Also in this case $W_o < eE$, because the bulk contribution to the energy density is such to stabilize the homeotropic orientation. To have a numerical estimation of the critical anchoring energies we use the analogy between the

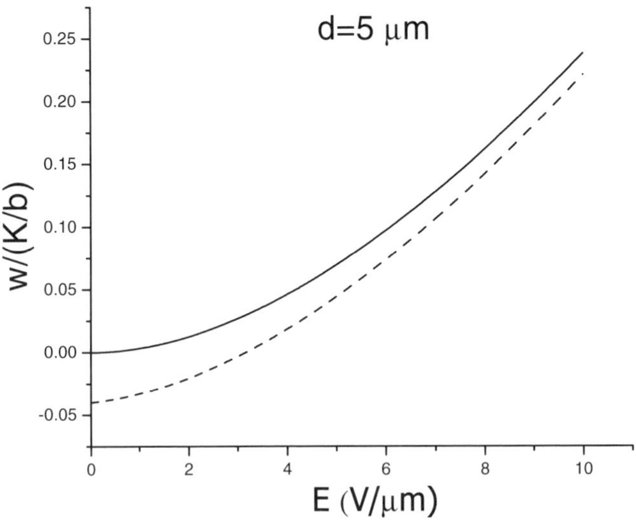

FIGURE 6.6
Critical lines $W_e/(K/b)$ (solid) and $W_o/(K/b)$ (dashed) for deformations even and odd in z deduced by means of the simple analysis reported in Section 6.2.2. For large E the curves are practically coincident. For a given anchoring energy W the critical value of the surface field for the even deformation in z is smaller than the one for the odd deformation. Hence, the actual deformation induced by the surface field is the even one. The curves are drawn for the case $eE > 0$, using the parameters $e = 1 \times 10^{-11}$ C/m, $K = 10^{-11}$ N, $b = 0.1\mu$m and $0 < E < 10^7$ V/m. Reprinted from Ref. [24].

linear term and the flexoelectric coupling, and assume $|e| \sim 10^{-11}$C/m [20, 25], $K \sim 10^{-11}$N [26], $\epsilon \sim 10\epsilon_0$ [26], and $b \sim 10^{-7}$m. The parameter b has been assumed of the order of Debye's screening length for commercial liquid crystals [21]. For the density of surface adsorbed charges we use the value estimated by Thurston *et al.* [21], $\sigma \sim 10^{-4}$C/m^2. With the values reported above, $E \sim \sigma/\epsilon \sim 10^6$V/m, and $eE \sim 10^{-5}$J/m^2. Hence, the phenomenon described by us is important all the time the extrapolation length $L = K/W$ is in the micron scale. In Figure 10.3 we show the critical curves $W_e = W_e(E)$, and $W_o = W_o(E)$ deduced by means of Eq. (6.52) and Eq. (6.54), respectively. In the limit of large d they are practically coinciding, as expected. In fact in this limit the deformation is localized in the surface layers of thickness b, and the bulk is homeotropically oriented. However $E_e < E_o$. This implies that for a given anchoring energy strength, the deformation characterized by a tilt angle which is an even function of z costs less energy than the one odd in z.

By means of the linearized analysis presented above it is not possible to obtain the full profile $\theta_e(z)$. But simple considerations show that the deformation is mainly localized in the surface layers. In fact, since $W < eE$, it follows that $\theta'_e(-d/2) = [(W - eE)/K]\theta(-d/2) < 0$. Furthermore, from Eq. (6.49) we have $\theta''_e(z) = \nu\theta_e(z) > 0$ for $-d/2 \leq z \leq d/2$. Consequently, the tilt angle is larger in the surface layers than in the bulk.

Let us consider now the case $eE < 0$. In this framework the bulk differential equation is

$$\theta''(z) + \frac{|eE|}{Kb}\theta(z) = 0. \tag{6.55}$$

The relevant boundary conditions are

$$\theta' - \frac{W + |eE|}{K}\theta = 0 \quad \text{and} \quad \theta' + \frac{W + |eE|}{K}\theta = 0, \tag{6.56}$$

for $z = -d/2$ and $z = d/2$, respectively. In this case the presence of the surface field reinforces the anchoring energy strength, because W is substituted by $W + |eE|$. Furthermore, it destabilizes the bulk, as follows from Eq. (6.55). A calculation similar to the one reported above shows that the eigenvalue for the even deformation is

$$\frac{W_e}{K} = -\frac{|eE|}{K} + \sqrt{\frac{|eE|}{Kb}}\tan\sqrt{\frac{|eE|b}{K}}, \tag{6.57}$$

whereas the eigenvalue for the odd deformation is given by

$$\frac{W_o}{K} = -\frac{|eE|}{K} + \sqrt{\frac{|eE|}{Kb}}\,\frac{\sqrt{|eE|b/K}\,[(d/2b) - 1]\tan\sqrt{|eE|b/K} - 1}{\sqrt{|eE|b/K}\,[(d/2b) - 1] + \tan\sqrt{|eE|b/K}}. \tag{6.58}$$

The critical curves $W_e = W_e(E)$ and $W_o = W_o(E)$ are reported in Figure 6.7. From this figure it follows that the deformation even in z is favored. It can take place also in the case of strong anchoring, where $W \to \infty$. The relevant critical field is found to be

$$E = \left(\frac{\pi}{2}\right)^2 \frac{K}{|e|b}. \tag{6.59}$$

Since in the case under consideration $\theta'(-d/2) = [(W + |eE|)/K]\theta(-d/2) > 0$, and $\theta''(z) = -\sqrt{(|eE|/Kb)}\theta(z) < 0$ for $-d/2 \leq z \leq d/2$, the deformation is larger in the bulk than in the surface layers.

6.2.3 Half-space approximation

The analysis presented in the previous section is very simple, and allows one to obtain explicit expressions for the surface fields inducing the instability due to the presence of the elastic term linear in the spatial derivative of the

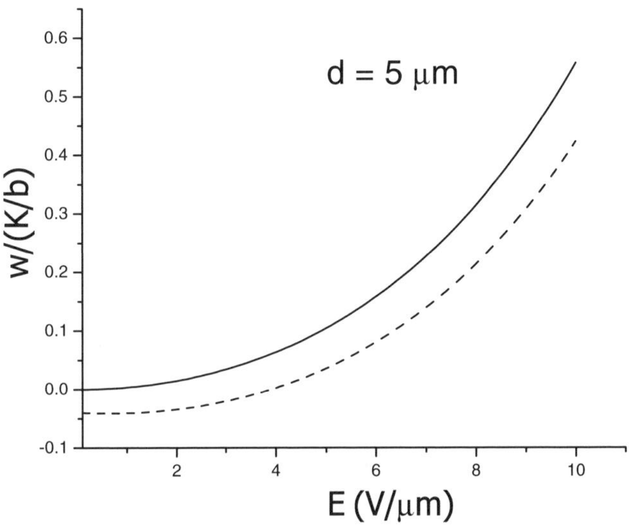

FIGURE 6.7
As in Figure 6.6, with $eE < 0$. In this case the flexoelectric contribution reinforces the anchoring energy, changing W in $W + |eE|$, and destabilizes the homeotropic orientation in the bulk. Now, even in the strong anchoring situation the bulk can be distorted by the surface field. As in the previous case the even deformation is favored. Parameters as in Figure 6.6, with $e = -1 \times 10^{-11}$ C/m. Reprinted from Ref. [24].

nematic director. However, it is connected with a very special z-dependence of the surface field. It is possible to generalize previous analysis if $d \gg b$. In this case the sample can be considered, for what concerns the surface field, as formed by two half spaces. In each half space the field can be assumed exponentially decreasing with a typical length b. The investigation is limited to the half space $z \geq 0$, and we assume that the surface field is

$$E(z) = Ee^{-z/b}. \tag{6.60}$$

By substituting (6.60) into (6.34) the bulk differential equation is now

$$\theta''(z) - \frac{eE}{Kb} e^{-z/b} \theta(z) = 0, \tag{6.61}$$

that has to be solved with the boundary condition

$$\theta' - \frac{W - eE}{K} \theta = 0, \tag{6.62}$$

at $z = 0$, and

$$\lim_{z \to \infty} \theta'(z) = 0. \tag{6.63}$$

Let us suppose first $eE > 0$. In this case the solution of Eq. (6.61), which remains finite for $z = 0$ and whose first z derivative tends to zero for $z \to \infty$, is

$$\theta(E, z) = \beta I_0 \left(2\sqrt{\frac{eEb}{K}} e^{-z/2b} \right), \tag{6.64}$$

where I_0 is modified Bessel's function of zero order and β is a constant. By means of (6.64) the critical field is found to be

$$\frac{W}{K} = \frac{eE}{K} + \left\{ \frac{\theta'(E, z)}{\theta(E, z)} \right\}_{z=0}. \tag{6.65}$$

In Figure 6.8, we show the critical line obtained by means of (6.65) and the one given by (6.52). As is evident from the figure, (6.52) is a good approximation of (6.65), and the agreement increases with d. Also in the case considered now, $W < eE$. Consequently, $\theta'(0) < 0$. Since $\theta''(z) > 0$ for $0 \leq z < \infty$, we deduce that $\theta(0) > \theta(\infty)$.

Let us suppose now that $eE < 0$. In this case the solution of the problem under consideration is

$$\theta(z) = \beta J_0 \left(-2\sqrt{\frac{|eE|b}{K}} e^{-z/2b} \right), \tag{6.66}$$

where J_0 is Bessel's function of zero order and β is a constant. By substituting (6.66) into (6.65) we obtain the critical field for the instability. Note that even

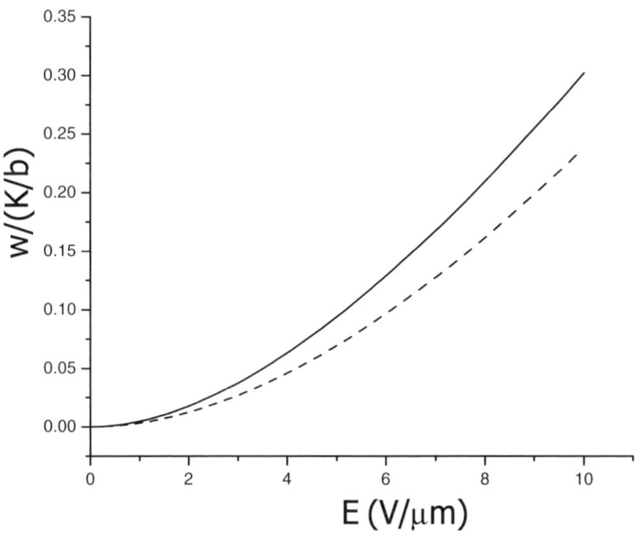

FIGURE 6.8
Critical line $W/(K/b)$(solid) deduced by means of the model reported in Section 6.2.3, where the surface field is assumed to be exponentially decreasing with a typical length b. Case $eE > 0$. The dashed line is the approximated curve $W_e/(K/b)$ obtained in Section 6.2.2. Physical parameters as in Figure 6.6. Reprinted from Ref. [24].

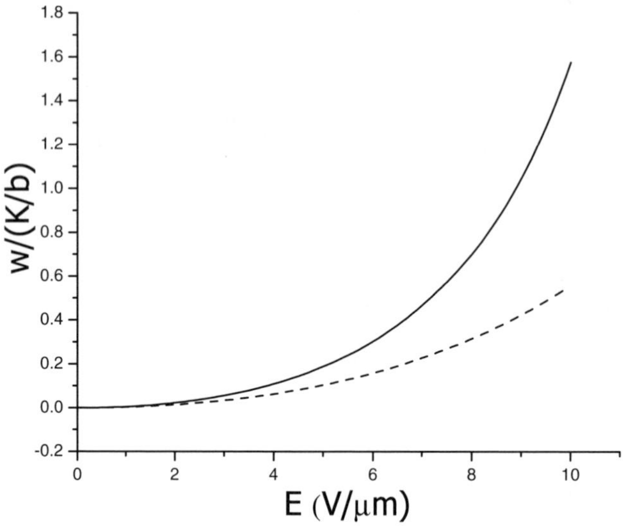

FIGURE 6.9
As in Figure 6.8 with $eE < 0$. The dashed line is the approximated curve $W_e/(K/b)$ obtained in Section 6.2.2. Physical parameters as in Figure 6.7. Reprinted from Ref. [24].

in the case of strong anchoring the instability exists. Its critical field is given by

$$J_0\left(-2\sqrt{\frac{|eE|b}{K}}\right) = 0, \tag{6.67}$$

from which we obtain

$$E^* = \left(\frac{j_0}{2}\right)^2 \frac{K}{|e|b}, \tag{6.68}$$

where $j_0 = 2.4048$ is the first zero of Bessel's function $J_0(x)$. In Figure 6.9 we show the critical line obtained with the present analysis by comparing with the one given by (6.57).

6.2.4 Finite sample limited by two identical surfaces

If d is comparable with b the half space approximation does not work any longer. In this case the field due to the limiting surfaces, assumed as identical, can be approximated by

$$E(z) = -E\frac{\sinh(z/b)}{\sinh(d/2b)}. \tag{6.69}$$

Note that from (6.69) $E(-d/2) = -E(d/2) = E$, and the surface field is localized in surface layers of thickness of the order of b. If $d \gg b$, from (6.69), we re-obtain the half-space approximation discussed in the previous section. By substituting (6.69) into (6.34) we obtain the bulk differential equation for the present problem in the form

$$\theta'' - \frac{eE}{Kb}\frac{\cosh(z/b)}{\sinh(d/2b)}\theta = 0, \tag{6.70}$$

which has to be solved with the boundary conditions (6.35). The even solution of Eq. (6.70) is

$$\theta_e(z) = \alpha_e \Psi_e(E, z), \tag{6.71}$$

and the odd solution of the same equation is

$$\theta_o(z) = \alpha_o \Psi_o(E, z), \tag{6.72}$$

where $\Psi_e(E, z)$ and $\Psi_o(E, z)$ are the modified Mathieu's function, usually indicated by [27]

$$\Psi_e(E, z) = \mathcal{C}\left(0, -2\frac{eEb}{K\sinh(d/2b)}, -i\frac{z}{2b}\right),$$
$$\Psi_o(E, z) = \mathcal{S}\left(0, -2\frac{eEb}{K\sinh(d/2b)}, -i\frac{z}{2b}\right). \tag{6.73}$$

Repeating previous calculations step by step we obtain that the critical surface field to induce an even or odd deformation is given by

$$\frac{W_i}{K} = \frac{eE}{K} + \left\{\frac{\Psi_i'(E, z)}{\Psi_i(E, z)}\right\}_{z=-d/2}, \tag{6.74}$$

where $i = e, o$. If $eE > 0$ the behaviour of W_e and W_o is the one reported in Figure 6.10. As before, the actual deformation is θ_e, because the relevant critical field is smaller than the one connected to θ_o. In the opposite case of $eE < 0$ the critical lines are shown in Figure 6.11. As emphasized above, in this case the instability is possible even in the case of strong anchoring. The relevant threshold field is obtained by the condition

$$\Psi_e(E, -d/2) = \mathcal{C}\left(0, -2\frac{eEb}{K\sinh(d/2b)}, i\frac{d}{4b}\right) = 0. \tag{6.75}$$

In Figure 6.12, we show the ratio between the critical field given by Eq. (6.75) and E^* obtained by means of (6.68) versus the thickness of the sample. As expected, if $d \gg b$, $E \to E^*$. The agreement is already very good for $d \sim 10b$.

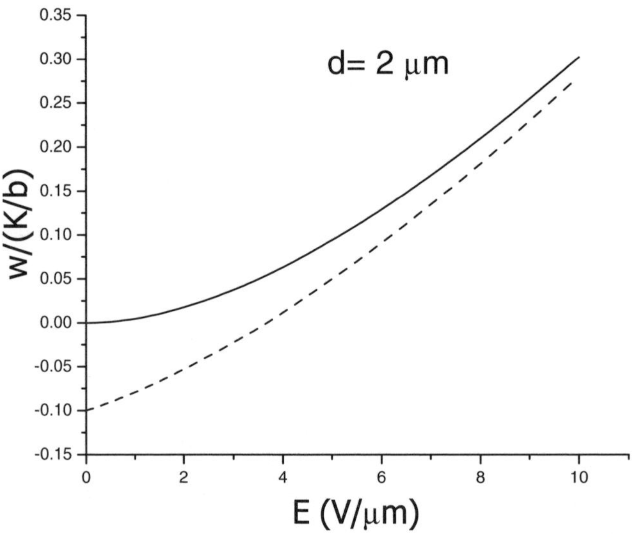

FIGURE 6.10
Critical lines $W_e/(K/b)$ (solid) and $W_o/(K/b)$ (dashed) deduced by means of analysis reported in Section 6.2.4. Physical parameters as in Figure 6.8. Reprinted from Ref. [24].

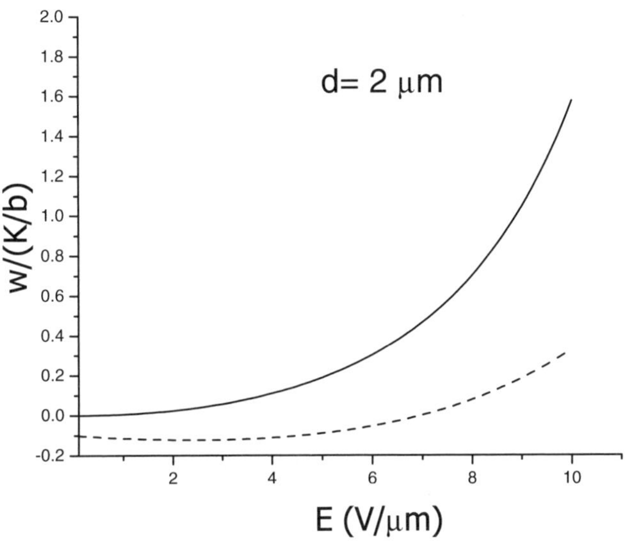

FIGURE 6.11
Critical lines $W_e/(K/b)$ (solid) and $W_o/(K/b)$ (dashed) deduced by means of the analysis reported in Section 6.2.4. Physical parameters as in Figure 6.9. Reprinted from Ref. [24].

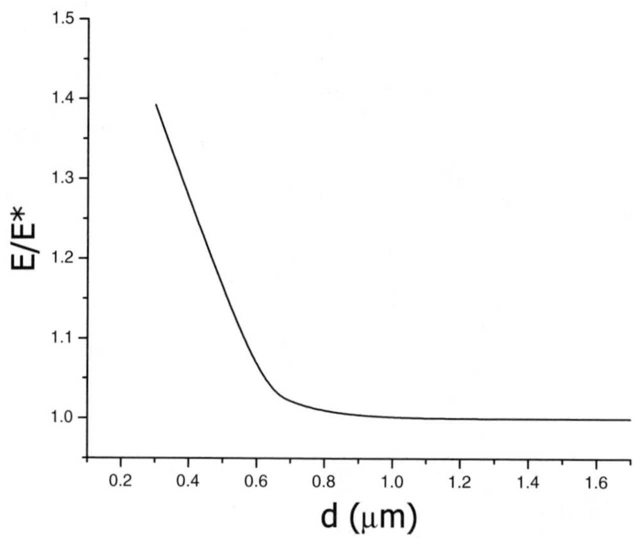

FIGURE 6.12
Reduced critical field E/E^*, where E^* is the critical field in the half-space approximation (defined in Eq. (6.68)) to have a surface instability in a sample of finite thickness, versus d. Physical parameters as in Figure 6.6. Reprinted from Ref. [24].

6.2.5 Influence of the dielectric anisotropy on the instability

Up to now we described the effect of the linear term in the spatial derivatives of the nematic director on the stable orientation as the coupling of an intrinsic electric field with the flexoelectric polarization, because the contributions to the energy density are of the same functional type. However, if the linear term is really connected to the coupling of a surface electric field due to the adsorption phenomenon, the analysis has to take into account also the electric term connected with the dielectric anisotropy [22, 23].

Let us consider a nematic sample limited by two identical substrates, where the surface electric field is such that $E(z) = -E(-z)$. For this case we have obtained different approximated solutions of the problem. As is evident from Figure 6.10 and Figure 6.11 the agreement between the half-space approximation and the case where d is of the same order of magnitude of b is very good already for $d = 10\,b$. Since $b \sim 0.1$ μm, for usual nematic samples used in the experiment our half-space approximation works very well. In this approximation we can investigate the influence of the dielectric anisotropy ϵ_a on the phenomenon under consideration. If $\epsilon_a \neq 0$ the bulk energy density of the nematic liquid crystal submitted to an electric field is, in the limit of $\theta \to 0$,

$$f = \frac{1}{2}K\theta'^2 + \frac{1}{2}\epsilon_a E^2(z)\theta^2 + eE(z)\theta\theta'. \tag{6.76}$$

The bulk equilibrium equation of the problem is, in the exponential approximation for the electric field,

$$K\theta'' - \left(\epsilon_a E^2 e^{-2z/b} + \frac{eE}{b}e^{-z/b}\right)\theta = 0, \tag{6.77}$$

which has to be solved with the boundary conditions (6.62) and (6.63), where E is supposed positive. If $e = 0$, Eq. (6.77) becomes

$$K\theta'' - \epsilon_a E^2 e^{-2z/b}\theta = 0, \tag{6.78}$$

whose solution is

$$\theta(z) = \alpha I_0\left(-\sqrt{\frac{\epsilon_a}{K}}Ebe^{-2z/b}\right), \tag{6.79}$$

where α is an integration constant. In this case the surface instability exists only if $\epsilon_a < 0$, and the corresponding critical field is

$$E = \frac{j_0}{b}\sqrt{\frac{K}{|\epsilon_a|}}. \tag{6.80}$$

Notice that $\sqrt{6} = 2.45 \sim j_0$, thus indicating the good agreement between the approximated result obtained in Ref. [22] and (6.80). The term connected with the dielectric anisotropy and the one connected with the flexoelectric polarization for the values of the physical parameters used above are comparable for $z = 0$, if $\epsilon_a \sim \epsilon_0$. In this situation, the dielectric anisotropy can play an important role in the instability we are analyzing.

Solution of Eq. (6.77), with the boundary conditions (6.62) and (6.63), is

$$\theta = \alpha \exp\left(-\mathrm{sign}(e)\sqrt{\frac{\epsilon_a}{K}}Ebe^{-z/b}\right)\mathcal{F}(E,z) \tag{6.81}$$

where $\mathcal{F}(E,z)$ is the hypergeometric function [27]:

$$\mathcal{F}(E,z) = F_1\left\{\frac{1}{2}\left(1+\sqrt{\frac{e^2}{K\epsilon_a}}\right), 1, 2\,\mathrm{sign}(e)\sqrt{\frac{\epsilon_a}{K}}Ebe^{-z/b}\right). \tag{6.82}$$

The eigenvalues for the present problem are obtained by substituting (6.81) into Eq. (6.65). In Figure 6.13 we show the critical curves relevant to the present case, for different values of the dielectric anisotropy. For small ϵ_a the influence of the dielectric term is negligible. However, for $\epsilon_a \sim \epsilon_0$, the dielectric contribution is such that the homeotropic configuration is stable for all surface fields, for reasonable values of the flexoelectric coefficient. On the contrary, for $\epsilon_a \sim -\epsilon_0$ the value of the surface field giving rise to the instability is strongly reduced for the destabilizing effect of the dielectric term.

As before, the critical field to have the instability in the case of strong anchoring is obtained by the condition

$$\mathcal{F}(E,0) = F_1\left\{\frac{1}{2}\left(1+\sqrt{\frac{e^2}{K\epsilon_a}}\right), 1, 2\,\mathrm{sign}(e)\sqrt{\frac{\epsilon_a}{K}}Eb\right) = 0. \tag{6.83}$$

In Figure 6.14, we show the critical field for the instability versus the dielectric anisotropy, for $e = \pm 1$. As expected, if $e > 0$, which means that the flexoelectric term stabilizes the homeotropic orientation in the bulk, the surface instability is possible only for $\epsilon_a < 0$. On the contrary, if $e < 0$, the instability can take place even with positive dielectric anisotropy. In this case, from Eq. (6.83), the maximum value of the dielectric anisotropy to have the instability is $\epsilon_a < e^2/K$, which is similar to the one obtained in approximated manner in Ref. [22].

6.2.6 Spontaneous Fréedericksz transition

We have discussed the possibility to observe spontaneous Fréedericksz transition induced by the surface field. This instability is actually a flexoelectric instability induced by the electric field due to the adsorption phenomenon. The analysis was devoted to the case in which the electric field is thickness independent [14]. However, it can be easily extended to take into account its

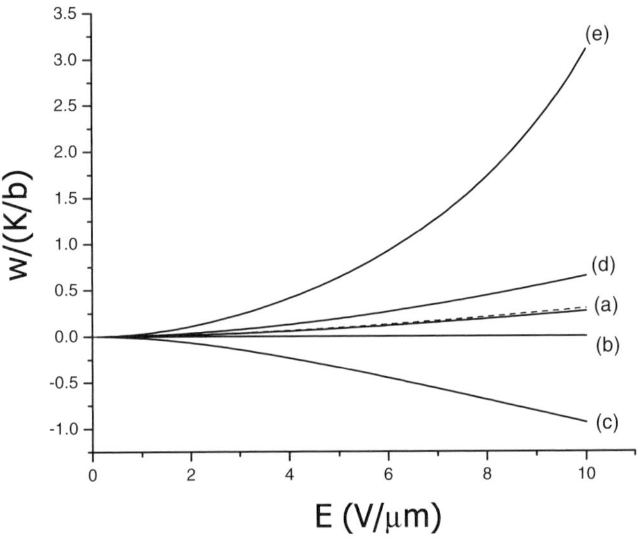

FIGURE 6.13
Critical line $W/(K/b)$ deduced by means of the analysis reported in Section 6.2.5. The dashed line is the critical curve obtained in Section 6.2.3 in which the dielectric anisotropy was neglected. Physical parameters as in Figure 6.8, with $\epsilon_a/\epsilon_0 = .1$ (a), 1 (b), 5 (c), -1 (d), and -5 (e). Reprinted from Ref. [24].

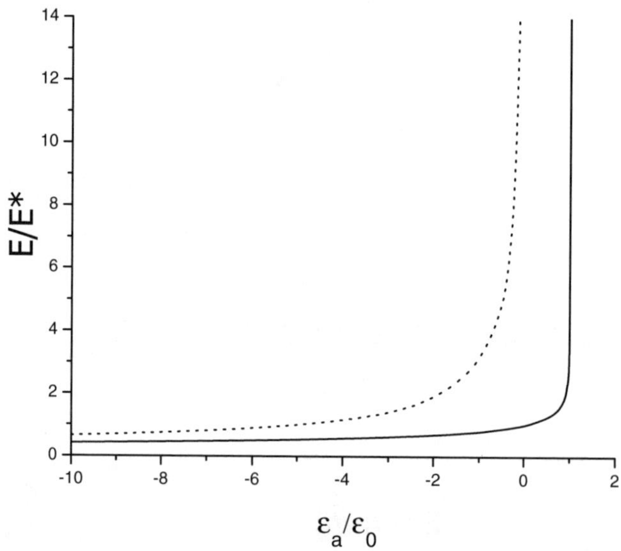

FIGURE 6.14
Reduced critical field E/E^*, where E^* is the critical field in the half-space approximation (defined by Eq. (6.68)) for a compensated nematic liquid crystal ($\epsilon_a = 0$), to have a surface instability versus ϵ_a. $e = 1$, dotted line, $e = -1$, solid line. Other physical parameters as in Figure 6.6. Reprinted from Ref. [24].

thickness dependence by means of the results reported before. In a first approximation, it is possible to assume, in the limit of large adsorption energies, that the surface field involved in the phenomenon can be given by

$$E = \frac{\sigma}{\epsilon} = \frac{\Sigma}{\epsilon} \frac{d}{d + 2\lambda_0}, \tag{6.84}$$

which corresponds to the Langmuir isotherm law (6.31). In (6.84), Σ is the saturation surface density of adsorbed charges, which depends on the adsorption energies. A more precise thickness dependence of the surface field will be obtained in Chapter 8. Since E is an increasing function of d, the phase diagrams (W, E) reported in Figures 6.6 through 6.12 are actually phase diagrams (W, d).

From the discussion reported above, a possible mechanism explaining the experimental data by [28, 29] could be connected to the existence of the flexoelectric contribution due to the ion adsorption. Of course, to observe the predicted instability in this case it is necessary that $W < |eE|$.

Conclusions

In this chapter we have analyzed the effect of surface electric field of ionic origin on homeotropically oriented samples of NLC. The surface electric field distribution was considered in the exponential form, i.e., consistent with the linear version of a Poisson-Boltzmann theory. Even in this approximation, it is shown that the ionic adsorption can be effectively responsible for experimentally detected effects. In fact it is often found that it is rather difficult to get a homeotropic alignment of materials with negative dielectric anisotropy. The preliminary results obtained in this chapter indicate that experimental results connected to the destabilization of the uniform pattern in NLC samples can be understood in the framework of a electrostatic model, that takes into account the explicit contribution of adsorbed ions to the final field distribution across the sample. As a special result, it is shown that the flexoelectric term can induce, in particular cases, a spontaneous Fréedericksz transition, where the control parameter is the thickness of the sample. In the next chapter we discuss the applicability of the exponential approximation for the electric field of ionic origin and, subsequently, we formulate the problem of obtaining the field distribution in general terms, without linearizing the fundamental equations governing its behavior.

[1] Blinov LM, Kabayenkov AY, and Sonin AA. Experimental studies of the anchoring energy of nematic liquid crystals–Invited lecture. *Liquid Crystals*, **5**, 645 (1989).

[2] Jerome B. Surface effects and anchoring in liquid crystals. *Reports on Progress in Physics*, **54**, 391 (1991).

[3] Gomes OA, Falcão RC, and Mesquita ON. Anomalous capillary length in cellular nematic-isotropic interfaces. *Physical Review Letters*, **86**, 2577 (2001).

[4] Strangi G, Versace C, and Scaramuzza N. Surface anchoring energy modulation in liquid crystal cells with mixed conductor boundary layers. *Applied Physics Letters*, **78**, 2455 (2001).

[5] Alexe-Ionescu AL, Barbero G, and Petrov AG. Gradient flexoelectric effect and thickness dependence of anchoring energy. *Physical Review E*, **48**, R1631 (1993).

[6] Barbero G and Durand G. On a possible mechanism for the spontaneous Fréedericksz effect. *Liquid Crystals*, **2**, 401 (1982).

[7] Barbero G and Durand G. Ion adsorpion and equilibrium distribution of charges in a cell of finite thickness. *Journal de Physique*, **51**, 281 (1990).

[8] Barbero G, Zvezdin AK, and Evangelista LR. Ionic adsorption and equilibrium distribution of charges in a nematic cell. *Physical Review E*, **59**, 1846 (1999).

[9] Nazarenko VG and Lavrentovich OD. Anchoring transition in a nematic liquid-crystal composed of centrosymetric molecules. *Physical Review E*, **49**, R990 (1994).

[10] Kühnau U, Petrov AG, Klose G, and Schmiedel H. Measurements of anchoring energy of a nematic liquid crystal, 4-cyano-4 '-n-pentylbiphenyl, on Langmuir-Blodgett films of dipalmitoyl phosphatidylcholine. *Physical Review E*, **59**, 578 (1999).

[11] Nazarenko VG, Pergamenshchik VM, Koval'chuk OV, Nych AB, and Lev BI. Non-Debye screening of a surface charge and a bulk-ion-controlled anchoring transition in a nematic liquid crystal. *Physical Review E*, **60**, 5580 (1999).

[12] Fazio VSU and Komitov L. Alignment transition in a nematic liquid crystal due to field-induced breaking of anchoring. *Europhysics Letters*, **46**, 38 (1999).

[13] Fazio VSU, Nannelli F, and Komitov L. Sensitive methods for estimating the anchoring strength of nematic liquid crystals on Langmuir-Blodgett monolayers of fatty acids. *Physical Review E*, **63**, 061712 (2001).

[14] Meister R and Jerome B. Influence of a surface electric field on the anchoring characteristics of nematic phases at rubbed polyimides. *Journal of Applied Physics*, **86**, 2473 (1999).

[15] Schmickler W. *Interfacial Electrochemistry.* Oxford University Press, Oxford, 1996.

[16] Meyer RB. Piezoelectric effects in liquid crystals. *Physical Review Letters*, **29**, 918 (1969).

[17] Derzhanski A and Petrov AG. Flexoelectricity in nematic liquid crystals. *Acta Physica Polonica A*, **55**, 747 (1979).

[18] Blinov LM and Kabaenkov AY. Temperature dependence and the "size effect" exhibited by the anchoring energy of a nematic with a planar orientation on a solid substrate. *Sovietic Physics–JETP*, **66**, 1002 (1987).

[19] Karat PP and Madhusudana NV. Elastic and optical properties of some 4'-n-alkyl-4-cyanobiphenyls. *Molecular Crystals and Liquid Crystals*, **36**, 51 (1976).

[20] Murthy PRM, Raghunathan VA, and Madhusudana NV. Experimental determination of the flexoelectric coefficients of some nematic liquid crystals. *Liquid Crystals*, **14**, 483 (1986).

[21] Thurston RN, Cheng J, Meyer RB, and Boyd GD. Physical mechanisms of DC switching in a liquid crystal bistable boundary layer display. *Journal of Applied Physics*, **56**, 263 (1984).

[22] Barbero G, Evangelista LR, and Madhusudana NV. Effect of surface electric field on the anchoring of nematic liquid crystals. *European Physical Journal B*, **1**, 337 (1998).

[23] Pereira HA and Evangelista LR. Surface electric field-induced instabilities in a homeotropically oriented nematic liquid crystal sample. *European Physical Journal E*, **3**, 123 (2000).

[24] Barbero G and Evangelista LR. Role of the linear elastic term in the spatial derivatives of the nematic director in a 1D geometry. *Liquid Crystals*, **30**, 633 (2003).

[25] Blinov LM, Barnik MI, Ozaki M, Shtykov NM, and Yoshino K. Surface and flexoelectric polarization in a nematic liquid crystal directly measured by a pyroelectric technique. *Physical Review E*, **62**, 8091 (2000).

[26] Vertogen G and de Jeu WH. *Thermotropic Liquid Crystals, Fundamentals*. Springer Verlag, Berlin, 1988.

[27] Gradshtein IS and Ryzhik IM. *Table of Integrals, Series and Products*. Academic Press, New York, 1965.

[28] Chuvyrov AN and Lachirov AV. Investigation of nature of homeotropic orientation of nematic liquid crystals molecules–feasibility of application in modulation spectroscopy. *Sovietic Physics–JETP*, **47**, 749 (1978).

[29] Chuvyrov AN. The influence of spontaneous polarization on the mechanical stability of nematic liquid crystals. *Kristallografiya*, **25**, 188 (1980).

7

EXPONENTIAL APPROXIMATION FOR THE ELECTRIC FIELD OF IONIC ORIGIN

In the first part of this chapter, we present a general model to describe the influence of the ion adsorption on the anisotropic part of the surface energy of a nematic liquid crystal in contact with a substrate. We show that in the limit of small adsorption energy, the exponential approximation for the electric field of ionic origin works well. In this limit, the dielectric and flexoelectric contributions to the surface energy are, respectively, quadratic and linear on the density of adsorbed ions. In the opposite limit of large adsorption energy, the exponential approximation for the electric field does not work, and the two contributions to the surface energy are found to depend both linearly on the surface density of adsorbed charges. Approximated formulae reported in literature are derived from the general equations as particular cases, and their limits discussed. An expression for the surface polarization in nematic liquid crystal due to the ion adsorption is also deduced.

In the second part of the chapter the influence of an external field on the effective anchoring energy of a nematic liquid crystal in contact with a substrate is theoretically analyzed in the hypothesis that the electrodes are perfectly blocking and that there is no selective ion adsorption. We discuss the ionic contribution to the surface energy in general terms, obtaining explicit formulae valid for the weak and strong adsorption energies. The analysis is performed in the framework of the Poisson-Boltzmann theory, where the ions are assumed dimensionless, and the problem is faced by a mean field approach. The theory predicts an effective anchoring energy dependent on the applied dc voltage. According to the sign of the dielectric anisotropy and of the flexoelectric coefficient the dependence of the anchoring energy strength with the bias can be monotonic or not. For large bias voltage the effective anchoring energy strength tends to a constant value.

7.1 Non-linear Poisson-Boltzmann equation

We consider a neutral liquid containing n_0 ions per unit volume in the thermodynamical limit (infinite sample) [1]. In this situation it is locally and globally neutral. When the liquid is limited by two surfaces of infinite area at a distance d apart, due to the selective ion adsorption phenomenon, the liquid will be locally charged. If d is very large with respect to the Debye's screening length λ_0, in the middle of the sample the liquid will be, practically, locally neutral [2]. In this section, the analysis is limited to this case, which represents a good approximation for ordinary samples ($d \sim 10\,\mu$m), made with commercial NLC (typically $\lambda_0 \sim 0.5\,\mu$m [3]). This case is known as the half-space approximation ($d/\lambda_0 \gg 1$), where it is possible to consider only one surface, placed at $z = 0$, with surface charge density $\sigma = n_{ad}q$, as we have done in Section 6.2.3. Here, n_{ad} is the surface density of adsorbed ions of charge q. We indicate by $V(z)$ the electrical potential, with respect to the reference state where the liquid is locally neutral (at $z \to \infty$), and by $V_T = k_\mathrm{B}T/q$ the thermal electrical potential ($V_T \sim 25$ mV at room temperature for monovalent ions). The quantity

$$\psi(z) = \frac{qV(z)}{k_\mathrm{B}T} = \frac{V(z)}{V_T}, \tag{7.1}$$

which is the electrostatic energy of the ion in $k_\mathrm{B}T$ units or the electrical potential measured in V_T units, will be called reduced potential. We assume in this chapter that only positive ions are adsorbed.

According to Maxwell-Boltzmann statistics the densities of positive and negative ions are given by Eq. (6.6), here rewritten as

$$n_\pm(z) = n_0 \exp[\mp\psi(z)]. \tag{7.2}$$

It follows that the net charge density, as given by Eq. (6.5), is

$$\rho(z) = q[n_+(z) - n_-(z)] = -2\, n_0\, q \sinh[\psi(z)] \tag{7.3}$$

and Poisson's equation $\nabla \cdot \mathbf{E} = \rho/\epsilon$, in the present case, is written as

$$\frac{d^2\psi(z)}{dz^2} = \frac{1}{\lambda_0^2}\sinh[\psi(z)], \tag{7.4}$$

where λ_0 is the Debye's screening length introduced in (6.9). The surface electric field (at $z = 0$) is $E = E(0) = \sigma/\epsilon = n_{ad}q/\epsilon$, from which we obtain

$$\psi'(0) = -\frac{1}{\ell}, \tag{7.5}$$

where

$$\ell = \frac{\epsilon k_B T}{\sigma q} = \frac{\epsilon k_B T}{n_{ad} q^2} \tag{7.6}$$

is a new length connected with the adsorption phenomenon, first introduced in [4]. Note that $\ell \propto n_{ad}^{-1}$. Hence it diverges for $n_{ad} \to 0$ (weak adsorption), and tends to zero for $n_{ad} \to \infty$ (strong adsorption). From Eq. (7.4) we obtain

$$\frac{1}{2}\left(\frac{d\psi}{dz}\right)^2 = \frac{1}{\lambda_0^2}\cosh[\psi(z)] + k, \tag{7.7}$$

where k is an integration constant. Since in the present case $\lim_{z\to\infty} \psi(z) = 0$ and $\lim_{z\to\infty} \psi'(z) = 0$, from Eq. (7.7) it follows that $k = -\lambda_0^{-2}$, and hence

$$\frac{d\psi(z)}{dz} = -\frac{2}{\lambda_0}\sinh\left[\frac{\psi(z)}{2}\right]. \tag{7.8}$$

Equation (7.8) can be further integrate to give

$$\psi(z) = 2\ln\left[\frac{1+\gamma\exp(-z/\lambda_0)}{1-\gamma\exp(-z/\lambda_0)}\right], \tag{7.9}$$

where $\gamma = \tanh(\psi_s/4)$, and

$$E(z) = 2\frac{V_T}{\lambda_0}\sinh\left[\frac{\psi(z)}{2}\right], \tag{7.10}$$

for the reduced potential and the electric field. One obtains, in particular,

$$E = E(0) = 2\frac{V_T}{\lambda_0}\sinh\left(\frac{\psi_s}{2}\right), \tag{7.11}$$

connecting the surface field with the surface potential, ψ_s. From boundary condition (7.5) and Eq. (7.11) we deduce

$$\sinh\left(\frac{\psi_s}{2}\right) = \Lambda, \tag{7.12}$$

where $\Lambda = \lambda_0/2\ell$ is a measure of the importance of the adsorption phenomenon on the electric potential distribution across the sample. The cases of weak and strong adsorption correspond to small and large Λ, respectively. By means of Eq. (7.12) the parameter γ defined above is

$$\gamma = \tanh\left[\frac{\psi_s}{4}\right] = \sqrt{\frac{(1+\Lambda^2)^{1/2}-1}{(1+\Lambda^2)^{1/2}+1}}. \tag{7.13}$$

and, therefore,

$$\psi_s = 2\ln(\Lambda + \sqrt{1+\Lambda^2}). \tag{7.14}$$

Equation (7.14) shows that in the half-space approximation the reduced surface potential depends only on Λ, the ratio between the two intrinsic lengths of the problem. For $\Lambda \sim 10$ [5] one obtains $\psi_s \sim 6$. Note that, if $\Lambda \ll 1$ then $\psi_s \ll 1$, whereas if $\Lambda \gg 1$ then $\psi_s \gg 1$. It is possible to write the inequalities $\Lambda \ll 1$ and $\Lambda \gg 1$, corresponding to the cases of weak and strong adsorption, respectively, as $n_{ad} \ll 4n_0\lambda_0$, and $n_{ad} \gg 4n_0\lambda_0$, if the definitions of Λ, λ_0 and ℓ are used. From the inequalities written above it follows that there is a weak adsorption when the number of adsorbed ions is small with respect to $n_0\lambda_0$, i.e., with respect to the surface density of ions contained in a layer whose thickness coincides with λ_0, in thermodynamical equilibrium.

By substituting $\psi(z)$, given by Eq. (7.9), into the expressions for $n_\pm(z)$ the bulk densities of positive and negative ions are found to be

$$n_\pm(z) = n_0 \left\{ \frac{1 \mp \gamma \exp(-z/\lambda_0)}{1 \pm \gamma \exp(-z/\lambda_0)} \right\}. \qquad (7.15)$$

Up to now the liquid was assumed isotropic. However, if the liquid is a nematic medium, in a first approximation it is possible to evaluate the ionic distribution as reported above. After that we can evaluate the coupling between the electric field of ionic origin, $E(z)$, with the anisotropic properties of the NLC. The coupling of $E(z)$ with the dielectric anisotropy and with the flexoelectric properties of the liquid crystal gives rise to a surplus of anchoring energy strength, of the kind

$$W_\mathrm{D} = -\frac{1}{2}\epsilon_a \int_0^\infty E^2(z)dz, \qquad (7.16)$$

and

$$W_\mathrm{Q} = \int_0^\infty eE'(z)dz. \qquad (7.17)$$

As shown in Section 7.3, Eqs. (7.16) and (7.17) hold in the hypothesis of constant nematic orientation over the spatial region where the electric potential is changing, as we have assumed in the analysis [see Eq. (6.17) and (6.20)] . Using Eq. (7.10), W_D and W_Q given by Eq. (7.16) and (7.17) become

$$W_\mathrm{D} = -4\frac{\epsilon_a V_T^2}{\lambda_0} \sinh^2\left(\frac{\psi_s}{4}\right) \qquad (7.18)$$

and

$$W_\mathrm{Q} = -2\frac{eV_T}{\lambda_0} \sinh\left(\frac{\psi_s}{2}\right). \qquad (7.19)$$

The electrostatic contribution of ionic origin to the anchoring energy strength, $W_\mathrm{E} = W_\mathrm{D} + W_\mathrm{Q}$, is then

$$W_{\rm E} = -2\frac{V_T}{\lambda_0}\left[2\epsilon_a V_T \sinh^2\left(\frac{\psi_s}{4}\right) + e\sinh\left(\frac{\psi_s}{2}\right)\right]. \qquad (7.20)$$

Equation (7.20) generalizes the formula reported in Section 6.1.3, for the same phenomenon, valid only in the case small adsorption. Using Eq. (7.12) it is possible to rewrite $W_{\rm D}$ and $W_{\rm Q}$ in terms of Λ. Simple calculations give

$$W_{\rm D} = -2\left(\frac{\epsilon_a V_T^2}{\lambda_0}\right)(\sqrt{1+\Lambda^2}-1), \qquad (7.21)$$

and

$$W_{\rm Q} = -2\frac{eV_T}{\lambda_0}\Lambda. \qquad (7.22)$$

The total contribution to the anchoring energy strength is then

$$W_{\rm E} = -2\frac{V_T}{\lambda_0}\left[\epsilon_a V_T(\sqrt{1+\Lambda^2}-1) + e\Lambda\right]. \qquad (7.23)$$

It follows that, for $\Lambda \to 0$ (*weak adsorption*),

$$W_{\rm E} = -2\frac{V_T}{\lambda_0}\left[\frac{1}{2}\epsilon_a V_T \Lambda^2 + e\Lambda\right] + \mathcal{O}(\Lambda^3). \qquad (7.24)$$

In the opposite limit, where $\Lambda \to \infty$ (*strong adsorption*), Eq. (7.23) gives

$$W_{\rm E} = -2\frac{V_T}{\lambda_0}(\epsilon_a V_T + e)\Lambda. \qquad (7.25)$$

These cases will be reconsidered later, to compare the results of the present analysis with the ones reported by other groups.

Before concluding this section we analyze when the adsorption phenomenon can play an important role on the renormalization of the anchoring energy strength. To this end we rewrite Eq. (7.23) as

$$W_{\rm E} = -W\left[\sqrt{1+\Lambda^2} - 1 + \left(\frac{e}{\epsilon_a V_T}\right)\Lambda\right], \qquad (7.26)$$

where $W = 2\epsilon_a V_T^2/\lambda_0$. By assuming $\epsilon_a \sim 10\,\epsilon_0$ and $e \sim 5\times 10^{-11}$ C/m, typical for a nematic liquid crystal like 5CB [6] and $0.1\,\mu$m$\leq \lambda_0 \leq 0.5\,\mu$m [3], we obtain 2×10^{-7} J/m$^2 \leq W \leq 10^{-6}$ J/m^2. In this case $W_{\rm E} \sim 10^{-6}$ J/m^2 if $0.2 \leq \Lambda \leq 1$, whereas $W_{\rm E} \sim 10^{-4}$ J/m^2 for $10 \leq \Lambda \leq 50$. Hence, we can conclude that in the limit of *weak adsorption* $W_{\rm E}$ can be important only if the interface nematic liquid crystal/solid substrate is characterized by weak anchoring energy. In the case where the bare anchoring energy strength of the interface is rather strong ($\sim 10^{-4}$ J/m^2), $10 \leq \Lambda \leq 50$ and $\ell = \lambda_0/2\Lambda$ is in the nanometer scale. The estimation reported above shows that the Debye's screening length plays an important role in surface effects of liquid crystals.

Consequently, in all experimental investigations devoted to the characterization of surface energy or flexoelectric coefficients, this parameter has to be measured carefully. Without information on this quantity, all discussions on the origin of the surface energy strength or on the surface polarization based on experimental investigation could be meaningless.

7.1.1 Limits of weak and strong adsorption

Let us analyze first the case of *weak adsorption*, already considered in Section 6.1.3 in another framework. In this limit $\psi_s \ll 1$ and hence $\gamma = \tanh(\psi_s/4) \sim \psi_s/4$. From Eq. (7.9) we obtain

$$\psi(z) = \psi_s \exp(-z/\lambda_0), \tag{7.27}$$

and

$$E(z) = E \exp(-z/\lambda_0), \tag{7.28}$$

where

$$E = V_T \frac{\psi_s}{\lambda_0} = \frac{V_T}{\ell}, \tag{7.29}$$

as follows from Eq. (7.11) and (7.12). The bulk densities of ions are then

$$n_\pm(z) = n_0[1 \mp \psi_s \exp(-z/\lambda_0)]. \tag{7.30}$$

In particular $n_\pm(0) = n_0(1 \mp \psi_s)$. As expected, if $\psi_s \ll 1$, the electrical equilibrium is only slightly perturbed. The electrostatic contributions to the anchoring energy strength are, in this limit,

$$W_{\rm D} = -(1/4)\epsilon_a \lambda_0 E^2 \quad \text{and} \quad W_{\rm Q} = -eE, \tag{7.31}$$

as follows from the expressions reported above for $W_{\rm D}$ and $W_{\rm Q}$ in terms of ψ_s and from Eq. (7.11). Hence

$$W_{\rm E} = -\frac{1}{4}\epsilon_a \lambda_0 E^2 - eE, \tag{7.32}$$

as reported in Section 6.1.3. In this approximation $W_{\rm D} \propto E^2$ and $W_{\rm Q} \propto E$.

Let us consider now the case of strong adsorption. In this framework, $\psi_s \gg 1$ and $\gamma = \tanh \psi_s \sim 1 - 2\exp(-\psi_s/2)$. Hence

$$\psi(z) = 2 \ln \left\{ \frac{1 + [1 - 2\exp(-\psi_s/2)] \exp(-z/\lambda_0)}{1 - [1 - 2\exp(-\psi_s/2)] \exp(-z/\lambda_0)} \right\}, \tag{7.33}$$

and the electric field is still given by Eq. (7.9). For $z/\lambda_0 \ll 1$, $\psi(z) \gg 1$. Consequently, $\sinh(\psi/2) \sim (1/2)\exp(\psi/2)$, that by using Eq. (7.33) in the limit $\psi_s \gg 1$ is equivalent to

$$\sinh\left(\frac{\psi}{2}\right) = \frac{\exp(\psi_s/2)}{2 + (z/\lambda_0)\exp(\psi_s/2)}. \tag{7.34}$$

In the considered limit of $\psi_s \gg 1$ from Eq. (7.10) we have

$$\exp(\psi_s/2) = (\lambda_0/V_T)E. \tag{7.35}$$

By substituting this expression into Eq. (7.34) and the result in Eq. (7.10) one obtains

$$E(z) = \frac{E}{1 + (z/2\ell)}. \tag{7.36}$$

In this limit ($\psi_s \gg 1$, $z/\lambda_0 \ll 1$), $n_\pm(z)$ are given by

$$n_\pm(z) = n_0 \left\{ \frac{1 \mp [1 - 2\exp(-\psi_s/2)](1 - z/\lambda_0)}{1 \pm [1 - 2\exp(-\psi_s/2)](1 - z/\lambda_0)} \right\}. \tag{7.37}$$

In particular

$$n_+(0) = n_0 \exp(-\psi_s/2) \sim 0 \quad \text{and} \quad n_-(0) = n_0 \exp(\psi_s/2). \tag{7.38}$$

Equation (7.36) has been deduced in Ref. [4] and presented as a general result, valid for all z. On the contrary, it is valid only for $z/\lambda_0 \ll 1$. For large z/λ_0 from Eq. (7.9) we obtain, for all ψ_s, $\psi(z) = 4\gamma\exp(-z/\lambda_0)$, showing that the reduced potential, and hence the electric field, are always exponentially decreasing, and Eq. (7.36) does not work. In this case we have, furthermore

$$n_\pm(z) = n_0\{1 \mp 2[1 - 2\exp(-\psi_s/2)]\exp(-z/\lambda_0)\}, \tag{7.39}$$

showing again that for $z \to \infty$, $n_\pm(z) \to n_0$, as expected.

In the limit under consideration ($\psi_s \gg 1$) from the expression of $W_{\rm D}$ and $W_{\rm Q}$ in terms of ψ_s we derive

$$W_{\rm D} = -\epsilon_a V_T E, \quad \text{and} \quad W_{\rm Q} = -eE. \tag{7.40}$$

These expressions show that, in this limit, $W_{\rm D}$ and $W_{\rm Q}$ are both linear in E. The total electrostatic contribution to the surface energy strength ($W_{\rm E} = W_{\rm D} + W_{\rm Q}$) is then

$$W_{\rm E} = -(\epsilon_a V_T + e)E. \tag{7.41}$$

It follows that, in the limit of large adsorption, the electrostatic contribution can stabilize or destabilize the planar orientation according to the sign of $(\epsilon_a V_T + e)$, independently of the number of adsorbed ions. The importance of the flexoelectric term was already pointed out in Section 6.1.3 using the exponential approximation for the electric field.

7.1.2 Charge distribution and thickness of the surface layer

The net charge density is $\rho(z) = -2n_0 q \sinh\psi(z)$, as discussed above [see Eq. (7.3)]. Using Eq. (7.8) it is easy to show that

$$\mathcal{Q}(\infty) = \int_0^\infty \rho(z)dz = -\sigma, \tag{7.42}$$

as expected since the system is globally neutral. The electric charge, per unit surface, contained in a surface layer of thickness b is, in analogy with Eq. (7.42),

$$\mathcal{Q}(b) = \int_0^b \rho(z)dz. \tag{7.43}$$

By rewriting Eq. (7.43) in the form

$$\mathcal{Q}(b) = \mathcal{Q}(\infty) - \int_b^\infty \rho(z)dz, \tag{7.44}$$

and using Eq. (7.8) and Eq. (7.9) we obtain

$$R = 1 - \frac{\mathcal{Q}(b)}{\mathcal{Q}(\infty)} = \frac{2}{\Lambda}\frac{\gamma\exp(-b/\lambda_0)}{1-\gamma^2\exp(-2b/\lambda_0)}, \tag{7.45}$$

where we have used Eq. (7.42) and the definitions of λ_0, ℓ and Λ. The meaning of R is evident from the definition. If $R \to 0$ the layer of thickness b contains practically all the counterions. In the opposite case of $R \to 1$, the counterions present in the surface layer are a negligible fraction of the total number .

In general, fixing R, from Eq. (7.45) it is possible to obtain the corresponding thickness of the surface layer b, as discussed in [2]. Straightforward calculations give

$$b = \lambda_0 \ln\left\{\frac{R\Lambda}{\sqrt{1+(R\Lambda)^2}-1}\sqrt{\frac{(1+\Lambda^2)^{1/2}-1}{(1+\Lambda^2)^{1/2}+1}}\right\}. \tag{7.46}$$

For $\Lambda \to 0$, $b \to -\lambda_0 \ln R$, i.e., $b \sim \lambda_0$, and for $\Lambda \to \infty$, $b \to (\lambda_0/R)\Lambda^{-1}$, as expected. For $R = 0.1$ and $\Lambda \sim 10$, $b \sim \lambda_0$. In this case $(9/10)$ of the ionic charge is contained in a surface layer of thickness λ_0 [5].

Using the simple results reported above we can now obtain the total dipole moment and the bulk polarization of the sample under consideration. If the sample is symmetric, i.e., it has the same adsorption energy on the two surfaces, the net dipole moment vanishes, for symmetry reasons. The dipole moment, per unit surface, of half sample is given by

$$P_S = \int_0^\infty z\rho(z)dz, \tag{7.47}$$

and the corresponding bulk polarization is $P_{\rm B}(z) = z\rho(z)$. The quantity $P_{\rm S}$ defined in Eq. (7.47) will be known also as "surface polarization". Taking into account that $\rho(z) = \epsilon\nabla \cdot \mathbf{E}$, and using $\psi(z)$ instead of $\mathbf{E}$ one obtains

$$P_{\rm B}(z) = -\epsilon V_T z \psi''(z), \tag{7.48}$$

where the prime indicates derivative with respect to z. In the half-space approximation $\psi(z)$ is given by Eq. (7.9), and $P_{\rm B}(z)$ can be written in the form

$$P_{\rm B}(z) = -8n_0 q z \gamma \exp(-z/\lambda_0)\frac{1+\gamma\exp(-2z/\lambda_0)}{[1-\gamma\exp(-2z/\lambda_0)]^2}. \tag{7.49}$$

If $\psi_s \ll 1$, Eq. (7.49) becomes

$$P_{\rm B}(z) = -2n_0 q z \psi_s \exp(-z/\lambda_0), \tag{7.50}$$

where $\psi_s \sim \lambda_0/\ell = 2\Lambda$, as follows from Eq. (7.12). In the opposite limit of $\psi_s \gg 1$, Eq. (7.49) gives

$$P_{\rm B}(z) = -8n_0 q z \exp(-z/\lambda_0)\frac{1+\exp(-2z/\lambda_0)}{[1-\exp(-2z/\lambda_0)]^2}. \tag{7.51}$$

Equations (7.50) and (7.51) show that $P_{\rm B}(z)$ is localized over a few ℓ or λ_0, according to the considered limit.

Using Eq. (7.47) it is possible to evaluate the surface polarization. Simple calculations give

$$P_{\rm S} = -\epsilon V_T \psi_s = -\epsilon V_{\rm S}. \tag{7.52}$$

By assuming $\epsilon \sim 10\,\epsilon_0$ and $\psi_s \sim 4$, which implies $\Lambda \sim 5$, we obtain for the surface polarization $P_{\rm S} \sim 10^{-11}$ C/m. If this formula is applied to a nematic liquid crystal ϵ has to be substituted with an average value, that in a first approximation is

$$\langle\epsilon\rangle = \frac{1}{3}(2\epsilon_\perp + \epsilon_\parallel). \tag{7.53}$$

Finally, we evaluate the polarization induced by the electric field of ionic origin on a nematic liquid crystal. The macroscopic polarizability tensor of a nematic medium is given by Eq. (1.71). If the nematic liquid crystal is submitted to an electric field the induced polarization is

$$P_i = \epsilon_0 \chi_{ij} E_j = \epsilon_0[\chi_a n_i(\mathbf{n}\cdot\mathbf{E}) + \chi_\perp E_i]. \tag{7.54}$$

Since $\mathbf{E}(z)$, which is parallel to the z-axis, is different from zero, practically, only in a surface layer of thickness ranging between λ_0 and a few ℓ, we conclude that $P_i(z)$ is localized near to the adsorbing substrate. The surface polarization of ionic origin in the nematic liquid is obtained by integrating $P_i(z)$ from z=0 to $z \to \infty$. Simple calculations give

$$P_{\mathrm{S\,i}} = \epsilon_0(\chi_a n_i n_z + \chi_\perp \delta_{iz}) V_{\mathrm{S}}. \tag{7.55}$$

If we assume that the substrate is isotropic and the easy axis is parallel to the z-axis (homeotropic orientation), then the NLC possesses cylindrical symmetry around the z-axis, and the average surface polarization is

$$\langle P_{\mathrm{S\,i}} \rangle = \epsilon_0(\chi_a n_z^2 + \chi_\perp) V_{\mathrm{S}}. \tag{7.56}$$

The net surface polarization P_{S} is obtained by adding to $\langle P_{\mathrm{S\,i}} \rangle$ the quantity P_{S} evaluated above, and it is found to be given by

$$P = \left\{ (\epsilon_a - \epsilon_0)\, n_z^2 - \left[\left(\frac{\epsilon_a}{3} \right) + \epsilon_0 \right] \right\} V_{\mathrm{S}}. \tag{7.57}$$

As expected, P depends on n_z^2. Its amplitude, by assuming $\epsilon_a \sim 10\epsilon_0$ is of the order of 10^{-11} C/m for $\psi_s \sim 4$, i.e., $\Lambda \sim 5$. This value is of the same order [7, 8], or larger [9] than the values reported in literature for the nematic surface polarization. From this result we conclude that, probably, the observed surface polarization in NLC has an ionic origin. This possibility was already mentioned in the literature [7]–[9], but no estimations were reported. It is different from the one discussed by Petrov and Derzhanski [10], which is linear in n_z, whose origin is connected with the different chemical affinities of the two extremities of the nematic molecules with the substrate.

7.2 External electric field effect on the nematic anchoring energy

In this section we analyze the influence of an external electric field on the nematic anchoring energy when the selective ion adsorption can be neglected [11]. The charge separation induced by the external field gives rise to an effect similar to the one connected to the selective ion adsorption. It is responsible for a bias-voltage dependence of the effective surface energy [12]. To perform the analysis we consider the Poisson-Boltzmann theory for an isotropic liquid. The results are then employed for an NLC sample.

7.2.1 Poisson-Boltzmann theory for a liquid submitted to an external field

Let us consider an isotropic liquid limited by two non-adsorbing surfaces, at a distance d apart. The z-axis is normal to the bounding surfaces, with the origin in the middle of the sample. The liquid is supposed to contain ions. In the absence of external electric field the liquid is globally and locally neutral, i.e., $n_+(z) = n_-(z) = n_0$. If U is the difference of potential across the sample due to the external power supply Eq. (7.2) can be rewritten as

$$n_\pm(z) = ne^{\mp\psi(z)}, \tag{7.58}$$

where, now, n is the density of ions where $\psi = 0$. Equation (7.58) has been obtained by assuming $V(\pm d/2) = \pm U/2$, i.e., $V(z) = -V(-z)$. In Eq. (7.58) $n = n_\pm(0)$ has to be determined by imposing the conservation of the number of ions. The bulk density of electric charge is the same as in (7.3), but with n in the place of n_0. A simple calculation shows that

$$\int_{-d/2}^{d/2} \rho(z)dz = -2nq \int_{-d/2}^{d/2} \sinh\psi(z)dz = 0, \tag{7.59}$$

as required. The conservation of the number of ions implies that

$$n_0 d = \int_{-d/2}^{d/2} n_+(z)dz. \tag{7.60}$$

The distribution of $\psi(z)$ across the sample is obtained by solving the Poisson's equation, that by taking into account Eq. (7.3), reads

$$\frac{d^2\psi}{dz^2} = \frac{1}{\lambda^2}\sinh\psi(z), \tag{7.61}$$

with the boundary conditions

$$\psi(\pm d/2) = \pm u = \pm\frac{U}{2V_T}. \tag{7.62}$$

Notice that in Eq. (7.61), λ is the effective Debye's screening length of the liquid when the bias voltage is U (it is not the same as in (7.4)). It is, in fact, given by

$$\lambda^2 = \frac{\epsilon k_B T}{2q^2 n} = \frac{n_0}{n}\lambda_0^2. \tag{7.63}$$

In the present problem, λ depends on the applied voltage, which was not the case in the previous section. By using Eq. (7.61) one easily obtains

$$\frac{d\psi}{dz} = \frac{\sqrt{2}}{\lambda}\sqrt{\cosh\psi + k}, \tag{7.64}$$

where k is again an integration constant, whose value is directly connected with the electric field at $z = 0$ by

$$k = \frac{\lambda^2}{2}\left(\frac{d\psi}{dz}\right)_0^2 - 1. \tag{7.65}$$

In particular, if $(d\psi/dz)_0 = 0$ then $k = -1$ and Eq. (7.64) can be easily integrated. This case will be considered in the next section. By means of

Eqs. (7.64) and (6.9), we can rewrite Eqs. (7.60) and (7.62), respectively, in the form

$$I(k,u)=\sqrt{2}\frac{d}{\lambda} \quad \text{and} \quad J(k,u)=\sqrt{2}\frac{d\lambda}{\lambda_0^2}, \tag{7.66}$$

where

$$J(k,u)=\int_{-u}^{u}\frac{e^{-\psi}}{\sqrt{\cosh\psi+k}}d\psi \quad \text{and} \quad I(k,u)=\int_{-u}^{u}\frac{1}{\sqrt{\cosh\psi+k}}d\psi. \tag{7.67}$$

Equations (7.66) imply that

$$I(k,u)J(k,u)=2\left(\frac{d}{\lambda_0}\right)^2, \tag{7.68}$$

which determines $k=k(U)$. When this quantity is known, the effective Debye's screening length is

$$\lambda=\sqrt{2}\frac{d}{I(k,u)}. \tag{7.69}$$

In Figure 7.1, $k=k(U)$ is reported for $d=10\mu\,\text{m}$ and $\lambda_0=0.36\mu\,\text{m}$, typical for a nematic cell made with commercial liquid crystal [3, 13]. As follows from this figure, for bias voltage in the range $0\leq U\leq 0.2\,\text{V}$, $(d\psi/dz)_0\approx 0$, which implies that the electric field for $z=0$ vanishes. This means that for small bias voltage, the ionic charges screen completely the field due to the external power supply. The liquid behaves as a conductor: the ions move until the electric field inside it vanishes. In this case the electric field is localized close to the bounding surfaces over a thickness of the order of $\lambda_0=\lambda(0)$.

Figure 7.2 shows $\lambda=\lambda(U)$. As is expected, for $U\to 0$, $\lambda\to\lambda_0$. On the contrary, for large U, $\lambda\to\infty$ because all the ions are pushed at the surfaces, and there are no more ions in the liquid, which becomes a true insulating material.

7.2.2 Limiting cases of small and large bias voltage

The reduced voltage across the sample $\psi(z)$ for arbitrary bias voltages can be obtained by integrating Eq. (7.64), in the form

$$\int_{-u}^{\psi(z)}\frac{d\psi}{\sqrt{\cosh\psi(z)+k}}=\frac{\sqrt{2}}{\lambda}\left(z+\frac{d}{2}\right). \tag{7.70}$$

Now we consider the particular cases of small and large bias voltage. In the case of a small applied voltage ($u=qU/2V_T\ll 1$), $|\psi(z)|\leq u\ll 1$. Consequently, the solution of Eq. (7.61) with the boundary conditions (7.62), is

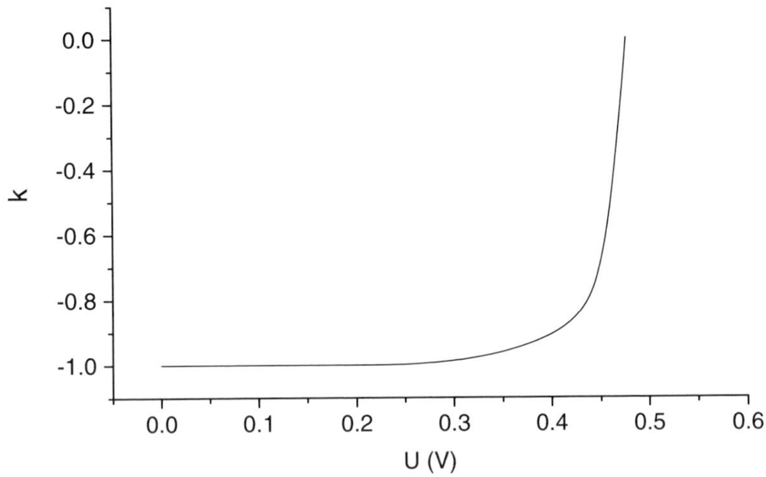

FIGURE 7.1
Integration constant k versus the applied voltage U. For small U, $k \to -1$. In this situation the electric field at $z = 0$ vanishes. For $d = 10\,\mu$m and $\lambda_0 = 0.36\,\mu$m, this happens for $U < 0.2\,$V. Reprinted with permission from Ref. [11]. Copyright (2002) by the American Physical Society.

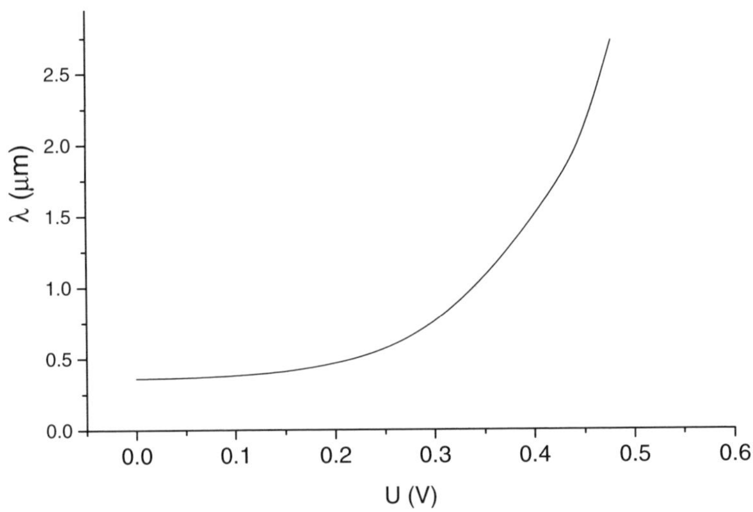

FIGURE 7.2
Effective Debye's screening length λ versus the applied voltage U. For $U \to 0$, $\lambda \to \lambda_0$. For large U all the ions are pushed at the surfaces, and the liquid behaves as a perfect insulator, for which $\lambda \to \infty$. The curve refers to $d = 10\,\mu$m and $\lambda_0 = 0.36\,\mu$m. Reprinted with permission from Ref. [11]. Copyright (2002) by the American Physical Society.

$$\psi(z) = u\frac{\sinh(z/\lambda)}{\sinh(d/2\lambda)}. \tag{7.71}$$

In this case

$$e^{-\psi(z)} \approx 1 - \psi(z) = 1 - u\frac{\sinh(z/\lambda)}{\sinh(d/2\lambda)} \tag{7.72}$$

and Eq. (7.60) gives $n \approx n_0$, which implies

$$\lambda = \lambda_0 + \mathcal{O}(u^2). \tag{7.73}$$

It follows that, in the considered limit, Eq. (7.71) reads

$$\psi(z) = u\frac{\sinh(z/\lambda_0)}{\sinh(d/2\lambda_0)}. \tag{7.74}$$

The charge distributions in the present case are found to be

$$n_\pm(z) = n_0\left[1 \mp u\frac{\sinh(z/\lambda_0)}{\sinh(d/2\lambda_0)}\right]. \tag{7.75}$$

As expected, the positive ions are collected near the negative electrode, and the negative ions near to the positive one, in surface layers of thickness of the order of λ_0. Another important particular case is the one in which k is very close to -1. As pointed out above, in this situation the electric field in the middle of the sample vanishes. By assuming $k = -1$, the solution of Eq. (7.61), with the boundary conditions (7.62), is

$$\psi(z) = \psi_-(z) + \psi_+(z), \tag{7.76}$$

where

$$\psi_\pm(z) = 2\ln\left[\frac{1 \pm \tanh(u/4)\, e^{(z\mp d/2)/\lambda_0}}{1 \mp \tanh(u/4)\, e^{(z\mp d/2)/\lambda_0}}\right]. \tag{7.77}$$

In this limit, the total electric potential $\psi(z)$ is obtained by adding the potential $\psi_-(z)$ and $\psi_+(z)$ created by the two surfaces separately. This means that the sample is considered as formed by two half-spaces. This approximation works well if $d/2\lambda_0$ is very large.

Let us consider, finally, the case of large applied voltage ($u = U/2V_T \gg 1$). In this situation practically all the ions are collected at the surfaces, and in the bulk the liquid can be considered as a perfect insulator. In a first approximation the electric potential across the sample is given by

$$\psi_0(z) \approx 2u\frac{z}{d}. \tag{7.78}$$

By substituting (7.78) into Eq. (7.60) we obtain

$$n = 2un_0 e^{-u}, \tag{7.79}$$

and, using Eq. (7.63), the effective Debye's screening length is

$$\lambda^2 = \lambda_0^2 \frac{e^u}{2\,u}, \tag{7.80}$$

showing that for $u \to \infty$, $\lambda \to \infty$ too. By means of Eq. (7.79) the charge distributions across the sample are found to be

$$n_{\pm}(z) = 2un_0 e^{[-u(1 \pm 2z/d)]}. \tag{7.81}$$

In particular, $n_+(-d/2) = n_-(d/2) = 2n_0 u$. Equation (7.81) shows that, in the considered limit of large bias voltage, the ionic charges are confined in a surface layer whose thickness is of the order of $d/2u$. Hence, for $d/2u \approx \lambda_0$, i.e., $u \approx d/2\lambda_0$, the thickness of the surface layer is comparable with the Debye's screening length.

By putting $\psi(z) = \psi_0(z) + \psi_1(z)$ into Eq. (7.61) and taking into account Eq. (7.80) with the boundary conditions (7.62) we obtain

$$\psi_1(z) = 2\frac{e^{-u}}{u}\left(\frac{d}{2\lambda_0}\right)^2 \sinh u \left[-2\frac{z}{d} + \frac{\sinh(2uz/d)}{\sinh u}\right]. \tag{7.82}$$

Consequently, since in the present limit $u \gg 1$ and, hence, $\sinh u \approx e^u/2$, we have for $\psi(z)$ the expression

$$\psi(z) = 2u\frac{z}{d}\left[1 - \frac{1}{u^2}\left(\frac{d}{2\lambda_0}\right)^2\right] + 2\frac{e^{-u}}{u}\left(\frac{d}{2\lambda_0}\right)^2 \sinh\left(2u\frac{z}{d}\right). \tag{7.83}$$

Therefore, we derive that, in the considered case, in the bulk, the electrical potential is approximately given by

$$\psi_{\rm B}(z) = 2u\frac{z}{d}\left[1 - \frac{1}{u^2}\left(\frac{d}{2\lambda_0}\right)^2\right]. \tag{7.84}$$

Equation (7.84) shows that the presence of the ionic charges reduces the effective potential. The trend of $\psi(z)$ differs from $\psi_{\rm B}(z)$ mainly close to the surfaces at $z = \pm d/2$, for a quantity

$$\Delta\psi(z) = 2\frac{e^{-u}}{u}\left(\frac{d}{2\lambda_0}\right)^2 \sinh\left(2u\frac{z}{d}\right). \tag{7.85}$$

As follows from (7.83) the analysis presented above holds for $(d/2\lambda_0 u)^2 \ll 1$, which implies $u \gg d/2\lambda_0$, or if $d \approx 10\,\mu$m and $\lambda_0 \approx 0.36\,\mu$m we have $U \gg 0.8\,$V.

In Figure (7.3), the potential across the sample, numerically evaluated by means of Eq. (7.64) with k and λ given by Eq. (7.68) and (7.69) respectively,

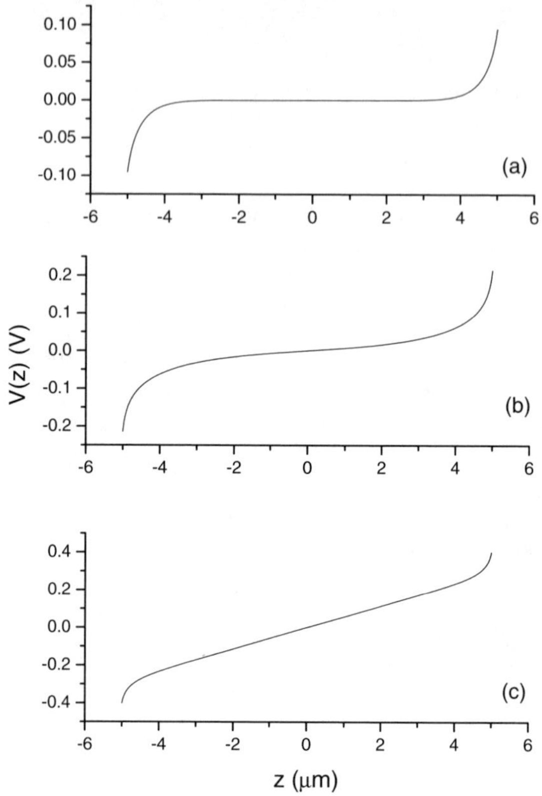

FIGURE 7.3
Voltage $V(z)$ versus z, for $d = 10\,\mu$m and $\lambda = 0.36\,\mu$m. For small U, E vanishes at $z = 0$. (a) $U = 0.2\,$V, (b) $U = 0.4\,$V, and (c) $U = 0.8\,$V. Reprinted with permission from Ref. [11]. Copyright (2002) by the American Physical Society.

is shown for different bias voltages. The trend agrees with the one discussed above.

In Figure (7.4), the electric field in the bulk $E_{\rm B} = E(0)$ for different λ_0 is reported. As expected, the presence of the surface charges of ionic origin reduces the bulk electric field. This reduction is, for commercial liquid crystals, rather important. In fact, for a bias voltage of the order of 1 V the actual bulk field is negligible with respect to the one in the absence of the ions. It follows that to identify the bulk electric field with the applied one is not correct.

To proceed further, we can do a kind of "experimental" physics by considering numerical results for the electric potential profile as the experimental data to be analyzed. It is possible to use an analytical expression, depending on two adjustable parameters, to fit these experimental data in a very good

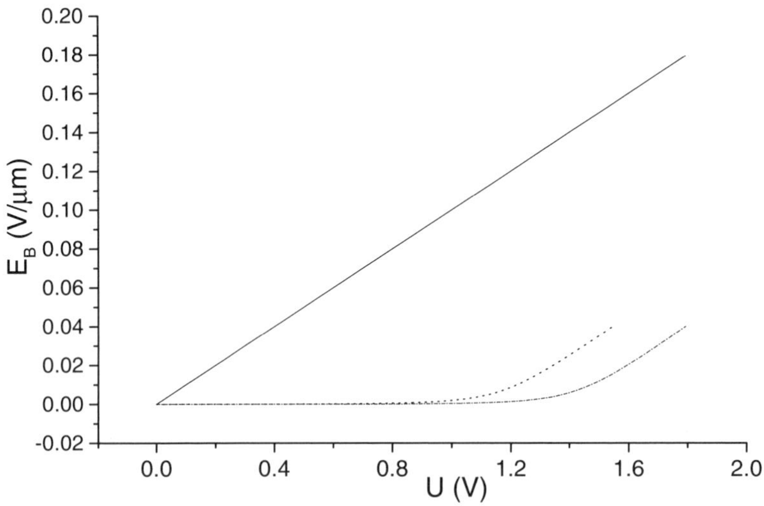

FIGURE 7.4
Bulk electric field $E_\mathrm{B} = E(0)$ versus the bias voltage U for $d = 10\ \mu$m, $\lambda_0 \to \infty$ (solid), $\lambda_0 = 0.36\ \mu$m (dotted), $\lambda_0 = 0.1\ \mu$m (dash-dotted). Reprinted with permission from Ref. [11]. Copyright (2002) by the American Physical Society.

manner. The expression is

$$V(z) = \left(\frac{U}{2} - v\right) 2\frac{z}{d} + v\frac{\sinh\left(z/\lambda_s\right)}{\sinh\left(d/2\lambda_s\right)}, \tag{7.86}$$

where v and λ_s are the adjustable parameters, v plays the role of a potential screening due to the ions whereas λ_s plays the role of a length over which this screening takes place. It is the effective screening length to measure the importance of the variations of the electrical potential near the surfaces. The the exact solution of Eq. (7.4), for the case of small applied voltage, $u \ll 1$, is given by (7.71); the one for large applied voltage, $u \gg 1$, is given by (7.78). In this manner, the general solution proposed in (7.86) embodies the two limiting solutions, while leaving two parameters free to be adjusted. Some examples of the good agreement of the numerical solutions with the approximated one are given in Figure 7.5. In fact, the agreement is good in the entire range of applied voltage, ranging from very small to large values.

To obtain the behavior of the adjustable parameters as a function of the applied voltage, we fix the thickness of the sample and the Debye's screening length. For each value of the applied voltage U a fit using (7.86) is performed in order to obtain the best fit. This gives the values of $\lambda_s(U)$ and $v(U)$ for that considered sample. Figure 7.6 shows the best fits giving the value of $\lambda_s(U)$ in units of λ_0. For large applied voltage, $\lambda_s(U)$ decreases but tends to zero only for very large voltage; for small applied voltage $\lambda_s(U) \to \lambda_0$, as expected. There is a plateau separating the regions of small and large voltages. This

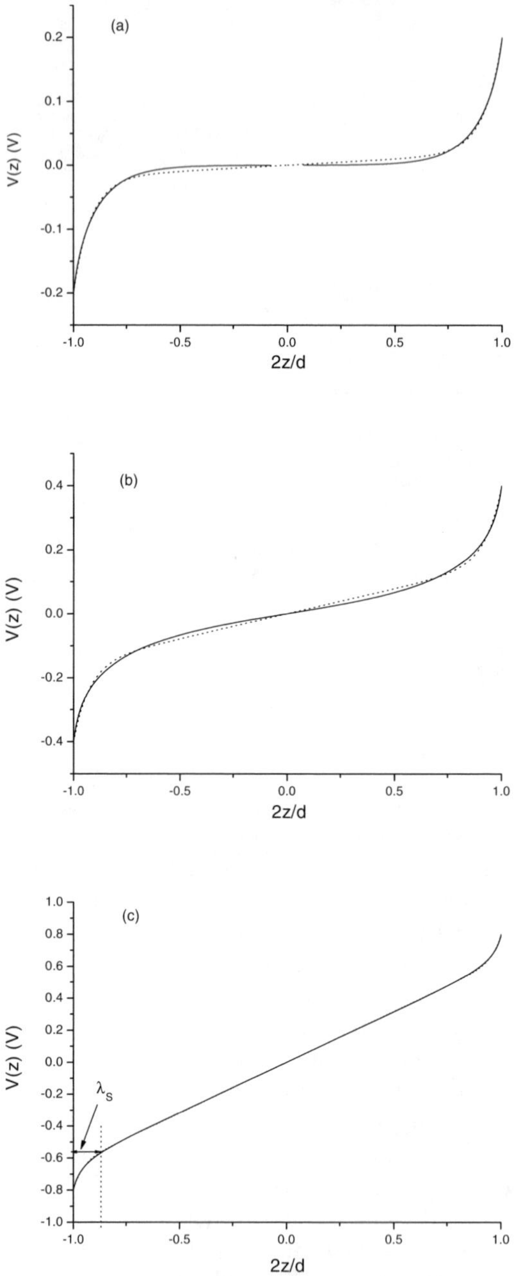

FIGURE 7.5

Electrical potential $V(z)$ versus $2z/d$, for $d = 10\,\mu$m and $\lambda_0 = 0.5\,\mu$m. Solid line corresponds to the numerical solution of Eq. (7.4); dotted line corresponds to the best fit obtained from solution (7.86). (a) $U = 0.2$ V, (b) $U = 0.4$ V, and (c) $U = 0.8$ V. In curve (c) there is an illustration of the effective screening length introduced in Eq. (7.86).

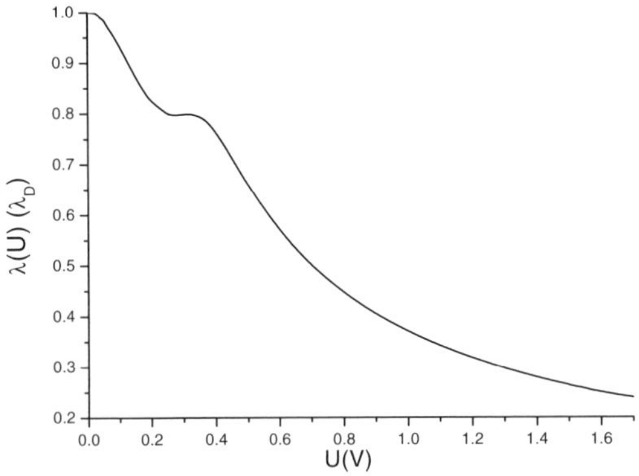

FIGURE 7.6
λ_s/λ_0 versus U for $d = 10\,\mu$m and $\lambda_0 = 0.5\,\mu$m. The results correspond to the best fit using solution (7.86) to approximate the numerical result of Eq. (7.4).

behavior can be understood by considering that for small applied voltage, the ionic charges (of a given sign) present in the liquid move towards the electrode of opposite sign forming a charged surface layer of density σ, that balances the surface density of charges furnished by the external power supply, Σ. There is a narrow region for which $\sigma \approx \Sigma$ characterizing the plateau. After that, the surface density σ cannot balance Σ any longer. When $\Sigma \gg \sigma$ the screening effect of internal charges becomes negligible. Therefore, the distribution of charges and fields in the sample is characterized by two regimes of low and high applied voltage, as we will show in Section 8.2.

Figure 7.7 shows the best fits giving $v(U)$. In the low-voltage region $v(U)$ is linear. It changes the behavior in a drastic manner in the region where $\lambda_s(U)$ presents a plateau. After that, it tends to zero, but remains always such that $v < U$.

7.3 Contribution to the surface energy of dielectric origin

In previous sections we have evaluated the influence of the ionic charge on the electric potential across a sample of an isotropic liquid. If the liquid is an anisotropic fluid, as an NLC, in a first approximation the electric potential is still given by the equations reported above. However, in this event the pres-

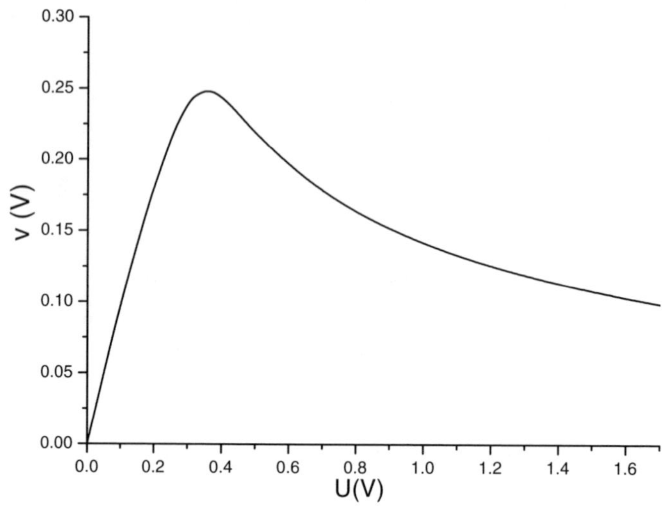

FIGURE 7.7
Trend of v versus U, for $d = 10\,\mu$m and $\lambda_0 = 0.5\,\mu$m. The curve was obtained by the same procedure as in Figure 7.6.

ence of the ionic charges gives rise to a surplus of surface energy characterizing the nematic liquid crystal-substrate interface. To evaluate the dielectric contributions to the surface energy we have to take into account the coupling of the external field with the dielectric anisotropy, $W_{\rm D}$, and with the flexoelectric properties of the liquid crystal, defined in Eq. (6.16) and (6.19), which are bulk energy densities.

Let us indicate by $E_{\rm B} = E(0)$ and by $E = E(d/2)$ the values of the electric field in the middle and at the surface of the sample. The dielectric energy, per unit surface, is

$$F_{\rm E} = \int_{-d/2}^{d/2} \left[-\frac{1}{2}\epsilon_a E(z)^2 \cos^2\theta + e\left(\cos^2\theta - \frac{1}{3}\right)\frac{dE}{dz}\right] dz, \tag{7.87}$$

by assuming the nematic orientation constant across the sample. In this case, as underlined before, the flexoelectric contribution to the dielectric energy reduces to the quadrupolar contribution. $F_{\rm E}$, given by (7.87), can be written as

$$F_{\rm E} = \int_{-d/2}^{d/2} \left[-\frac{1}{2}\epsilon_a [E(z)^2 - E_{\rm B}^2] \cos^2\theta + e\left(\cos^2\theta - \frac{1}{3}\right)\frac{dE}{dz}\right] dz$$

$$+\int_{-d/2}^{d/2}\left[-\frac{1}{2}\epsilon_a E_{\rm B}(z)^2\cos^2\theta\right]dz. \tag{7.88}$$

Taking into account that $E(z)$ is changing very rapidly close to the limiting surfaces we can put (7.88) in the form

$$F_{\rm E}=f_1+f_2+\int_{-d/2}^{d/2}\left[-\frac{1}{2}\epsilon_a E_{\rm B}(z)^2\cos^2\theta+e\left(\cos^2\theta-\frac{1}{3}\right)\frac{dE_{\rm B}}{dz}\right]dz, \tag{7.89}$$

where

$$f_1=-\frac{1}{2}\epsilon_a\cos^2\theta_1\int_{-d/2}^{0}[E^2(z)-E_{\rm B}^2]dz-e\left(\cos^2\theta_1-\frac{1}{3}\right)(E-E_{\rm B}), \tag{7.90}$$

and

$$f_2=-\frac{1}{2}\epsilon_a\cos^2\theta_2\int_{0}^{d/2}[E^2(z)-E_{\rm B}^2]dz+e\left(\cos^2\theta_2-\frac{1}{3}\right)(E-E_{\rm B}), \tag{7.91}$$

with $\theta_1=\theta(-d/2)$ and $\theta_2=\theta(d/2)$. f_1 and f_2 are the dielectric contributions, due to the ions, to the surface energy. The relevant anchoring energy strengths, coinciding with the coefficient of $\cos^2\theta_i$ $(i=1,2)$, are then

$$W_{\rm D}=-\frac{1}{2}\epsilon_a\int_0^{d/2}\left[E^2(z)-E_{\rm B}^2\right]dz, \tag{7.92}$$

and

$$W_{\rm Q}=\pm e(E-E_{\rm B}), \tag{7.93}$$

where $\pm$ refers to $z=\pm d/2$. Using now Eq. (7.64) and boundary conditions (7.62), $W_{\rm D}$ and $W_{\rm Q}$ can be written as

$$W_{\rm D}=-\frac{1}{2}\epsilon_a\left(\frac{k_{\rm B}T}{q}\right)^2\frac{\sqrt{2}}{\lambda}\int_0^u\frac{\cosh\psi}{\sqrt{\cosh\psi+k}}d\psi \tag{7.94}$$

and

$$W_{\rm Q}=\pm e\frac{k_{\rm B}T}{q}\frac{\sqrt{2}}{\lambda}\left(\sqrt{\cosh u+k}-\sqrt{1+k}\right). \tag{7.95}$$

In Figure (7.8), we report the quantity $W_{\rm E}=W_{\rm D}+W_{\rm Q}$ versus the bias voltage U for $\epsilon_a=10$ and typical values of e relevant to the surface at $z=d/2$ where $V(d/2)=U/2$.

As is evident from Figure (7.8), $W_{\rm E}$ tends to a saturation voltage for large bias, as experimentally observed [12]. In the limit of small voltage, where $k\approx-1$, using Eq. (7.76) and Eq. (7.91), we obtain (for $z=d/2$)

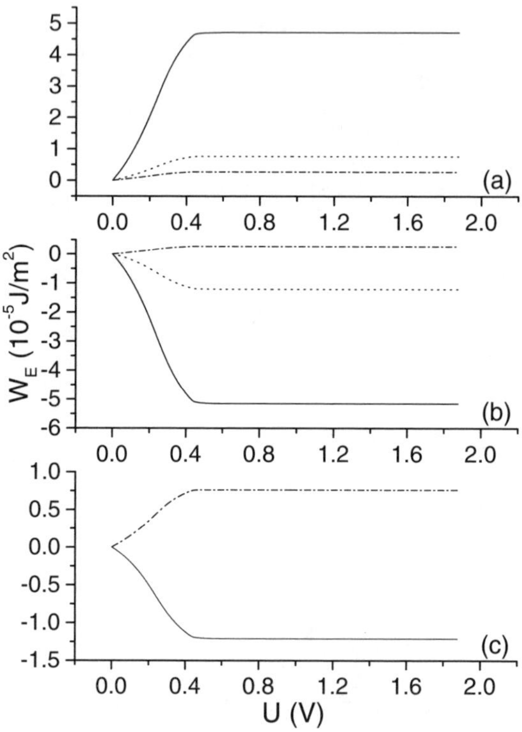

FIGURE 7.8

Dielectric contribution to the surface energy $W_{\rm E}$ versus the applied voltage U for $\epsilon_a = 10\epsilon_0$, $d = 10\,\mu$m, $\lambda_0 = 0.36\,\mu$m and (a) $e = 5 \times 10^{-11}\,$C/m (solid), $e = 10^{-11}\,$C/m (dotted) and $e = 5 \times 10^{-12}\,$C/m (dashed), (b) $e = -5 \times 10^{-12}\,$C/m (solid), $e = -10^{-11}\,$C/m (dotted)and $e = -5 \times 10^{-11}\,$C/m (dashed), and (c) comparison of $W_{\rm E}$ versus U for $e = 5 \times 10^{-11}\,$C/m (dashed) and $e = -5 \times 10^{-11}\,$C/m (solid). Reprinted with permission from Ref. [11]. Copyright (2002) by the American Physical Society.

$$W_{\rm E} = -2\frac{k_{\rm B}T}{q\lambda_0}\left[2\epsilon_a\frac{k_{\rm B}T}{q}\sinh^2\left(\frac{u}{4}\right) + e\sinh\left(\frac{u}{2}\right)\right]. \tag{7.96}$$

This equation shows that according to the value of e/ϵ_a non-monotonic behaviors of $W_{\rm E}(U)$ are possible. For large bias voltage, using Eq. (7.83) we obtain (for $z = d/2$)

$$W_{\rm E} = -\frac{1}{2}\left(\epsilon_a\frac{k_{\rm B}T}{q} + e\right)\frac{k_{\rm B}T}{q}\frac{d}{\lambda_0^2} + \mathcal{O}(u^{-1}). \tag{7.97}$$

It is now possible to estimate the order of magnitude of the renormalization of the bare anchoring energy due to the ions dissolved in the liquid crystal. If $\epsilon_a = 10\epsilon_0$, $e = 10^{-11}$ C/m [7]–[9], $\lambda_0 = 0.36\,\mu$m [3, 13], and $d = 10\,\mu$m, the saturation value for $W_{\rm E}$ is of the order of 10^{-5} J/m^2. However, if $\lambda_0 = 0.1\,\mu$m, $W_{\rm E}$ tends to 10^{-4} J/m^2. In other words, the saturation value of $W_{\rm E}$ strongly depends on λ_0.

If we use (7.86) for the electric potential distribution, the "exact" expression for the anchoring energy becomes, respectively,

$$\begin{aligned} W_{\rm D} &= -\frac{\epsilon_a}{2}\int_0^{d/2}\left[E(z)^2 - E_{\rm B}^2\right]dz \\ &= \epsilon_a\left\{\frac{v^2 - vU}{d} + \frac{vU - 2v^2}{2\lambda_s\sinh(d/\lambda_s)}\right. \\ &\left. + \frac{1}{8\lambda_s}\left[\frac{dv^2}{\lambda_s} - v^2\sinh\left(\frac{d}{2\lambda_s}\right)\right]\frac{1}{\sinh^2[d/(2\lambda_s)]}\right\} \end{aligned} \tag{7.98}$$

for the dielectric part and

$$\begin{aligned} W_{\rm Q} &= e\left[E_s - E_B\right] \\ &= -\frac{e}{\lambda_s}\frac{v}{d}\tanh\left(\frac{d}{4\lambda_s}\right) \end{aligned} \tag{7.99}$$

for the flexoelectric part.

Conclusions

The analysis of the influence of the ion adsorption on the anisotropic part of the surface energy of an NLC, in contact with a solid substrate, shows that the exponential approximation for the electric field of ionic origin works well only in the limit of weak adsorption. In this case, the dielectric contribution to the anchoring strength is quadratic, whereas the flexoelectric contribution is linear, in the surface electric field. In the opposite limit of strong adsorption, both contributions are linear in the surface field. On the other hand, it indicates that the surface polarization, experimentally observed in nematic samples, can have an ionic origin. These conclusions have been obtained under the assumption that the adsorbed charge density is a given quantity. As is well known, it depends on the actual surface potential, but this aspect of the problem will be analyzed in details in the next chapters.

In this chapter we have also analyzed the effect of a bias voltage on the ionic distribution in an isotropic liquid sample. The results can be applied to the case in which the liquid is an NLC, showing that the non-homogeneous distribution of the electric field can be described by means of a surface energy of dielectric origin. This surface energy is due to the coupling of the electric field with the dielectric anisotropy and with the quadrupolar properties of the liquid crystal. The surface energy of dielectric origin depends on the applied voltage and tends to a saturation value for large applied voltages. Its value ranges, for typical values of ϵ_a and e, from $5 \times 10^{-5}\mathrm{J/m^2}$ to $10^{-4}\mathrm{J/m^2}$, for $0.1\mu\mathrm{m} \leq \lambda_0 \leq 0.36\mu\mathrm{m}$. This means that in the presence of a dc voltage the renormalization of the anchoring energy due to the bias could be important. To neglect it can be the source of mistakes in the determination of the flexoelectric coefficients.

[1] Barbero G and Olivero D. Ions and nematic surface energy: Beyond the exponential approximation for the electric field of ionic origin. *Physical Review E*, **65**, 031701 (2002).

[2] Bohinc K, Kralj-Iglic V, and Iglic A. Thickness of electrical double layer. Effect of ion size. *Electrochimica Acta*, **46**, 3033 (2001).

[3] Thurston RN, Cheng J, Meyer RB, and Boyd GD. Physical mechanisms of DC switching in a liquid crystal bistable boundary layer display. *Journal of Applied Physics*, **56**, 263 (1984).

[4] Kühnau U, Petrov AG, Klose G, and Schmiedel H. Measurements of anchoring energy of a nematic liquid crystal, 4-cyano-4'-n-pentylbiphenyl, on Langmuir-Blodgett films of dipalmitoyl phosphatidylcholine. *Physical Review E*, **59**, 578 (1999).

[5] Nazarenko VG, Pergamenshchik VM, Koval'chuk OV, Nych AB, and Lev BI. Non-Debye screening of a surface charge and a bulk-ion-controlled anchoring transition in a nematic liquid crystal. *Physical Review E*, **60**, 5580 (1999).

[6] Murthy PRM, Raghunathan VA, and Madhusudana NV. Experimental determination of the flexoelectric coefficients of some nematic liquid crystals. *Liquid Crystals*, **14**, 483 (1986).

[7] Blinov LM, Barnik MI, Ozaki M, Shtykov NM, and Yoshino K. Surface and flexoelectric polarization in a nematic liquid crystal directly measured by a pyroelectric technique. *Physical Review E*, **62**, 8091 (2000).

[8] Mazzulla A, Ciuchi F, and Sambles JR. Optical determination of flexoelectric coefficients and surface polarization in a hybrid aligned nematic cell. *Physical Review E*, **64**, 021708 (2001).

[9] Blinov LM, Barnik MI, Ohoka H, Ozaki M, and Yoshino K. Separate measurements of the flexoelectric and surface polarization in a model nematic liquid crystal p-methoxybenzylidene-p'-butylaniline: Validity of the quadrupolar approach. *Physical Review E*, **64**, 031707 (2001).

[10] Petrov AG and Derzhanski AI. Flexoelectricity and surface polarization. *Molecular Crystals and Liquid Crystals*, **41**, 41 (1977).

[11] Olivero D, Evangelista LR, and Barbero G. External electric field effect on nematic anchoring energy. *Physical Review E*, **65**, 031721 (2002).

[12] Strangi G, Versace C, and Scaramuzza N. Surface anchoring energy modulation in liquid crystal cells with mixed conductor boundary layers. *Applied Physics Letters*, **78**, 2455 (2001).

[13] Thurston RN. Equilibrium distributions of electric field in a cell with adsorbed charge at the surfaces. *Journal of Applied Physics*, **55**, 4154 (1984).

8

ADSORPTION PHENOMENON AND EXTERNAL FIELD EFFECTS: GENERAL APPROACH

In this chapter, the steady state distribution of ionic charges in a liquid, in the presence of surface adsorption, is determined in the framework of a general model. In the first part of the chapter, the effect of an electric field applied by means of blocking electrodes is considered. It is shown that the surface adsorption of ions dissolved in the liquid is responsible for an asymmetry in the electric field distribution. In the model the liquid is supposed dielectric, but containing impurities. These impurities, by means of a chemical reaction, can give rise to ions. The theory takes into account the activation energy for the ionization chemical reaction and the adsorption energy of the ions at the surface. The model is built for isotropic liquids, but the results are used to analyze the behavior of liquid crystalline systems. A set of four non-linear equations governing the charge and field distribution in a typical sample is deduced. These equations come from the condition of conservation of the number of particles, from first integrals of the Poisson-Boltzmann equation and from the appropriate boundary conditions for the electric field at the surfaces limiting the sample. The analysis of this set of equations permits us to establish the existence of the regimes of low and high external voltage, according to the importance of the adsorbed (internal) charges as compared with the charges supplied by an external power supply. In the remaining part of the chapter, the influence of the adsorption energy on the anchoring properties of an NLC sample is investigated for significant values of the material parameters characterizing the phase and by using the electric field distribution calculated in the framework of the model developed in the first part of the chapter. Likewise, the destabilizing effect of the surface electric field is reanalyzed, thus establishing the dependence of the threshold fields on the adsorption energy as well as on the thickness of the sample on more general grounds.

8.1 Motivation

In Chapters 6 and 7 the role of the ion adsorption as the main mechanism to explain the thickness dependence as well as the bias-voltage dependence of the anisotropic part of the anchoring energy connected to the NLC-substrate interface has been emphasized. By means of detailed calculations of the threshold field values required to destabilize the uniform pattern of a typical nematic sample, it was demonstrated that the ion adsorption is responsible for the presence of a surface electric field which can induce destabilization in the NLC orientation.

As we have seen, the adsorption of ionic impurities produces a counterion cloud forming a diffuse electric double layer in the liquid [1], which gives rise to an internal field. The asymmetry in the electric field distribution that results when an external field is combined with the internal field, due to the surface adsorption, has been used as the mechanism to explain the dc switching of a liquid crystal display [2]. The problem has been solved in the situation of a static equilibrium field distribution under constant voltage, for a cell whose thickness is d, for which the ratio λ_0/d is not small [3].

In the pioneer analysis of the problem, presented by Thurston [3], it is assumed that there are equal numbeers of positive and negative ionic charges per unit area in the liquid, but that some of these charges stuck at the walls by electrochemical forces. The immobilized ions at the surfaces are supposed of negative sign and represented by a surface charge density. It is supposed to be independent of the electrical potential difference applied by an external power supply to the cell [3]. However, as is well known, the adsorption phenomenon of ions from a solid surface depends on the actual electrical potential, due to the electric charges of external origin and to the adsorbed charges. Hence, the hypothesis that the surface density of adsorbed charges can be used as a control parameter to describe the charge and field distribution when intrinsic and voltage-induced charge layers are combined is questionable, as we have indicated at the end of Section 7.3.

For this reason, in this chapter, we discuss a more general model to determine the equilibrium distributions of charges and fields in an isotropic liquid containing ions, submitted to an external field, removing this simplifying hypothesis [4]. The liquid is assumed dielectric, but containing impurities able to give rise to ions by means of a chemical dissociation reaction. The activation energy of this chemical reaction enters in the model. We suppose that the electrodes are perfectly blocking, and the surfaces are assumed, in a first moment, to adsorb positive ions with a given adsorption energy. The model is built in the framework of the Poisson-Boltzmann theory [5]. The basic assumptions of this theory are: (1) the ions can be considered as dimensionless point charges; (2) the surface charge is assumed to be uniformly smeared over the surface; (3) the electrolyte solution is described as a continuum with uniform dielectric constant. In the absence of external field it is possible to determine, by means of the model, the chemical potential and the electric potential profile in the sample. For this situation one observes that, once an adsorption energy is fixed, in the limit of small thickness the surface charge of adsorbed particles presents a linear behavior with the thickness d. For very large values of d the surface density of charges tends to a saturation value [5], thus recovering the simplified result obtained in the Langmuir approximation as particular cases.

8.2 The model

We consider again a cell in the shape of a slab of thickness d, filled with a liquid characterized by a dielectric constant ϵ, but containing impurities which, by means of a chemical reaction, are the source of the ions. The chemical reaction $X \to B^+ + C^-$, where X is a molecule of the impurity and B^+ and C^- the ions resulting from its dissociation, has an activation energy $E_{\rm ac}$. The liquid is globally neutral and we consider first the general case where the surfaces are not identical, i.e., the adsorption energy on one surface is different from the one relevant to the other suface.

As we have done in preceding chapters, we use a Cartesian reference frame whose z-axis is normal to the limiting walls, located at $z = \pm d/2$, and assume that all the physical quantities entering in the model are only z dependent. The distribution of charges produced by the ion adsorption gives rise to a liquid which is locally charged, but globally neutral. For simplicity, we assume that only positive ions are adsorbed, and, in Section 8.5, a possible extension of the model, for the case in which both positive and negative ions are adsorbed, is discussed.

Now, differently from what we have done in the previous chapter, we denote by n_0 the bulk density of impurities (in an infinite sample) which will be the source of the ions, due to the chemical reaction mentioned above (internal charges). The equilibrium distribution of the bulk density of non-dissociated impurities (i.e., which remain neutral) is given by

$$n_b = n_0 e^{\mu}, \tag{8.1}$$

where μ is the chemical potential in $k_{\rm B}T$ units. Likewise, the bulk densities of positive and negative ions are given by

$$n_{\pm}(z) = n_0 e^{\mu - \Delta \mp \psi(z)}, \tag{8.2}$$

where $\Delta = E_{\rm ac}/k_{\rm B}T$ is the activation energy and $\psi(z)$ is the reduced potential introduced in (7.1). In the analysis we assume that the liquid is locally neutral in an infinite sample, in the absence of the adsorption phenomenon. This condition fixes the zero of the potential. In the Maxwell-Boltzmann distribution of ionic charges in the sample, when the adsorption phenomenon is present and the sample is submitted to an external field, the potential is measured with respect to this ground state. The activation energy $E_{\rm ac}$ can be identified with the electrostatics interaction energy between the ions B^+ and C^- resulting from the dissociation of the molecule X. It is

$$E_{\rm ac} = \frac{1}{\epsilon_1}\frac{q^2}{(r_+ + r_-)}, \tag{8.3}$$

where r_- and r_+ are the radii of the negative and positive ion, and ϵ_1 is the dielectric constant of the liquid. The surface densities of (internal) adsorbed charges are given by

$$\sigma_{\rm i} = N_i e^{\mu - A_i - \psi_i}, \quad \text{for} \quad i = 1, 2, \tag{8.4}$$

where $\psi_1 = \psi(z = -d/2)$ and $\psi_2 = \psi(z = d/2)$ are the values of the reduced surface potentials, and A_1 and A_2 are the adsorption energies of the surfaces, measured in $k_{\rm B}T$ units. Furthermore, N_i is the surface density of sites where the ions can be adsorbed. The adsorption energy $E_{\rm ad}$, in a first approximation, can be identified with the electrostatics energy of an adsorbed ion with its image in the substrate (physical adsorption). It is given by [6]

$$E_{\rm ad} = \frac{q^2}{2r_+}\frac{\epsilon_1 - \epsilon_2}{\epsilon_1(\epsilon_1 + \epsilon_2)}, \tag{8.5}$$

where ϵ_2 is the dielectric constant of the substrate. Note that the adsorption phenomenon takes place, for electrostatics reasons, only if $\epsilon_2 > \epsilon_1$, as we will assume.

Let us now establish the fundamental equations of the model. We consider that only the internal charges can move to the surface. The external charges supplied to the system remain in the surface and are separated from the liquid by the blocking electrodes. In this manner the surface densities of charges will have both the internal and the external contribution, which, as we shall show later, gives rise to an asymmetry in the surface density of charges. The first requirement to be satisfied by the system is the conservation of the number of particles per unit surface, namely

$$\frac{N_+ + N_-}{2} + N_{\rm B} + \frac{\sigma_1 + \sigma_2}{2} = n_0 d, \tag{8.6}$$

where

$$N_\pm = \int_{-d/2}^{d/2} n_\pm(z)dz \quad \text{and} \quad N_{\rm B} = \int_{-d/2}^{d/2} n_b(z)dz = n_b d. \tag{8.7}$$

Using for $n_\pm(z)$ and σ_i, (8.2) and (8.4), it is possible to rewrite Eq. (8.6) in the form

$$e^{\mu}\left\{n_0 e^{-\Delta}\int_{-d/2}^{d/2}\cosh\psi(z)dz + n_0 d + \frac{N_1 e^{-A_1-\psi_1}+N_2 e^{-A_2-\psi_2}}{2}\right\} = n_0 d. \tag{8.8}$$

In this case the chemical potential is given by

$$e^{-\mu} = 1 + \frac{1}{2n_0 d}(N_1 e^{-A_1-\psi_1} + N_2 e^{-A_2-\psi_2}) + \frac{e^{-\Delta}}{d}\int_{-d/2}^{d/2}\cosh\psi(z)dz. \tag{8.9}$$

This equation connects the chemical potential μ with the surface potentials ψ_1 and ψ_2.

We are interested in the final equilibrium distribution of charges and fields when the applied voltage is held constant, i.e., no transients are considered. In the framework of the Poisson-Boltzmann theory, in the steady state the charge distribution and the electrical potential are related by Poisson's equation

$$\frac{d^2V}{dz^2} = -\frac{q}{\epsilon}[n_+(z) - n_-(z)], \tag{8.10}$$

which can be easily written as

$$\frac{d^2\psi}{dz^2} = \frac{1}{L^2}e^{\mu-\Delta}\sinh\psi, \tag{8.11}$$

where

$$L = \left(\frac{\epsilon k_B T}{2n_0 q^2}\right)^{1/2} \tag{8.12}$$

is an intrinsic length of the problem. This length is connected to the Debye's screening length λ_0 through the relation [5] $\lambda_0 = Le^{\Delta/2}$. A first integration of Eq. (8.11) can be easily performed giving

$$\frac{1}{2}\left(\frac{d\psi}{dz}\right)^2 = \frac{e^{\mu-\Delta}}{L^2}\left[\cosh\psi(z) + c\right], \tag{8.13}$$

where c is an integration constant to be determined by the boundary conditions. The electric field is

$$E(z) = -\frac{dV}{dz} = -\frac{k_B T}{q}\frac{d\psi}{dz}.$$

In the absence of external field its surface values are

$$E(-d/2) = q\frac{\sigma_1}{\epsilon} \qquad \text{and} \qquad E(d/2) = -q\frac{\sigma_2}{\epsilon},$$

and outside the slab the field is zero because the system is globally neutral. Notice that if the adsorption energies (A_1 and A_2) are different, or if the densities of sites on the surfaces (N_1 and N_2) are different, the electrical potential is not symmetric with respect to $z = 0$ also in the absence of external applied voltage. When an external field is applied the above conditions become

$$E(\mp d/2) = -\frac{k_B T}{q}\left(\frac{d\psi}{dz}\right)_{\mp d/2} = \pm q\,\frac{\sigma_{1,2} \mp \Sigma}{\epsilon}, \tag{8.14}$$

where Σ is the surface density of external charges. Equations (8.14) are written by assuming that the surface at $z = -d/2$ is connected with the negative pole of the external power supply. The requirement that the system is globally neutral can be expressed as

$$\sigma_1 + \sigma_2 + \int_{-d/2}^{d/2} n_+(z)dz = \int_{-d/2}^{d/2} n_-(z)dz, \tag{8.15}$$

which, with the help of Eq. (8.2) can be written in the form

$$\sigma_1 + \sigma_2 = 2n_0 e^{\mu-\Delta}\int_{-d/2}^{d/2} \sinh\psi(z)dz. \tag{8.16}$$

A simple calculation shows that Eq. (8.16) is an identity, if Eqs. (8.14) are taken into account. In fact, from Eq. (8.11) we have

$$\sinh\psi = L^2 e^{(-\mu+\Delta)}\frac{d^2\psi}{dz^2}.$$

By substituting this result into Eq. (8.16) we obtain

$$\sigma_1 + \sigma_2 = 2n_0 L^2\left[\left(\frac{d\psi}{dz}\right)_{-d/2} - \left(\frac{d\psi}{dz}\right)_{d/2}\right],$$

which is an identity, taking into account the definition of L and the boundary conditions (8.14). In order to solve the problem under consideration, we consider the two separated cases in which $\sigma_1 - \Sigma > 0$ and $\sigma_2 - \Sigma < 0$, defining, respectively, the regimes of low and high external voltage.

8.2.1 Low-voltage region

We limit first the analysis to the case of low external voltage, where $\sigma_1 - \Sigma \geq 0$. Since in our hypotheses $E(z = -d/2) > 0$ and $E(z = d/2) < 0$, the electrical potential has a minimum at some point z^* in the sample, where the electric field vanishes. A simple inspection shows that $\psi(z)$ can have only one extremum in $-d/2 \leq z \leq d/2$. In fact, from Eq. (8.13) $d\psi/dz = 0$ implies $\cosh\psi + c = 0$, which has only one solution. In the low voltage regime, since $d\psi/dz < 0$ for $z = -d/2$ and $d\psi/dz > 0$ for $z = d/2$, this point corresponds to a minimum. On the contrary, in the high voltage region, where $d\psi/dz < 0$ for $z = \pm d/2$, the $\psi(z)$ profile has to be monotonic. In fact, if not, $\psi(z)$ has to present at least two extrema, which is impossible for Eq. (8.13).

It follows that $(d\psi/dz)_{z=z^*} = 0$, and in Eq. (8.13) the integration constant c can be written in terms of $\psi^* = \psi(z^*)$ as $c = -\cosh\psi^*$. From Eq. (8.13) we obtain

$$\frac{d\psi}{dz} = \mp\frac{\sqrt{2}}{L}e^{(\mu-\Delta)/2}\sqrt{\cosh\psi - \cosh\psi^*}, \tag{8.17}$$

where the sign $-$ refers to the region $-d/2 \le z \le z^*$, and $+$ to the region $z^* \le z \le d/2$. Equations (8.17) can be integrated to give

$$\int_{\psi^*}^{\psi_2}\frac{d\psi}{\sqrt{\cosh\psi - \cosh\psi^*}} - \int_{\psi_1}^{\psi^*}\frac{d\psi}{\sqrt{\cosh\psi - \cosh\psi^*}} = \sqrt{2}\frac{d}{L}e^{(\mu-\Delta)/2}. \tag{8.18}$$

Furthermore, by using Eqs. (8.14) and (8.17), we have

$$\begin{aligned} \frac{\sqrt{2}k_{\rm B}T}{q^2L}e^{(\mu-\Delta)/2}\sqrt{\cosh\psi_1 - \cosh\psi^*} &= \frac{\sigma_1 - \Sigma}{\epsilon} \quad \text{and} \\ \frac{\sqrt{2}k_{\rm B}T}{q^2L}e^{(\mu-\Delta)/2}\sqrt{\cosh\psi_2 - \cosh\psi^*} &= \frac{\sigma_2 + \Sigma}{\epsilon}. \end{aligned} \tag{8.19}$$

The fundamental equations of the model are (8.9), (8.18) and (8.19). We have to solve this system of four equations to obtain μ, ψ_1, ψ_2 and ψ^*. Once this system of equations is solved, it is straightforward to obtain the surface charge densities $\sigma_{\rm i}$ by means of Eq. (8.4). As follows from these equations, the surface charge densities depend on the external charges at the surface through the chemical potential and the electric potentials at the surfaces.

8.2.2 High-voltage region

The border separating the two regimes is defined by $\sigma_1(\Sigma_c) - \Sigma_c = 0$, where Σ_c is the critical surface density of external charges. For $\Sigma = \Sigma_c$, $\psi^*(\Sigma_c) = \psi_1(\Sigma_c)$, as follows from Eq. (8.19). In the high-voltage regime the adsorbed charge, at $z = -d/2$, is then smaller than the one sent by the power supply on the electrode. From Eqs. (8.14) we now have that $E(-d/2) < 0$ and $E(d/2) < 0$. The electrical potential is a monotonic function of z and, consequently, the electric field never vanishes for $-d/2 \le z \le d/2$. In this case, from Eq. (8.13) we obtain

$$\int_{\psi_1}^{\psi_2}\frac{d\psi}{\sqrt{\cosh\psi + c}} = \sqrt{2}\frac{d}{L}e^{(\mu-\Delta)/2}, \tag{8.20}$$

connecting c to ψ_1 and ψ_2. By using Eqs. (8.13) and (8.14) we deduce that the boundary conditions read

$$\frac{k_B T}{q}\frac{\sqrt{2}}{L}\sqrt{\cosh\psi_1 + c} = q\frac{\Sigma - \sigma_1}{\epsilon} \quad \text{and} \quad \frac{k_B T}{q}\frac{\sqrt{2}}{L}\sqrt{\cosh\psi_2 + c} = q\frac{\Sigma + \sigma_2}{\epsilon}. \tag{8.21}$$

In this regime, the fundamental equations are (8.9), (8.20) and (8.21). These equations give μ, ψ_1, ψ_2 and c in terms of Σ and d.

8.3 The charge and field distributions

In this section, we shall consider the particular situation in which the adsorption energy is the same at both surfaces, and the system is submitted to an external field. The purpose is to obtain the charge and field distributions in the presence of an applied voltage, taking into account the phenomenon of ion adsorption. The problem is analyzed in the framework of the model presented in the preceding section. The solution of the system of four non-linear equations (8.9), (8.18) and (8.19) in the low-voltage regime and (8.9), (8.20) and (8.21) in the high-voltage regime, is searched numerically. To do this, we start with an estimation of the parameters entering in the model. We assume $\Delta = 8.0$, $N_1 = N_2 = N$ and $A_1 = A_2 = -0.1$. We work at a fixed thickness $d/2L = 200$, which implies that $\lambda_0 \approx 0.6\,\mu$m, and $d \approx 4\,\mu$m. Using these estimations we can determine the electrical potential and the field profiles in the regimes of low ($\sigma_1 > \Sigma$) and high ($\sigma_1 < \Sigma$) external potential, at a fixed $n_0 d/N = 0.85$.

In Figure 8.1, the difference $\sigma_1 - \Sigma$ versus Σ, where Σ represents the external charge density sent by the power supply on the limiting surfaces, is reported. This quantity decreases rapidly with Σ, and vanishes for $\Sigma = \Sigma_c$. For $\Sigma > \Sigma_c$, $\sigma_1 - \Sigma < 0$. Notice that for $\Sigma = 0$, σ_1 coincides with the adsorbed charges density in the absence of external field. It depends on A, Δ and d. For very large Σ, $\sigma_1 \to n_0 d$. In this limit, all positive ions are adsorbed on the surface at $z = -d/2$, connected with the negative electrode of the power supply.

Figure 8.2 shows the surface density of adsorbed charges versus Σ at the surface at high potential. This quantity is a decreasing function of Σ. The trends of σ_i versus Σ can be easily understood taking into account that σ_i are given by Eq. (8.4), in which enter the surface electrical potentials ψ_i.

The dependencies of ψ_i versus Σ are reported in Figure 8.3. The potential ψ_1 is a monotonic decreasing function of Σ, whereas ψ_2 is a monotonic increasing function of Σ. Notice that for $\Sigma \to \Sigma_c$, ψ_1 vanishes. For $\Sigma > \Sigma_c$, ψ_1 is negative. In Figure 8.3, the minimum value of ψ, $\psi^* = \psi(z^*)$, versus Σ is also reported. It presents a maximum for $\Sigma/N \approx 0.25$ and it decreases abruptly to zero for $\Sigma \to \Sigma_c$. The analysis shows that $\psi^* \to \psi_1$ for $\Sigma \to \Sigma_c$. However, $\psi_1 \to 0$ in a monotonic manner, whereas ψ^*, for $\Sigma \leq \Sigma_c$ increases

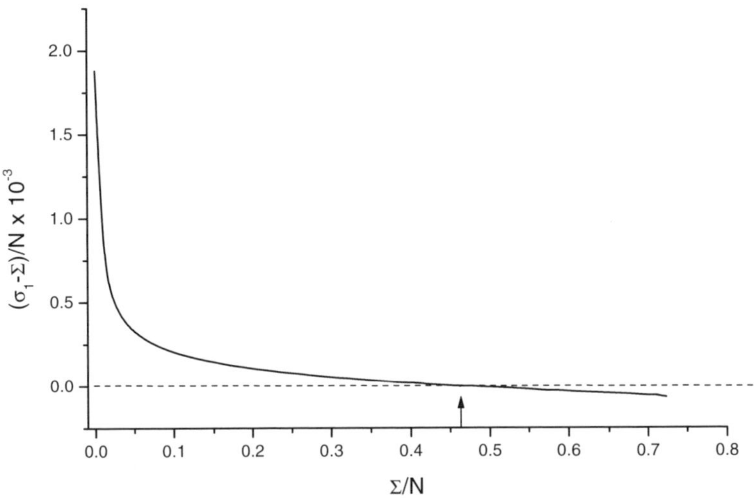

FIGURE 8.1
The behavior of the surface charge density σ_1 as a function of the external charge density Σ. $(\sigma_1 - \Sigma)/N$ versus Σ/N is reported in the two regimes of low and high applied voltage. The border between the two regimes is $\Sigma_c/N \approx 0.46$, where this difference changes its sign. The curves have been depicted for $\Delta = 8.0$ and $A = -0.1$. Reprinted with permission from Ref. [4]. Copyright (2001) by the American Physical Society.

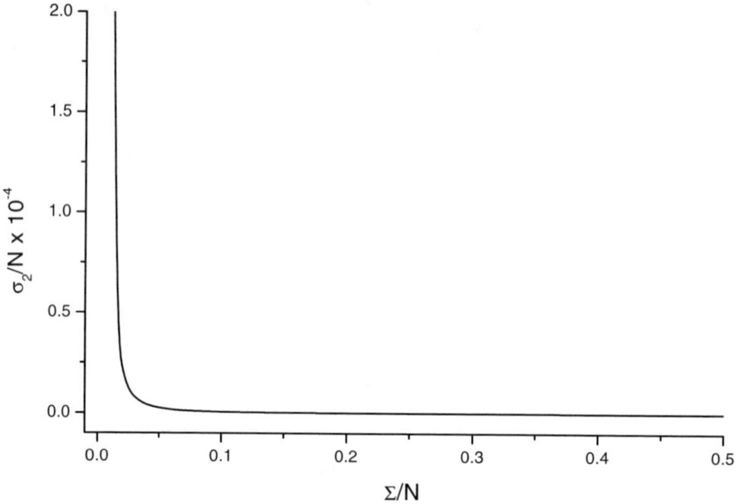

FIGURE 8.2
The behavior of the surface charge densities σ_2 as a function of the external charge density Σ. σ_2/N versus Σ/N is shown. The curves have been depicted for $\Delta = 8.0$ and $A = -0.1$. Reprinted with permission from Ref. [4]. Copyright (2001) by the American Physical Society.

up to a maximum value, and then it decreases to zero.

In Figure 8.4, the actual voltage difference across the sample $\psi_2 - \psi_1$ versus Σ is shown. As expected, for Σ very large it tends to be proportional to Σ.

The chemical potential μ versus Σ is displayed in Figure 8.5. As is evident from this figure, in the absence of an external field (i.e., for $\Sigma = 0$) $\mu(\Sigma = 0)$ is rather small, in absolute values and depends on A, Δ and d. μ changes very much with Σ, and tends to $-\infty$ for $\Sigma \to \infty$. In fact, in this limit all the positive ions are adsorbed at the surface at lower potential, and an exchange of particles between the surface and the bulk becomes impossible.

The electrical potential profiles for various Σ are shown in Figure 8.6. The dashed curve corresponds to the case in which the electrical potential is due only to the adsorption phenomenon ($\Sigma = 0$). In the symmetric case under consideration ($A_1 = A_2$, $N_1 = N_2$), $\psi(z)$ is an even function of z, with respect to the middle of the sample (at $z = 0$). The other curves correspond to $\Sigma/N \approx 0.4$ (low-voltage regime) and $\Sigma/N \approx 0.7$ (high-voltage regime). For the case we are illustrating here, the border between the two regimes is $\Sigma_c/N \approx 0.46$.

Finally, in Figure 8.7 the electric field profiles for various Σ are reported. Notice that, as pointed out before, for $\Sigma < \Sigma_c$ the electric field changes sign across the sample, whereas for $\Sigma > \Sigma_c$ it has the same direction everywhere.

As stated above, Σ_c is the border separating the two regimes. It is possible to obtain Σ_c by operating on the fundamental equations of the model. Let us

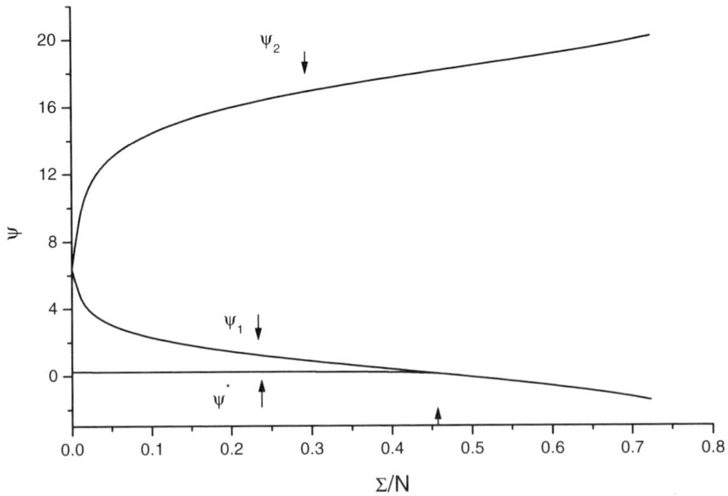

FIGURE 8.3
Surface electrical potentials ψ_1 and ψ_2 and the minimum value of $\psi(z)$, ψ^* versus the external charge density Σ/N. Reprinted with permission from Ref. [4]. Copyright (2001) by the American Physical Society.

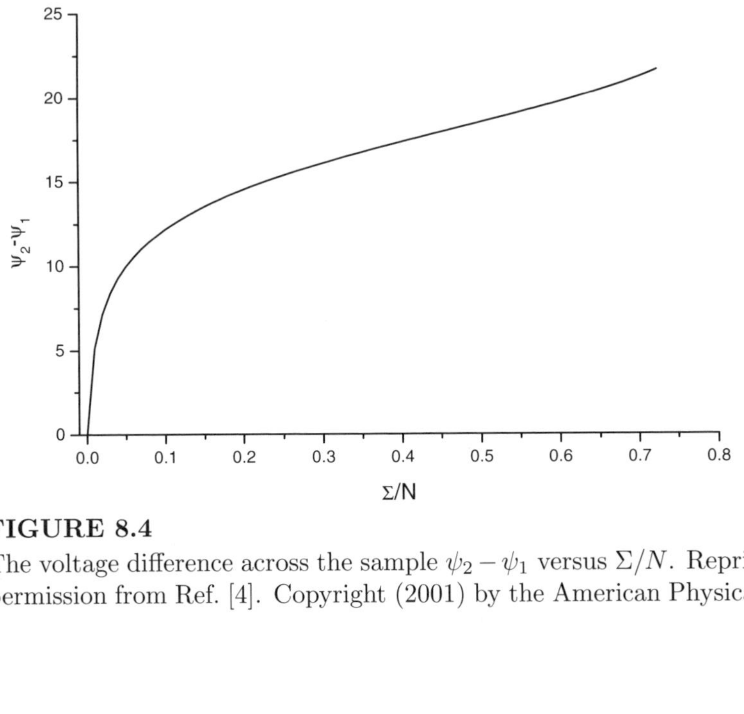

FIGURE 8.4
The voltage difference across the sample $\psi_2 - \psi_1$ versus Σ/N. Reprinted with permission from Ref. [4]. Copyright (2001) by the American Physical Society.

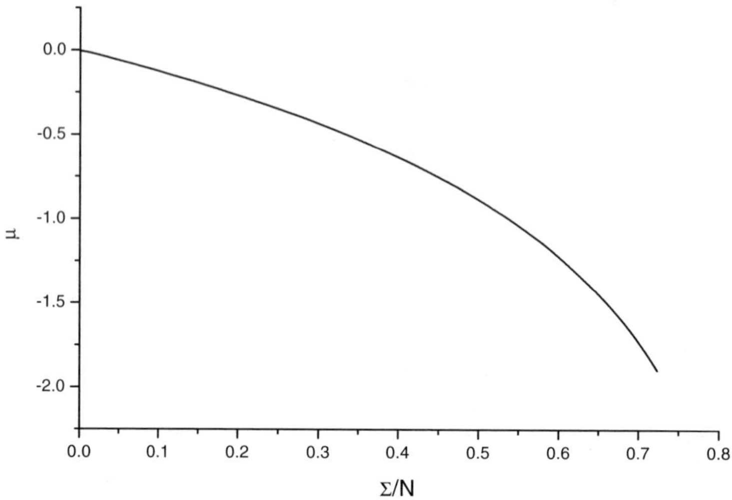

FIGURE 8.5
Chemical potential μ versus Σ/N. Reprinted with permission from Ref. [4]. Copyright (2001) by the American Physical Society.

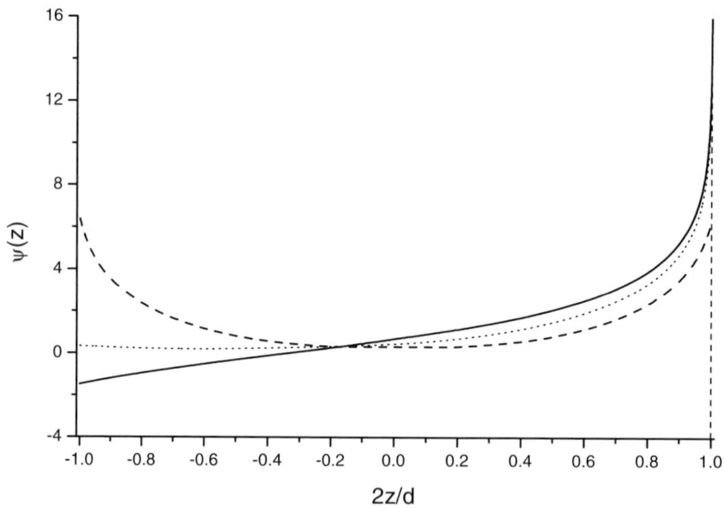

FIGURE 8.6
Electrical potential $\psi(z)$ versus $2z/d$. Dashed curve corresponds to $\Sigma/N = 0.0$ (no external potential), dotted curve to $\Sigma/N \approx 0.41$ and solid curve to $\Sigma/N \approx 0.72$. Reprinted with permission from Ref. [4]. Copyright (2001) by the American Physical Society.

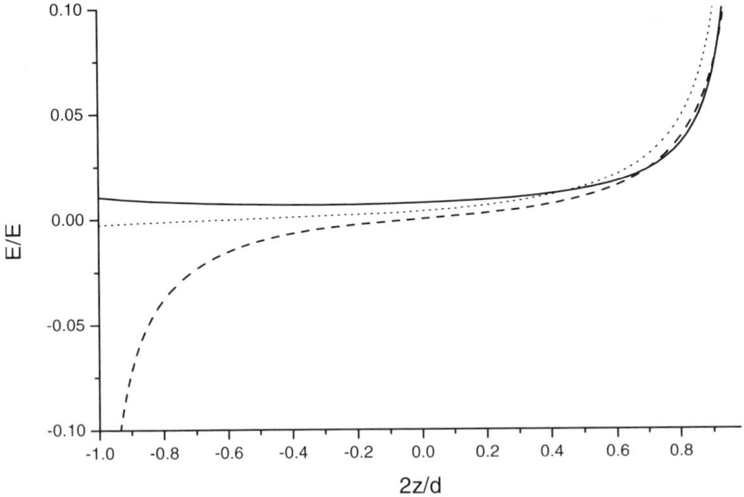

FIGURE 8.7
Reduced electric field E/E^*, where $E^* = \sqrt{2}V_T/L$ versus $2z/d$ in correspondence to the values of Σ/N considered in Figure 8.6. Reprinted with permission from Ref. [4]. Copyright (2001) by the American Physical Society.

consider these equations in the high-voltage regime. We can rewrite Eq. (8.9) and Eq. (8.20), respectively, in the forms

$$e^{-\mu} = 1 + \frac{1}{2n_0 d}\left[N_1 e^{-A_1-\psi_1} + N_2 e^{-A_2-\psi_2}\right] + \frac{L}{\sqrt{2}d} e^{-(\mu+\Delta)/2} J[\psi_1, \psi_2; c] \tag{8.22}$$

and

$$I[\psi_1, \psi_2; c] = \frac{\sqrt{2}d}{L} e^{(\mu-\Delta)/2}, \tag{8.23}$$

where

$$J[\psi_1, \psi_2; c] = \int_{\psi_1}^{\psi_2} \frac{\cosh\psi}{\sqrt{\cosh\psi + c}} d\psi, \quad \text{and} \quad I[\psi_1, \psi_2; c] = \int_{\psi_1}^{\psi_2} \frac{d\psi}{\sqrt{\cosh\psi + c}}. \tag{8.24}$$

In this manner, by using Eq. (8.24) it is possible to rewrite Eq. (8.22) as

$$e^{-\mu} = 1 + \frac{1}{2n_0 d}\left[N_1 e^{-A_1-\psi_1} + N_2 e^{-A_2-\psi_2}\right] + e^{-\Delta} \frac{J[\psi_1, \psi_2; c]}{I[\psi_1, \psi_2; c]}. \tag{8.25}$$

The definitions of the surface charge densities, Eqs. (8.4), give $\sigma_1(\Sigma) = e^{\mu-A_1-\psi_1}$, i.e.,

$$\sigma_1(\Sigma) = \frac{N_1 e^{-A_1-\psi_1}}{1 + [N_1 e^{-A_1-\psi_1} + N_2 e^{-A_2-\psi_2}]/2n_0 d + e^{-\Delta} J[\psi_1, \psi_2; c]/I[\psi_1, \psi_2; c]}. \tag{8.26}$$

For $\Sigma \to \Sigma_c^+$, $c \to -1$, $\psi_1 \to 0$, and $\psi_2 \to \psi_{2c} \gg 1$, as can be deduced from the numerical calculations reported above. Consequently,

$$\sigma_1(\Sigma_c) = \Sigma_c = \frac{N_1 e^{-A_1}}{1 + N_1 e^{-A_1}/2n_0 d + R}, \tag{8.27}$$

where

$$R = e^{-\Delta} \frac{J[0, \psi_{2c}; -1]}{I[0, \psi_{2c}; -1]}. \tag{8.28}$$

Numerical calculations show that

$$R \approx \frac{N_1 e^{-A_1}}{2n_0 d}. \tag{8.29}$$

Therefore, Σ_c is found to be

$$\Sigma_c \approx \frac{N_1 e^{-A_1}}{1 + N_1 e^{-A_1}/n_0 d}. \tag{8.30}$$

Notice that for $N_1 \to 0$ or $A_1 \to \infty$, $\Sigma_c \to 0$, as expected. In the opposite limit of $N_1 \to \infty$ or $A_1 \to -\infty$, $\Sigma_c \to n_0 d$, which corresponds to the case in which all the positive ions are adsorbed on the surface at $z = -d/2$. This behavior is illustrated in Figure 8.8 where Σ_c is depicted versus the adsorption energy A_1 as comes from the numerical calculations. Notice that the curve is quite well represented by Eq. (8.30) given above.

We have presented most of the results of the numerical calculations for an adsorption energy which is relatively low, just to emphasize the crucial role played by this quantity in a real sample. However, if we consider higher values for the adsorption energy, the magnitude of the quantities reported above is shifted, as expected, but the global behavior of the system remains the same.

8.4 Ionic adsorption in the absence of external field: Equilibrium distribution of charges

Let us consider in this section a particular case, dealing with the same system, but in the absence of external applied voltage ($\Sigma = 0$) [5]. We assume that the limiting surfaces are identical and adsorb selectively positive ions. In this case, we will consider the following approximations: $A_1 = A_2 = A$,

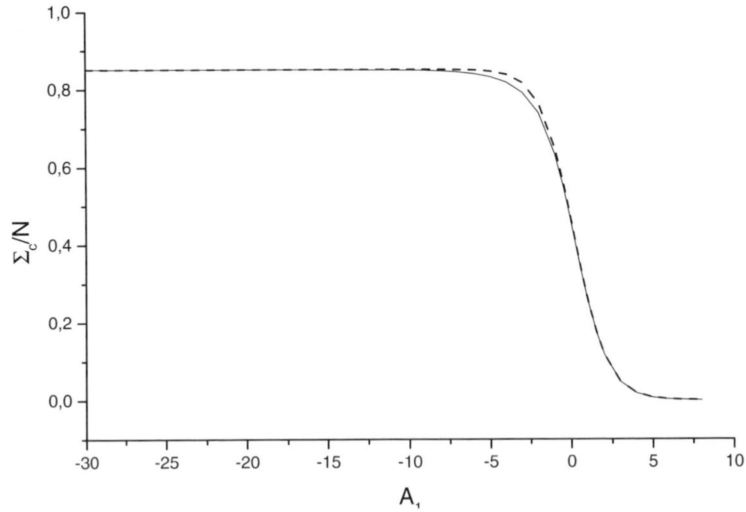

FIGURE 8.8
Critical external density of charge Σ_c versus adsorption energy A_1. Solid line corresponds to the numerical calculations; dashed line is the approximated relation represented by Eq. (8.30). Reprinted with permission from Ref. [4]. Copyright (2001) by the American Physical Society.

$N_1 = N_2 = N$, and, consequently, the surface density of charges reduces to $\sigma = \sigma_1 = \sigma_2$, and, furthermore, $\psi_1 = \psi_2 = \psi_s$. The surface adsorbed charges and the diffuse layer of oppositely charged mobile ions that they attract constitute the Debye double layer [1]. Since we do not consider the external electric field, there is a similar double layer at each wall. Likewise, we consider that $\psi_0 = \psi(0)$. The fundamental equations, in this case, are obtained as limiting case from the low-regime equations (considering that $\psi^* \to \psi_0$). The fundamental equation connecting the chemical potential with the electrical potential, Eq. (8.9), reduces to

$$e^{-\mu} = 1 + \frac{N}{n_0 d} e^{-A-\psi_s} + e^{-\Delta} \frac{1}{d} \int_{-d/2}^{d/2} \cosh\psi(z) dz. \tag{8.31}$$

The equation connecting ψ_s and ψ_0 with the chemical potential can be obtained from Eq. (8.18), and is written as

$$\int_{\psi_0}^{\psi_s} \frac{d\psi}{\sqrt{\cosh\psi - \cosh\psi_0}} = \frac{\sqrt{2}d}{2L} e^{\frac{\mu-\Delta}{2}}. \tag{8.32}$$

Finally, the equations coming from the boundary conditions can be obtained from Eq. (8.19) which now reduces to a single equation in the form

$$e^{\mu} = 2\left(\frac{\epsilon k_B T}{NLq^2}\right)^2 e^{-\Delta+2(A+\psi_s)}(\cosh\psi_s - \cosh\psi_0). \tag{8.33}$$

The fundamental equations of this particular case are now Eqs. (8.31), (8.32) and (8.33). They connect ψ_0, ψ_s and μ. When these equations are solved, as before, the surface density σ, which is due to the adsorption phenomenon, can be calculated from the expression

$$\sigma_1 = \sigma_2 = \sigma = Ne^{\mu-A-\psi_s}, \tag{8.34}$$

which is easily obtained from Eq. (8.4) for this particular case.

8.4.1 Limits of small and large thickness

Let us assume, as usual, $\Delta \gg 1$, i.e., $E_{ac} \gg k_B T$. We consider first the limit of small thickness ($d \to 0$). This implies, as it will be verified *a posteriori*, $\psi_s \gg 1$, $\psi_0 \gg 1$, and $\psi_s - \psi_0 =$ small. In this special case from the general formulae (8.31), (8.32), and (8.33) we obtain

$$\psi_s - \psi_0 \approx (d/2L)^2 e^{\mu-\Delta+\psi_0}, \tag{8.35}$$

and

$$\psi_s \approx \frac{\Delta - A}{2} + \frac{1}{2}\ln\left(\frac{N}{n_0 d}\right) + \frac{1}{2}\ln 2 + \mathcal{O}(d^2). \tag{8.36}$$

Equations (8.35) and (8.36) show that, in the considered limit of large Δ and small d, both ψ_s and ψ_0 are large quantities, such that $\psi_s - \psi_0 = \mathcal{O}(d^2)$, as previously assumed. We have furthermore for the chemical potential μ

$$\mu = \frac{A+\Delta}{2} - \frac{1}{2}\ln\left(\frac{N}{n_0 d}\right). \tag{8.37}$$

Equations (8.35), (8.36) and (8.37) solve the adsorption problem in the limit of small thickness. By substituting these equations in the expression for σ given above, we obtain for the surface charge density the expression

$$\sigma = q\,\frac{(Nn_0 d/2)^{1/2} e^{-\frac{\Delta+A}{2}}}{1 + (2N/n_0 d)^{1/2} e^{-\frac{\Delta+A}{2}}}, \tag{8.38}$$

that, in the limit of $|-(\Delta + A)/2| \gg 1$ tends to

$$\sigma = q\frac{n_0 d}{2}. \tag{8.39}$$

Let us consider now the limit $d \to \infty$. In this limit ψ_0 and ψ_s are expected to tend to a constant value. Consequently, from the general equations

(8.31), (8.32), and (8.33) we have that $\lim_{d\to\infty}\psi_0 = 0$. In this framework, by assuming again $\psi_s \gg 1$, we get

$$\psi_s = \frac{2}{3}\ln\left(\sqrt{2}\frac{N}{n_0 L}\right) + \frac{\Delta}{3} - \frac{2}{3}A. \tag{8.40}$$

By substituting Eq. (8.40) into the expression for σ we obtain

$$\sigma = Nq\left(\frac{n_0 L}{N}\right)^{2/3} e^{-\frac{\Delta+A}{3}}, \tag{8.41}$$

which is thickness independent. Equations (8.39) and (8.41) show that in the limit of small d the surface charge density σ is proportional to d, whereas in the opposite case of large d it tends to a constant value.

8.4.2 Arbitrary thickness

Let us consider now the solution of the adsorption problem for arbitrary thickness of the sample. The solution has to be searched by numerically solving the set of three coupled non-linear equations (8.31), (8.32) and (8.33) for a significant set of parameters Δ, A, N, and n_0. We assume $q = 2e$, where e is the modulus of the electronic charge, $r_+ = 10\,\text{Å}$, $r_- = 30\,\text{Å}$, $\epsilon_1 \approx 4$, which is typical for organic liquid, and $\epsilon_2 \approx 6$, which refers to a glass. A representative estimation for the parameters entering in the model can be obtained by considering a typical nematic liquid crystal sample of slab shape limited by two glasses. The typical dimension of a molecule is $R \simeq 40\text{Å}$ and $n_0 \approx 1/(4/3)\pi R^3$. Furthermore $N \approx 1/\pi R^2$. From the definition of L, given by Eq. (8.12), one obtains $L \simeq 30\,\text{Å}$. In this framework $A = E_{\text{ad}}/k_B T \approx -6$ and $\Delta = E_{\text{ac}}/k_B T \approx 18$. Note that if $\epsilon_2 \to \infty$, i.e., if the substrate is a metal, the adsorption energy tends to $A = -28.8$. By means of these values the Debye's screening length λ_0 is found to be of the order of the micron, which is consistent with the values reported in literature. More precise estimations can be performed, but the general results do not change in a significant manner, as we have pointed out before. In Figure 8.9, the chemical potential μ versus the thickness of the sample d is reported.

In the limit of small d, μ presents a logarithmic divergence, in agreement with Eq. (8.37). In the opposite limit of large d, μ tends to zero, as expected. The electrical potentials at the surface, ψ_s, and in the middle of the sample, ψ_0, versus d are shown in Figure 8.10. For small d, ψ_s and ψ_0 are large, but $\psi_s - \psi_0 \to 0$ as $d \to 0$. For large d, $\psi_0 \to 0$ and ψ_s tends to a constant value.

In Figure 8.11, the trend of the surface charge density σ versus the thickness d is shown. The global behavior is the expected one from the limiting cases discussed above. There is a linear behavior for small d and a clear saturation for large d. The general trend of $\sigma = \sigma(d)$ recalls the Langmuir isotherm [7]. However, as we discussed in Section 5.2.3, in the Langmuir problem of adsorption, the mutual interaction among the particles is neglected. On the contrary,

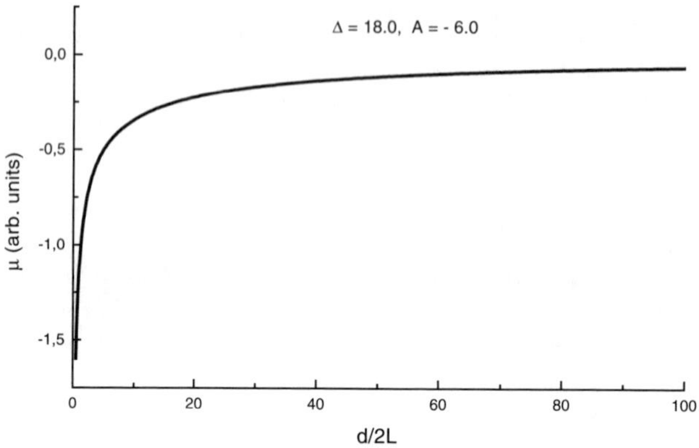

FIGURE 8.9
Chemical potential μ versus the thickness of the sample d. Reprinted with permission from Ref. [5]. Copyright (1999) by the American Physical Society.

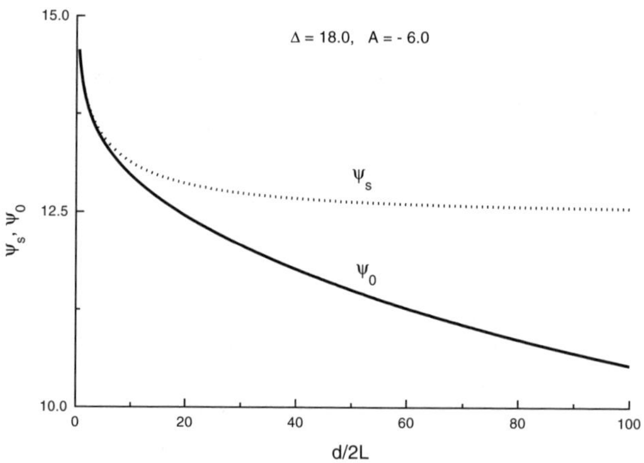

FIGURE 8.10
Electrical potential at the surface, ψ_s, and in the middle of the sample, ψ_0, versus the thickness of the sample d. Reprinted with permission from Ref. [5]. Copyright (1999) by the American Physical Society.

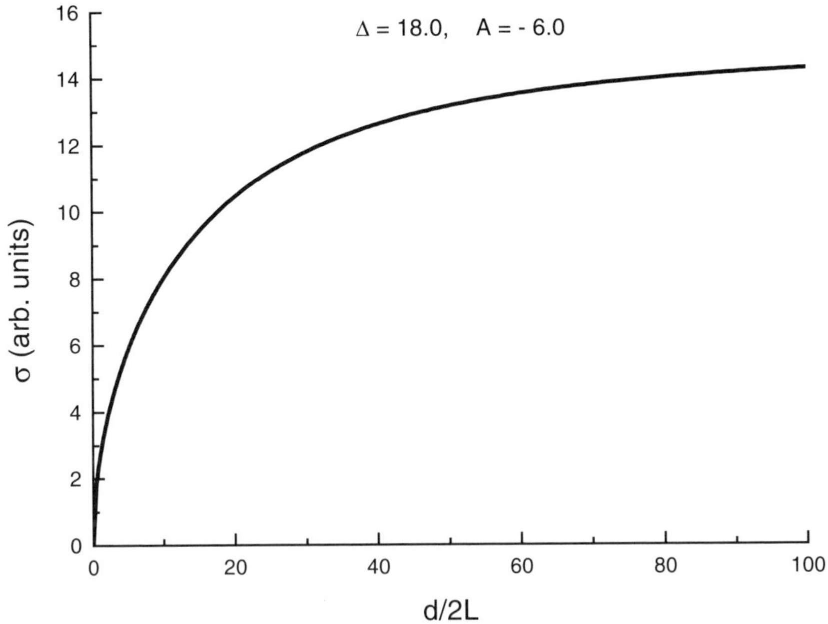

FIGURE 8.11
Surface density of adsorbed positive ions, σ, versus the thickness of the sample d. Reprinted with permission from Ref. [5]. Copyright (1999) by the American Physical Society.

in the model this interaction is explicitly taken into account by means of the electrical potential entering in the definitions of the charge distributions.

Before we proceed, it is convenient to briefly discuss some particular cases which can be analyzed in the framework of the proposed model. The first one is the situation in which the adsorption energies are different on the surfaces $A_1 \neq A_2$ (asymmetric case), but the external voltage is absent. The fundamental equations of the model are easily obtained from the ones presented before by taking the limit $\Sigma \to 0$. A practical situation where this model has revealed its usefulness is the asymmetric electro-optical response in a nematic liquid crystal cell [8, 9]. Another limiting case to be considered refers to the situation in which the adsorption phenomenon is absent, but the sample is submitted to an external voltage. In this case we have $A_1 = A_2 \to \infty$. Consequently, $\sigma_{\rm i} \to 0$, as given by (8.4). In this case, as it follows from Eq. (8.21), $\psi_1 = -\psi_2 = \Psi$. Again, the fundamental equations of the model, now reduced to three coupled equations, can be easily obtained from the previous ones. They will connect μ, Ψ and c with Σ and d.

8.5 Extensions of the model

In the preceding analysis, we have considered a selective ion adsorption phenomenon in which only the positive ions are supposed to be adsorbed on the surface. This implies that the adsorption energy for the negative ones is infinite. Let us now consider the case in which both positive and negative ions can be adsorbed by the surfaces. We suppose, for simplicity, that the surfaces are identical, and A_+ and A_- are the adsorption energies for both species of ions. The surface density of adsorbed ions of a given sign is given by a natural extension of Eq. (8.4):

$$\sigma_{i,\pm} = N_\pm e^{\mu - A_\pm \mp \psi_i}, \tag{8.42}$$

where $i = 1, 2$ refers to the surfaces and $N_\pm$ are the surface densities of sites for + and − ions. We assume, furthermore, that $N_+ = N_- = N$. The actual surface density of adsorbed ions is given by

$$N_i = N e^{\mu}(e^{-A_+ - \psi_i} + e^{-A_- + \psi_i}), \tag{8.43}$$

whereas the actual surface charge density is

$$Q_i = q(\sigma_{i,+} - \sigma_{i,-}) = q\sigma_{\rm i}. \tag{8.44}$$

The equation representing the conservation of the number of particles, Eq. (8.6), is slightly modified, giving for the chemical potential

$$\begin{aligned} e^{-\mu} &= 1 + \frac{N}{2n_0 d}\left[e^{-A_+}\left(e^{-\psi_1} + e^{-\psi_2}\right) + e^{-A_-}\left(e^{\psi_1} + e^{\psi_2}\right)\right] \\ &+ \frac{e^{-\Delta}}{d}\int_{-d/2}^{d/2} \cosh\psi(z) dz. \end{aligned} \tag{8.45}$$

The basic equations of the model are then (8.18), (8.19), and (8.45) for the low-voltage regime, and (8.20), (8.21), and (8.45) for the high-voltage regime, if we consider in these equations $\sigma_{1,2}$ as defined in (8.44).

8.6 Effect of the adsorption energy on the anchoring energy of dielectric origin

In this section we shall consider two particular situations to investigate the direct effect of the adsorption energy on the electric field distributions in the sample and, consequently, on the anchoring energy of dielectric origin [10]. The basic equations of the model are numerically solved to obtain μ, ψ_1, ψ_2 and ψ^* (low-voltage regime) or c (high-voltage regime). Once these quantities are determined it is possible to establish the profile of $E(z)$ for different values of Σ and $A_\pm$ and also d as we have done in the preceding sections. We consider again two particular cases

8.6.1 Case I: $A_+ = A \neq 0$, $A_- \to \infty$, $\Sigma = 0$

This case refers to a situation in which only positive charges are adsorbed, in the absence of external voltage. This is the case we have considered in details in Section 8.4, where the behavior of μ, ψ_s and ψ_0 as a function of the thickness of the sample d was investigated. Here, our attention will be devoted to the calculation of the anchoring energy. This simplified case is of particular importance because it permits us to evaluate the direct effect of an adsorption energy on the anchoring energy of dielectric origin in the absence of external charges.

In Figure 8.12, the behavior of $W = W_{\rm E} = W_{\rm D} + W_{\rm Q}$ as a function of the thickness of the sample d is shown. The trend is in good agreement with the data from Ref. [11] as discussed in Section 6.1.3. However, two approximations were made. The first one considered an exponential decreasing distribution for the electric field in the sample; the second one considered an approximated expression for the surface density of charges as a function of the thickness of the sample. The present model removes these simplifying hypotheses.

In Figure 8.13, the behavior of the anchoring energy of dielectric origin $W = W_{\rm E}$ as a function of the adsorption energy for positive charges is shown. When the adsorption energy is not very high the order of magnitude of W agrees with the ones usually found ($W \approx 10^{-2}$ erg/cm^2). When the adsorption energy is very high, W tends to a constant value, independent of the value of A. This value is of the order of few erg/cm^2 and corresponds practically to a strong anchoring situation. This result indicates again that the ion adsorption can play a fundamental role in establishing the correct order of magnitude of W.

8.6.2 Case II: $A_+ = A \neq 0$, $A_- \to \infty$, $\Sigma \neq 0$

In this case, we have again only adsorption of positive charges, but now in the presence of an external voltage. In Figure 8.14, an illustrative result when the external density of charge is $\Sigma/N = 0.6$, is presented. The figure shows the behavior of $W = W_{\rm E}$ as a function of d. This situation has to be compared with the one depicted in Figure 8.12, where the external charges are absent. Notice that the effect of an external electric field strongly affects the magnitude of the anchoring energy. This result is in complete agreement with the ones established in Section 7.2, where it was demonstrated that the anchoring energy is bias-voltage dependent.

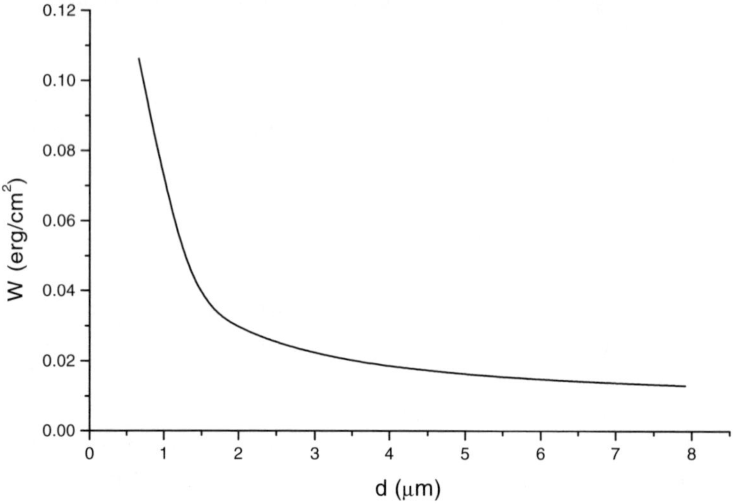

FIGURE 8.12
Anchoring energy $W = W_\mathrm{E} = W_\mathrm{D} + W_\mathrm{Q}$ versus the thickness of the sample d, in the absence of external applied voltage $\Sigma = 0$. The curve was plotted for $\epsilon_a = 14\epsilon_0$, $e = 4 \times 10^{-11}$ C/m [12], $\lambda_0 = 0.6\ \mu$m [3], and $A = -0.3$. Reprinted from Ref. [10] with the permission of Sociedade Brasileira de Física.

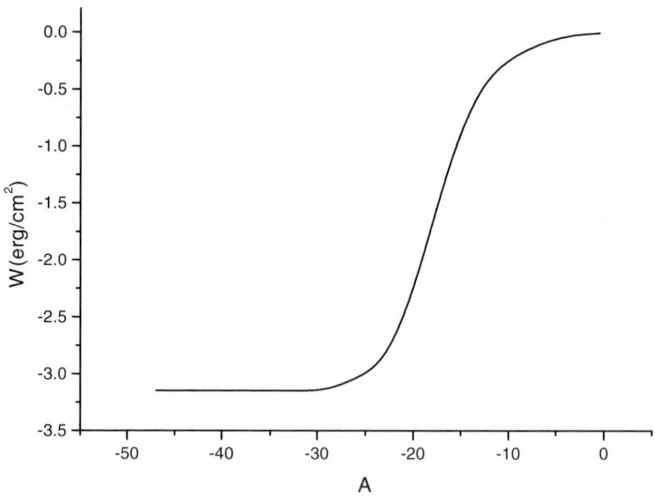

FIGURE 8.13
Anchoring energy W_E versus the adsorption energy of positive charges $A = A_+$ in the absence of external applied voltage $\Sigma = 0$. The material parameters are the same as in Figure 8.12. Reprinted from Ref. [10] with the permission of Sociedade Brasileira de Física.

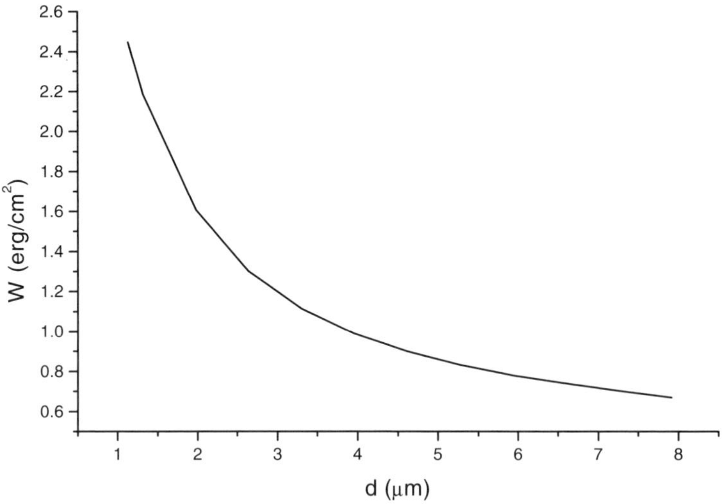

FIGURE 8.14
The same as in Figure 8.12, in the presence of external applied voltage $\Sigma/N = 0.6$. The material parameters are the same as in Figure 8.12. Reprinted from Ref. [10] with the permission of Sociedade Brasileira de Física.

In Figure 8.15, the quantity $W = W_{\rm E}$ is plotted as a function of the adsorption energy A, also in the presence of an external charge density $\Sigma/N = 0.6$. This figure illustrates the combined effect of an external voltage and an adsorption energy in the behavior of $W = W_{\rm E}$. Both effects act to increase the magnitude of W. Again we have a saturation value for W for large values of A, corresponding to a situation of strong anchoring (when $A \to -\infty$).

Finally, just to show the effect of the adsorption of negative charges on the net surface charge density, this quantity is exhibited as a function of the thickness of the sample in two cases in the absence of external applied voltage. In Figure 8.16, $\sigma/N = \sigma_1/N = \sigma_2/N$ is shown as function of the d for the case $A_+ = -0.4$ and $A_- \to \infty$. In this case, only the adsorption of positive charges is considered. One observes that the behavior of σ is linear with d, for small d, and tends to a value which is independent of d, for very large values of d, as discussed in Section 8.2.

In Figure 8.17, the same quantity is exhibited as a function of d for the case $A_+ = -0.4$ and $A_- = -1.0$. The global behavior is similar, in the sense that there is a linear behavior for small d and a saturation value for large d. However, as expected, the order of magnitude of the density in this case is less than in the preceding figure, because the net charge density is given by $\sigma = \sigma_+ - \sigma_-$ [see Eq. (8.44)]. Furthermore, in this situation σ tends to a saturation value only for very large values of d, as compared with the previous case where the saturation is abrupt.

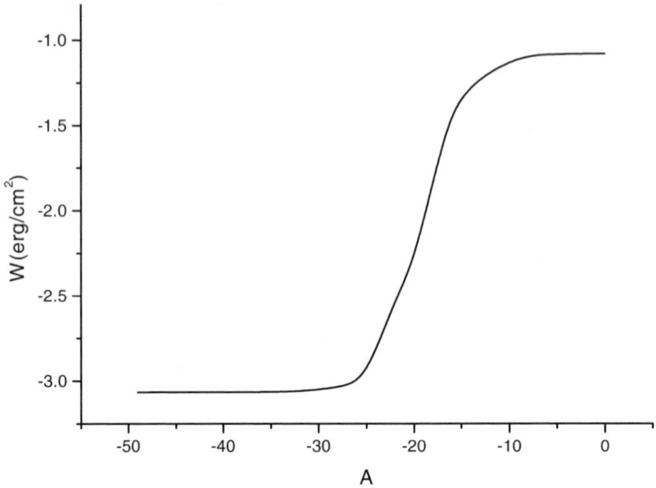

FIGURE 8.15
The same as in Figure 8.13, in the presence of external applied voltage $\Sigma/N = 0.6$. The material parameters are the same as in Figure 8.12. Reprinted from Ref. [10] with the permission of Sociedade Brasileira de Física.

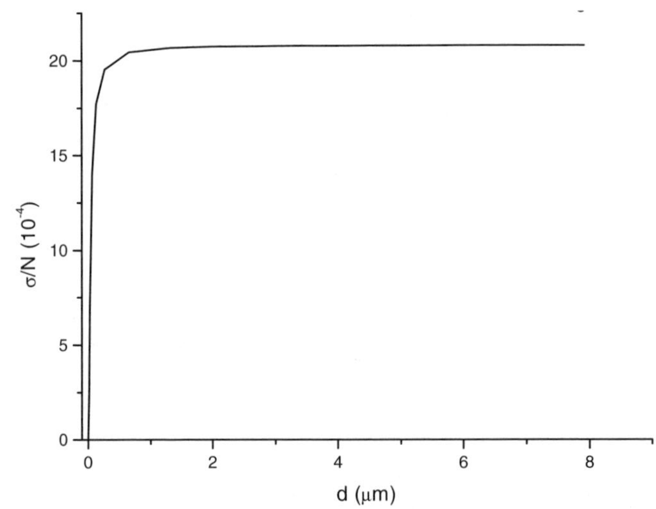

FIGURE 8.16
Surface charge density σ versus the thickness of the sample in the case of adsorption of only positive charges $A_+ = -0.4$ ($A_- \to \infty$). The material parameters are the same as in Figure 8.12. Reprinted from Ref. [10] with the permission of Sociedade Brasileira de Física.

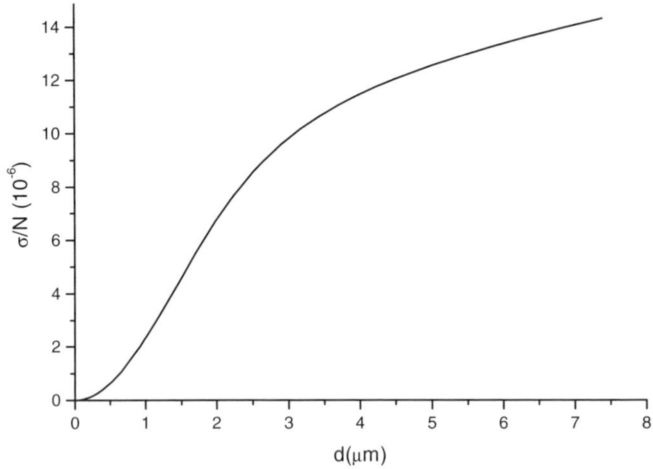

FIGURE 8.17
Surface charge density σ versus the thickness of the sample in the case of adsorption of positive (with $A_+ = -0.4$) and negative (with $A_- = -1.0$) charges. The material parameters are the same as in Figure 8.12. Reprinted from Ref. [10] with the permission of Sociedade Brasileira de Física.

8.7 Contribution of the ion adsorption phenomenon to the effective anchoring energy

In this section, the effect of the surface electric field, produced by selective ion adsorption, on the effective anchoring energy of an NLC sample is analyzed by assuming that the majority of the positive charges is adsorbed [1]. In Section 8.2, we have established the set of equations governing the spatial behavior of the electric field distribution whose origin is due to the ion adsorption by a surface. This set of equations is considerably simplified if one considers the limit in which practically all the positive charges are adsorbed at the surfaces [13]–[15]. This simplified analysis has also the advantage to establish, by means of the boundary conditions on the electric field used to solve the electrostatic problem, an effective non-Debye's screening length. It is similar to the one presented in Refs. [16]–[18], but here it is explicitly written in terms of the chemical potential and the adsorption energy characterizing the surface. The immediate consequence of the non-Debye screening is that the flexoelectric contribution to the anchoring energy can be the dominant one.

We consider that the sample is limited by identical surfaces, in the absence of external electric field. In this case, the equations governing the problem are the ones established in Section 8.4, namely Eqs. (8.31), (8.32) and (8.33).

Notice that from the condition

$$E(-d/2) = q\,\frac{\sigma}{\epsilon}, \tag{8.46}$$

rewritten in terms of ψ, and using Eq. (8.34), we have

$$\left(\frac{d\psi}{dz}\right)_{z=-d/2} = -\frac{N\,q}{\epsilon\,V_T} = -\frac{1}{\lambda_s}e^{-\psi_s}. \tag{8.47}$$

The quantity

$$\lambda_s = \left(\frac{\epsilon V_T}{Nq}\right) e^{-\mu+A}$$

has the dimension of a length, and coincides with the length introduced in Refs. [16, 18], and defined in (7.6), but now taking into account the chemical potential and the adsorption energy. The above set of equations has to be solved to obtain the electric field distribution across the sample [4]. Numerical calculations have shown that the curves for several different values of A can be well represented by a function of the form

$$E(Z) = \alpha \tan\left[\frac{\beta\pi Z}{2}\right], \tag{8.48}$$

for $-1 \leq Z(= 2z/d) \leq 1$, where α and β are, in principle, depending on the adsorption energy characterizing the surfaces, and on the thickness of the sample (see below). Therefore, for the cases we are dealing with here, the solution (8.48) can be considered as the exact one, as we shall demonstrate.

The Poisson-Boltzmann equation (8.11) can be approximated by

$$\frac{d^2\psi}{dz^2} = \frac{1}{L^2}e^{\mu-\Delta}\sinh\psi \approx \frac{1}{2L^2}e^{\mu-\Delta}\,e^{\psi}, \tag{8.49}$$

when we consider $\psi = V/V_T \gg 1$. This means that the number of positive charges remaining in the bulk (not adsorbed) is very small, as can be seen from the definition of $n_\pm(z)$ introduced before. Equation (8.49) can be integrated to give

$$\frac{d\psi}{dz} = \mp\frac{1}{L}\,e^{(\mu-\Delta)/2}\sqrt{e^{\psi} - e^{\psi_0}}, \tag{8.50}$$

where "$-$" refers to the interval $-d/2 \leq z < 0$ and "$+$" to $0 < z \leq d/2$. The electric potential can be obtained from (8.50) by further integration, giving

$$t[\psi, \psi_0] = -\frac{1}{2\,L}e^{(\mu-\Delta+\psi_0)/2}\left(z + \frac{d}{2}\right) + t[\psi_s, \psi_0], \tag{8.51}$$

where $t[a, b] = \tan^{-1}\left[\sqrt{e^{a-b} - 1}\right]$. From Eq. (8.51) one obtains (at $z = 0$)

$$t[\psi_s, \psi_0] = \frac{d}{4L} e^{(\mu-\Delta+\psi_0)/2}. \tag{8.52}$$

By using (8.52) into (8.51), one explicitly obtains the electric potential as

$$\psi(z) = \psi_0 + \ln\left\{1 + \tan^2\left[\frac{z}{2L} e^{(\mu-\Delta+\psi_0)/2}\right]\right\}. \tag{8.53}$$

In this manner, one can also determine the electric field distribution

$$E(z) = -\frac{4V_T}{d} t[\psi_s, \psi_0] \tan\left[\frac{2z}{d} t[\psi_s, \psi_0]\right], \tag{8.54}$$

where (8.52) has been used. Comparison of (8.54) with (8.48) yields:

$$\alpha = -\frac{4V_T}{d} t[\psi_s, \psi_0] \quad \text{and} \quad \beta = \frac{2}{\pi} t[\psi_s, \psi_0]. \tag{8.55}$$

To complete the solution of the entire electrostatics problem in the approximation in which most of the positive charges are adsorbed, we have to take into account (8.46). By using (8.33) and (8.54) we obtain another relation connecting μ, ψ_0 and ψ_s in the form

$$e^{\mu-A-\psi_s} = \frac{4\epsilon k_B T}{q^2 N d} t[\psi_s, \psi_0] \tan t[\psi_s, \psi_0]. \tag{8.56}$$

Finally, the last expression is obtained from (8.31) which reduces to

$$e^{-\mu} = 1 + \frac{N}{n_0 d} e^{-A-\psi_s} + e^{\psi_0-\Delta} \frac{\tan t[\psi_s, \psi_0]}{2\, t[\psi_s, \psi_0]}. \tag{8.57}$$

The set formed by the Eq. (8.52), (8.56) and (8.57) permits us to determine μ, ψ_0 and ψ_s as a function of the adsorption energy A and of the thickness of the sample d, in equilibrium at a temperature T. The sensibility of ψ_s with respect to changes in d is not very significant, for a fixed value of the adsorption energy, because, in the limit we are considering, the surface density of adsorbed charges given by Eq. (8.34) reaches a saturation, i.e., tends to a value independent of d [5]. On the contrary, ψ_0 changes in a substantial manner. The system, in any case, remains neutral, but an overwhelming adsorption of positive charges takes place. The situation is such that now ψ_s changes in a substantial manner with A for a fixed thickness, whereas ψ_0 is less sensible to the changes in A. Notice that in this limit $\mu \to 0^-$. In fact, by using Eq. (8.52) and (8.56), it is possible to rewrite (8.57) in the form

$$e^{-\mu} = 1 + \frac{2N}{n_0\, d} e^{-A-\psi_s},$$

which shows that when $d \to \infty$, $\mu \to 0^-$, in agreement with the behavior presented in Figure 8.9.

To analyze the influence of the ion adsorption on the anchoring energy of the sample we have to remember that, according to the analysis presented above, to the surface field of adsorption origin are connected the two contributions W_D and W_E to the surface energy. By using Eq. (8.54), (8.55), and (8.56), with the definitions of L and λ_s, in (7.92) and (7.93), we obtain, respectively, for the dielectric and flexoelectric contributions to the anchoring strengths, the expressions

$$W_{\rm D} = \frac{1}{2}\epsilon_a V_T^2 \left[\left(\frac{n_0 d}{N\lambda_s} \right) e^{A-\Delta+\psi_0} - \frac{2e^{-\psi_s}}{\lambda_s} \right] \quad \text{and} \quad W_{\rm Q} = -\frac{e}{\lambda_s} V_T \, e^{-\psi_s}. \tag{8.58}$$

Equations (8.58) show that the effective anchoring energy of a sample can be thickness dependent as found in many experimental results [11]. Moreover, these equations also show that the anchoring energy is renormalized by the contribution coming from the electrical properties whose origin is connected with the presence of ions in the sample. This dependence is here represented also by the adsorption energy. Equations (8.58) generalize Eqs. (7.94) and (7.95) for the case in which the adsorption energy is taken into account.

In Figure 8.18, the dependence of $W_{\rm E} = W_{\rm D} + W_{\rm Q}$ on the adsorption energy A is shown for two typical NLC samples. Changing the thickness of the sample the trend of $W_{\rm E}$ does not change in a significant manner. However, the absolute value of the anchoring strength changes in a substantial manner with A. Changing A one order of magnitude implies changing the anchoring strength by the same amount. Moreover, the order of magnitude of $|W_{\rm E}|$ indicates that this contribution cannot be neglected because it can be of the same order as the "bare" anchoring strength [10].

8.8 Destabilizing effect of a surface electric field generated by ion adsorption on the molecular orientation of nematic liquid crystals

By using the expression for the electric field (8.54) valid in the approximation discussed above, we can now evaluate the critical value of the surface field (or adsorbing energy) for which the initial homeotropic alignment becomes unstable, as we have done in Section 6.2. We consider a nematic liquid crystal sample limited by two adsorbing surfaces located at $z = \pm d/2$. The director $\mathbf{n}$ is defined by the tilt angle $\theta(z) = \cos^{-1}(\mathbf{n}\cdot\mathbf{k})$, where $\mathbf{k}$ is the normal to the bounding plates, and coincides with the easy direction, $\mathbf{n}_0$, for homeotropic orientation. We consider a general electrostatic problem incorporating the dielectric and flexoelectric properties of the medium, when only splay-bend

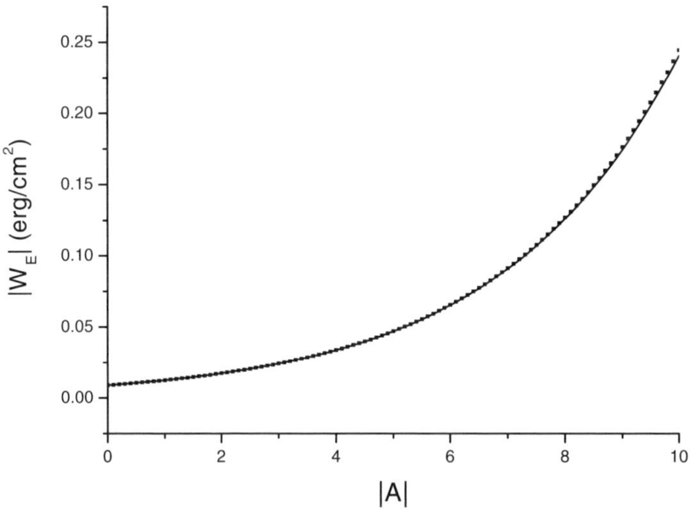

FIGURE 8.18
Anchoring energy of dielectric origin, $|W_{\mathrm{E}}|$ versus the adsorption energy (absolute value) for a sample of thickness $d = 8\,\mu$m (solid line) and $d = 15\,\mu$m (dotted line). The material parameters are (in SI units) $K = 4 \times 10^{-11}$, $e = 10^{-11}$ and $\epsilon_a = \epsilon_0$. Reprinted with permission from Ref. [13]. Copyright (2003) by the American Physical Society.

distortions are allowed in the system [19, 20]. In this framework the bulk free energy density in the one-constant approximation is given by

$$f = \frac{1}{2} K \theta'^2(z) - \frac{\epsilon_a}{2} E^2(z) \cos^2 \theta(z) + \frac{e}{2} \sin 2\theta(z) \theta'(z) E(z), \tag{8.59}$$

which, in the limit $\theta \to 0$, reduces to (6.76). The contribution of the surface to the free energy density for identical surfaces is written, by means of (2.9), in the form

$$f_S = -\frac{1}{2} W \cos^2 \theta_1 - \frac{1}{2} W \cos^2 \theta_2 \approx \frac{1}{2} W \theta_1^2 + \frac{1}{2} W \theta_2^2,$$

where, as before, $\theta_1 = \theta(z = -d/2)$, $\theta_2 = \theta(z = d/2)$. The total energy per unit area is given by $F = \int_{-d/2}^{+d/2} f\, dz + f_S$. Minimization of the total energy yields the differential equation governing the equilibrium profile represented by $\theta(z)$. Therefore, we have to find solutions to the following Euler-Lagrange equation:

$$K\theta''(z) - \epsilon_a E^2(z)\theta(z) + eE'(z)\theta(z) = 0. \tag{8.60}$$

Equation (8.60) has to be solved by satisfying the (also linearized) boundary conditions (6.35), i.e.,

$$\left[\frac{d\theta}{dz}\right]_{z=\mp d/2} \mp \left(\frac{W}{K} + \frac{e\, E(\mp d/2)}{K} \right) \theta(\mp d/2) = 0. \tag{8.61}$$

It is now possible to study the stability of the homeotropic pattern in the presence of the electric field distribution presented above. We have to search for solutions of Eq. (8.60) by taking into account that the field distribution is given by Eq. (8.54). Equation (8.60) can be rewritten in terms of the variable $Z = 2z/d$ in the form:

$$\theta'' + \left[R_2 - \frac{R_1 + R_2}{\cos^2 t\, Z} \right] \theta = 0, \tag{8.62}$$

where

$$R_1 = -2V_T \frac{e\, t^2}{K} \quad \text{and} \quad R_2 = V_T \frac{\epsilon_a t^2}{\pi\, K}. \tag{8.63}$$

By means of the transformation $y = \sin^2 t\, Z$, Eq. (8.62) can be identified with the Schrödinger equation for the Pöschl-Teller potential hole [21]

$$y(1-y)\theta'' + \left(\frac{1}{2} - y \right) \theta' + \frac{1}{4} \left[\frac{k^2}{t^2} - \frac{\lambda(\lambda - 1)}{1 - y} \right] \theta = 0, \tag{8.64}$$

where $k^2 = R_2$ and $\lambda = 1/2[1 + \sqrt{1 + 4(R_1 + R_2)/t^2}]$. If we now perform a further transformation, in the form

$$\theta = (1 - y)^{\lambda/2} u(y), \tag{8.65}$$

Eq. (8.64) becomes the hypergeometric differential equation [22]

$$y(y-1)u'' - t^2 \left[1 - 2(\lambda+1)y\right] u' - \frac{1}{2}\left[k^2 - \lambda^2 t^2\right] u = 0. \tag{8.66}$$

The general solution, Eq. (8.65), has the form [22]

$$\theta(Z) = (1-y)^{\lambda/2} \left[C_1 F(a,\, b,\, c,\, y) + i\, C_2 \sqrt{y}\, F(a-c+1,\, b-c+1,\, 2-c,\, y)\right], \tag{8.67}$$

where $F(a,b,c;y) = {}_2F_1(a,b,c;y)$ is the hypergeometric function, with

$$a = \frac{1}{2}\left(\lambda - \frac{k}{t}\right), \quad b = \frac{1}{2}\left(\lambda + \frac{k}{t}\right), \quad \text{and} \quad c = \frac{1}{2}.$$

Since we are interested in real solutions, without loss of generality, we can choose $C_2 = 0$, because in the interval $-1 < Z < 1$ the second term is imaginary. In this manner, the general problem represented by Eq. (8.62) has been solved in an exact manner, for an electric field distribution which is exact when the amount of adsorbed positive charges is very high. This solution can be used now in the boundary conditions to give the threshold values of the surface electric field. In fact, since we are considering identical surfaces, the boundary condition at $Z = -1$, i.e., the first of Eq. (8.61), gives now the threshold field $E(-d/2) = E_{\rm th}$. It can be put in the form

$$E_{\rm th} = \frac{4K}{e\,d}\left[\frac{\theta'(Z)}{\theta(Z)}\right]_{Z=-1} - \frac{W}{e}. \tag{8.68}$$

Equation (8.68), by taking into account $\theta(Z)$ as given by Eq. (8.67), is of utmost importance in this context. It establishes, in closed analytic form, the dependence of the threshold field on the thickness of the sample and on the adsorption energy, by means of the dependencies on these quantities that are found in ψ_S and ψ_0. It represents the generalization of the Rapini-Papoular expression for the threshold field in the Fréedericksz transition for the case of weak anchoring [19] and taking into account the flexoelectric effect. An expression of this kind can be the key to explain the observed spontaneous Fréedericksz transition induced by surface field, which is, in fact, a flexoelectric instability induced by the electric field due to the adsorption phenomenon.

In Figure 8.19, the dependence of the critical field on A is shown in the situation of weak anchoring for $e = \pm 5 \times 10^{-11}$ C/m. Due to the dependence

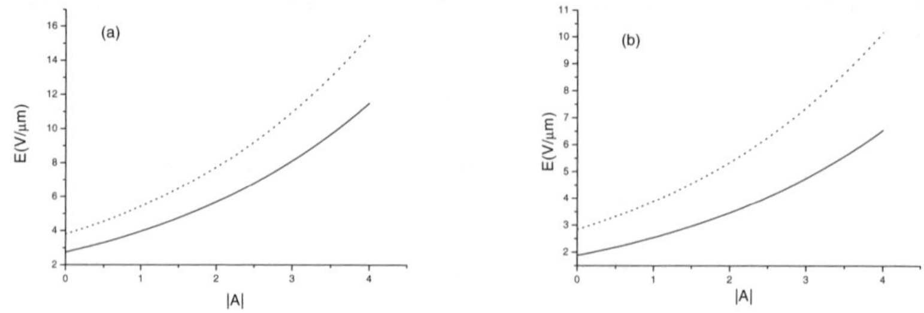

FIGURE 8.19
Critical field, E_{th}, as predicted by Eq. (8.68) versus the adsorption energy $|A|$ for a sample whose thickness is $d = 8\,\mu$m. Curve (a) refers to $e = 5 \times 10^{-11}$ C/m and (b) to $e = -5 \times 10^{-11}$ C/m. In both figures solid lines were calculated for the "bare" anchoring energy , i.e., $W = W_0 = 10^{-5}$ J/m^2 and dotted lines for the effective anchoring energy, given by $W_{\text{eff}} = W_0 + W_{\text{D}} + W_{\text{Q}}$ [see Eq. (8.58)]. The other parameters are $K = 10^{-11}$ N, $\epsilon_a = 10\epsilon_0$, $\Delta = 8.0$ and $\lambda_0 = 0.58\,\mu$m.

of ψ_{S} and ψ_0 on A, E_{th} is expected to be also dependent on A. The solid lines refer to a fixed value for the ("bare") anchoring energy $W_0 = 10^{-5}$ J/m^2. Actually, the anchoring energy is renormalized by the surface electric field generated by the adsorption of ions, giving rise to an effective anchoring energy that depends also on the adsorption energy characterizing the surface. The effective energy is given by $W_{\text{eff}} = W_0 + (W_{\text{D}} + W_{\text{Q}})$, where W_{D} and W_{Q} are given by Eqs. (8.58). These expressions show that the anchoring energy is not localized on the surface, but penetrates into the bulk, along a distance represented by λ_s. The results shown in the dotted lines of Figure 8.19 are consistent with the fact that the threshold field increases for increasing A.

[1] Israelachvili J. *Intermolecular and Surface Forces.* Academic Press, London, 1998.

[2] Thurston RN, Cheng J, Meyer RB, and Boyd GD. Physical mechanisms of DC switching in a liquid crystal bistable boundary layer display. *Journal of Applied Physics*, **56**, 263 (1984).

[3] Thurston RN. Equilibrium distributions of electric field in a cell with adsorbed charge at the surfaces. *Journal of Applied Physics*, **55**, 4154 (1984).

[4] Evangelista LR and Barbero G. Adsorption phenomenon and external field effect on an isotropic liquid containing impurities. *Physical Review E*, **64**, 021101 (2001).

[5] Barbero G, Zvezdin AK, and Evangelista LR. Ionic adsorption and equi-

librium distribution of charges in a nematic cell. *Physical Review E*, **59**, 1846 (1999).

[6] Landau L and Lifchitz EI. *Electrodymamique des Milieux Continus*. MIR, Moscow, 1969.

[7] Kubo R. *Statistical Mechanics*. North Holland, Amsterdam, 1967.

[8] Strangi G, Lucchetta DE, Cazzanelli E, Scaramuzza N, Versace C, and Bartolino R. Asymmetric electro-optical response in a liquid crystal cell containing a layer of amorphous tungsten trioxide. *Applied Physics Letters*, **74**, 534 (1999).

[9] Barbero G, Evangelista LR, and Olivero D. Asymmetric ionic adsorption and cell polarization in liquid crystals. *Journal of Applied Physics*, **87**, 2646 (2000).

[10] Pereira HA, Evangelista LR, Olivero D, and Barbero G. Influence of the adsorption energy on the dielectric contribution to the anchoring energy of nematic liquid crystals. *Brazilian Journal of Physics*, **32**, 2B, 584 (2002).

[11] Blinov LM, Kabayenkov AY, and Sonin AA. Experimental studies of the anchoring energy of nematic liquid crystals–Invited lecture. *Liquid Crystals*, **5**, 645 (1989).

[12] Barbero G, Chuvyrov AN, Krekhov AP, and Scaldin OA. Surface polarization and flexoelectric effect. *Journal of Applied Physics*, **69**, 6343 (1991).

[13] Pereira HA, Batalioto F, and Evangelista LR. Contribution of the ionic adsorption phenomenon to the effective anchoring energy of a nematic liquid crystal sample. *Physical Review E*, **68**, 040701(R) (2003).

[14] Stojmenovik G, Vermael S, Neyts K, van Asselt R, and Verchueren ARM. Dependence of the lateral ion transport on the driving frequency in nematic liquid crystal displays. *Journal of Applied Physics*, **96**, 3601 (2004).

[15] Gomes OA, Falcão RC, and Mesquita ON. Anomalous capillary length in cellular nematic-isotropic interfaces. *Physical Review Letters*, **86**, 2577 (2001).

[16] Kühnau U, Petrov AG, Klose G, and Schmiedel H. Measurements of anchoring energy of a nematic liquid crystal, 4-cyano-4'-n-pentylbiphenyl, on Langmuir-Blodgett films of dipalmitoyl phosphatidylcholine. *Physical Review E*, **59**, 578 (1999).

[17] Nazarenko VG, Pergamenshchik VM, Koval'chuk OV, Nych AB, and Lev BI. Non-Debye screening of a surface charge and a bulk-ion-controlled

anchoring transition in a nematic liquid crystal. *Physical Review E*, **60**, 5580 (1999).

[18] Nazarenko VG, Pergamenshchik VM, Koval'chuk OV, and Lev BI. Non-Debye charge screening and adsorbed-ion-induced anchoring transition in a nematic liquid crystal. *Molecular Crystals and Liquid Crystals*, **352**, 1 (2000).

[19] Barbero G and Evangelista LR. *An Elementary Course on the Continuum Theory for Nematic Liquid Crystals.* World Scientific, Singapore, 2001.

[20] Pereira HA, Batalioto F, and Evangelista LR. Destabilizing effect of a surface eletric field generated by ion adsorption on the molecular orientation of nematic liquid crystals. *European Physical Journal E*, **16**, 267 (2005).

[21] Flügge S. *Practical Quantum Mechanics.* Springer-Verlag, Berlin, 1994.

[22] Morse PM and Feshbach H. *Methods of Theoretical Physics.* McGraw-Hill, New York, 1953.

9

A FERMI–LIKE DESCRIPTION OF THE ADSORPTION PHENOMENON

The aim of this chapter is to present a Fermi-like description for the adsorption of charged or neutral particles at a surface. This theory accounts, in a natural way, for the saturation phenomenon found in real samples. In fact, the Fermi-like distribution–owing to the exclusion principle–naturally takes into account the occupation of the adsorption sites without the need to consider steric potentials with artificial cut-offs. For this reason, the saturation in the coverage of the surface by the adsorbed particles can be found also in the perfect gas approximation. Furthermore, this approach removes some limitations of the classical one, based on a Maxwell-Boltzmann (MB) description. As an application, the steady state distribution of ionic impurities in a sample of an isotropic fluid, whose limiting surfaces are supposed to adsorb positive charged particles, is investigated having as the classical counterpart the theory developed in the first part of Chapter 8. The fundamental equations of the problem can be numerically solved, thus permitting the determination of the electrical potential distribution in the sample, and the full thickness dependence of the surface density of adsorbed charges. In the "classical limit" the results of the MB approach are recovered.

9.1 The Fermi-like model

To start we have to re-consider the basic equations of the MB distribution of neutral non-interacting particles in the presence of a surface with a given adsorption energy. These equations have been introduced in Section 5.2.4 to investigate the phenomenon of the adsorption of magnetic grains. The fundamental equations are (5.48), (5.49) and (5.50). They give the number of particles in the bulk, η_B, and in the surface, η_S, in terms of the number of particles of the system, $\mathcal{N}$, the number of sites in the bulk, $\mathcal{N}_B$, the number of sites in the surface, $\mathcal{N}_S$, and the adsorption energy A. In the limit $A \to -\infty$, we obtain $\eta_B = 0$ and $\eta_S = \mathcal{N}$, as expected. However, this is not a good result because all the particles will be adsorbed no matter the number of adsorption sites is. In fact, if $\mathcal{N} < \mathcal{N}_S$ the result makes sense, but if $\mathcal{N} > \mathcal{N}_S$

(which is always true for a relatively large system) the result is surely wrong. As emphasized before, if the interaction between adsorbed particles is taken into account, a saturation in the coverage can be reached by an intrinsic mechanism. Otherwise, it is difficult to justify this approach for a real system, except in the case in which the concentration of potentially adsorbed particles is very small (which means that $\mathcal{N} \ll \mathcal{N}_\mathrm{S}$).

To better illustrate this point, let us consider again the problem of adsorption of magnetic grains in a lyotropic nematic sample doped with ferrofluid. A system of this kind was considered in Section 5.2.4. For a sample of thickness $d = 0.5\,\mathrm{mm}$, with magnetic grains of spherical shape, of radius $R \simeq 5$ nm, the bulk concentration was $c = 3 \times 10^{15}$ grains/cm^3. In the case of complete adsorption the surface density of magnetic grains is $cd \sim 1.5 \times 10^{14}$ grains/cm^2. The maximum density of occupation sites, however, is $\mathcal{N}_\mathrm{S} = 1/(\pi R^2) \simeq 10^{12}$ sites/cm^2. This result indicates that the coverage of adsorption particles for this system should be saturated since $\mathcal{N} \gg \mathcal{N}_\mathrm{S}$. Therefore, the adsorption phenomenon in a system like this one cannot be satisfactorily described in the framework of the "classical" approach.

Let us now consider the same problem by using a Fermi-like distribution for the particles [1]. A discussion of this possibility has been recently reported in [2], assuming for the chemical potential the one valid for a perfect gas, where the interaction between the particles is neglected. Here, as before, μ will be determined by imposing the conservation of the number of particles $\mathcal{N}$. The choice of the Fermi-Dirac distribution has to be considered as a phenomenological assumption which makes use of the special occupation number properties of the fermionic statistics. These properties are the key mechanism to take into account, in a natural way, for the saturation phenomenon found in real samples. In this framework, η_B and η_S have to be replaced, respectively, by

$$\eta_\mathrm{B} = \frac{\mathcal{N}_\mathrm{B}}{1+e^{-\mu}} \qquad \text{and} \qquad \eta_\mathrm{S} = \frac{\mathcal{N}_\mathrm{S}}{1+e^{A-\mu}}. \tag{9.1}$$

The conservation of the number of particles yields

$$e^{-\mu} = -\alpha + \sqrt{\alpha^2+\beta}, \tag{9.2}$$

where

$$\alpha = \frac{(\mathcal{N}-\mathcal{N}_\mathrm{B}) + (\mathcal{N}-\mathcal{N}_\mathrm{S})e^{-A}}{2\mathcal{N}} \quad \text{and} \quad \beta = \frac{\mathcal{N}_\mathrm{S}+\mathcal{N}_\mathrm{B}-\mathcal{N}}{\mathcal{N}}e^{-A}. \tag{9.3}$$

In the limit in which $A \to \infty$, no adsorption is expected. In fact, we obtain from (9.2) and (9.3),

$$e^{-\mu} = -\frac{\mathcal{N}-\mathcal{N}_\mathrm{B}}{2\mathcal{N}} + \frac{|\mathcal{N}-\mathcal{N}_\mathrm{B}|}{2\mathcal{N}}, \tag{9.4}$$

which gives $\eta_{\rm B} = \mathcal{N}_{\rm B}$ if $\mathcal{N} > \mathcal{N}_{\rm B}$, and $\eta_{\rm B} = \mathcal{N}$ if $\mathcal{N} < \mathcal{N}_{\rm B}$. In the opposite limit, in which $A \to -\infty$, we obtain

$$e^{-\mu} = -\frac{(\mathcal{N} - \mathcal{N}_{\rm S})}{2\mathcal{N}} e^{-A} + \frac{|\mathcal{N} - \mathcal{N}_{\rm S}|}{2\mathcal{N}} e^{-A} + \frac{|\mathcal{N} - \mathcal{N}_{\rm S}|}{(\mathcal{N} - \mathcal{N}_{\rm S})^2} (\mathcal{N}_{\rm S} + \mathcal{N}_{\rm B} - N). \quad (9.5)$$

Consequently, if $\mathcal{N} < \mathcal{N}_{\rm S}$, $\eta_{\rm B} = 0$ and $\mathcal{N}_{\rm S} = \mathcal{N}$ (i.e., all the particles are adsorbed, as expected). However, if $\mathcal{N} > \mathcal{N}_{\rm S}$, $\eta_{\rm B} = \mathcal{N} - \mathcal{N}_{\rm S}$ and $\eta_{\rm S} = \mathcal{N}_{\rm S}$, i.e., the number of adsorbed particles coincides with the number of adsorption sites. Notice that in this case, in contrast to the MB approach considered before, the adsorption tends naturally to a saturation, even in this situation of perfect gas approximation. In the absence of an adsorption energy ($A = 0$), we obtain

$$e^{-\mu} = \frac{(\mathcal{N}_{\rm B} + \mathcal{N}_{\rm S} - \mathcal{N})}{\mathcal{N}},$$

which gives

$$\eta_{\rm B} = \frac{\mathcal{N}\mathcal{N}_{\rm B}}{\mathcal{N}_{\rm B} + \mathcal{N}_{\rm S}} \qquad \text{and} \qquad \eta_{\rm S} = \frac{\mathcal{N}\mathcal{N}_{\rm S}}{\mathcal{N}_{\rm B} + \mathcal{N}_{\rm S}}.$$

Therefore, we have $\eta_{\rm B}/\eta_{\rm S} = \mathcal{N}_{\rm B}/\mathcal{N}_{\rm S}$, which is a well known result, according to which in the absence of adsorption phenomenon, the number of particles on the surface and in the bulk are proportional to the respective number of sites.

Consider now the situation in which the number of potentially adsorbed particles ($\mathcal{N}$) is very small with respect to $\mathcal{N}_{\rm B}$ and $\mathcal{N}_{\rm S}$. From Eq. (9.2) we obtain

$$e^{-\mu} = \frac{\mathcal{N}_{\rm B} + \mathcal{N}_{\rm S} e^{-A}}{\mathcal{N}},$$

which coincides with (5.48). In the same manner, $\eta_{\rm B}$ and $\eta_{\rm S}$, given by (9.1), are reduced, respectively, to (5.49) and (5.50). Therefore, in this limit, the Fermi-like distribution can be replaced by the Boltzmann one.

Very often, in practical situations, the sample is a slab of thickness d and area S, such that $\sqrt{S} \gg d$. In this case the equations reported above allow analysis of the adsorption phenomenon by considering the thickness of the sample as a control parameter. In this framework, $\mathcal{N}_{\rm B} = N_{\rm B} S d$, $\mathcal{N} = NSd$, and $\mathcal{N}_{\rm S} = 2N_{\rm S}S$, where $N_{\rm B}$, N and $N_{\rm S}$ are the bulk density of sites, the bulk density of particles in the absence of adsorption, and the surface density of sites, respectively. The factor 2 in $\mathcal{N}_{\rm S}$ appears because there are two surfaces, each of area S, limiting the slab. It is interesting to analyze the covering ratio θ versus $\mathcal{N}/\mathcal{N}_{\rm S} = (N/2N_{\rm S})d$ for different $\mathcal{N}_{\rm B}/\mathcal{N} = N_{\rm B}/N$. Small values of $N_{\rm B}/N$ correspond to a liquid phase, whereas very large $N_{\rm B}/N$ is relevant to a vapor or gas phase. The value of A ranges from a few units, which corresponds to the region of physical adsorption, to a few tens of units, which corresponds to a region of chemisorption [3].

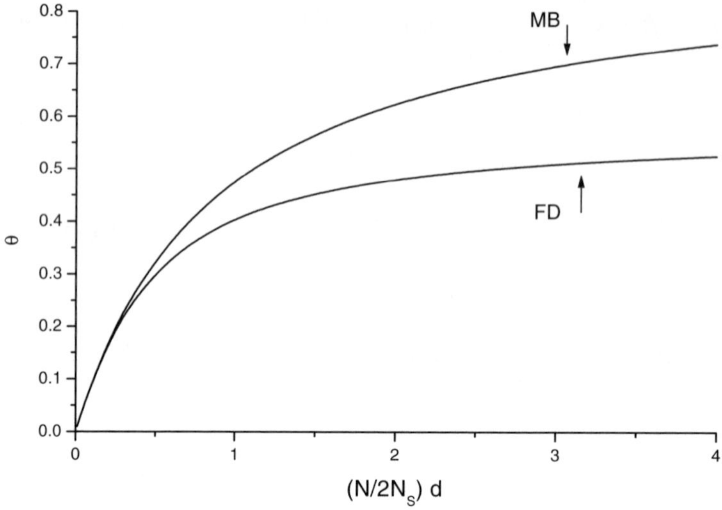

FIGURE 9.1
The trend of the coverage θ as a function of $N/(2N_{\rm S})d$ for a practical case in which the adsorption energy is $A = -1.0$ and $N_{\rm B}/N = 3$, both in the MB and Fermi-Like (FD) distributions. Reprinted from Ref. [1] with permission from Elsevier.

Notice that, according to Eq. (5.50) for large $\mathcal{N}/\mathcal{N}_{\rm S} = 2Nd/N_{\rm S}$ the covering ratio $\theta^{\rm MB} = \eta_{\rm S}/\mathcal{N}_{\rm S}$ tends to

$$\theta^{\rm MB} = \frac{N}{N_{\rm B}} e^{-A}, \tag{9.6}$$

whereas in the same limit the proposed FD approach yields

$$\theta^{\rm FD} = \left[1 + \left(\frac{N_{\rm B}}{N} - 1\right) e^{A}\right]^{-1}. \tag{9.7}$$

On the other hand, for small $\mathcal{N}/\mathcal{N}_{\rm S}$, the covering ratios are

$$\theta^{\rm MB} = \theta^{\rm FD} = \frac{N}{2N_{\rm S}} d, \tag{9.8}$$

i.e., they are proportional to the thickness of the sample. As is evident from (9.6) and (9.7), for $A \to -\infty$, $\theta^{\rm MB}$ diverges, as pointed out above, whereas $\theta^{\rm FD} \to 1$, as expected. From (9.6) one also concludes that the MB approach can work well only if $\theta^{\rm MB} < 1$, which implies $e^{-A} < N_{\rm B}/N$. The covering ratios for $N_{\rm B}/N = 3$, according to the two approaches, are reported in Figure 9.1 for $A = -1.0$, and in Figure 9.2 for $A = -10.0$ as a function of $(N/2N_{\rm S})d$.

Let us consider the case of small adsorption energy, where the two approaches predict reasonable values for the saturation of θ. In this case, it is

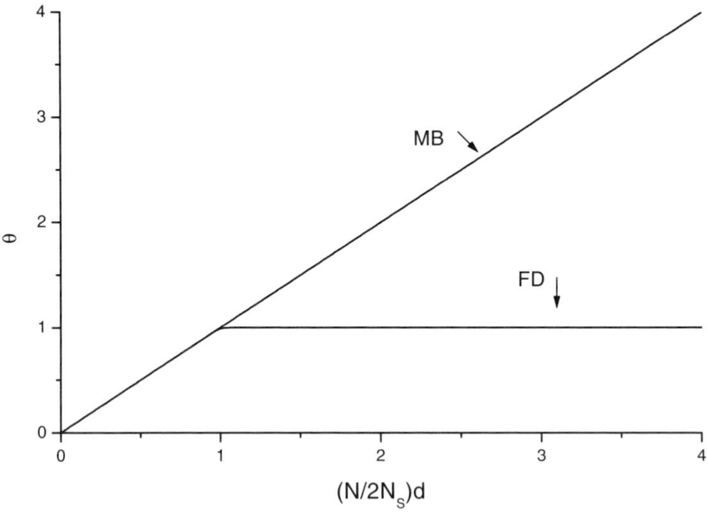

FIGURE 9.2
The same as in Figure 9.1 but for $A = -10.0$. Reprinted from Ref. [1] with permission from Elsevier.

possible to introduce an intrinsic length, D, connected with the saturation phenomenon. It can be defined as

$$\theta_{\text{sat}} = \left(\frac{d\theta}{d\,d}\right)_{d=0} D. \tag{9.9}$$

For the statistics considered here, using for θ the expressions reported above, we have

$$D^{\text{MB}} = 2\frac{N_{\text{S}}}{N_{\text{B}}} e^{-A} \tag{9.10}$$

and

$$D^{\text{FD}} = 2\frac{N_{\text{S}}}{N_{\text{B}}} \left[1 + \left(\frac{N_{\text{B}}}{N} - 1\right) e^{A}\right]^{-1}. \tag{9.11}$$

It follows that D^{MB} is independent of the density of potentially adsorbed particles, which is at least a strange result. On the contrary, D^{FD} depends, correctly, on this quantity. Furthermore, for $A < 0$, which is the case of interest in the adsorption phenomenon, D^{FD} is always smaller than D^{MB}. Finally, we mention that if $e^{A} \ll N_{\text{B}}/N$, which corresponds to a relatively large adsorption energy, $D^{\text{FD}} \sim 2N_{\text{S}}/N$, as expected.

In Figures 9.3 and 9.4, μ is evaluated by assuming again $N_{\text{B}}/N \approx 3$ (liquid phase) for different A. For small values of A (Figure 9.3) the behavior of μ is similar for both statistics. By increasing A, the chemical potential relevant to

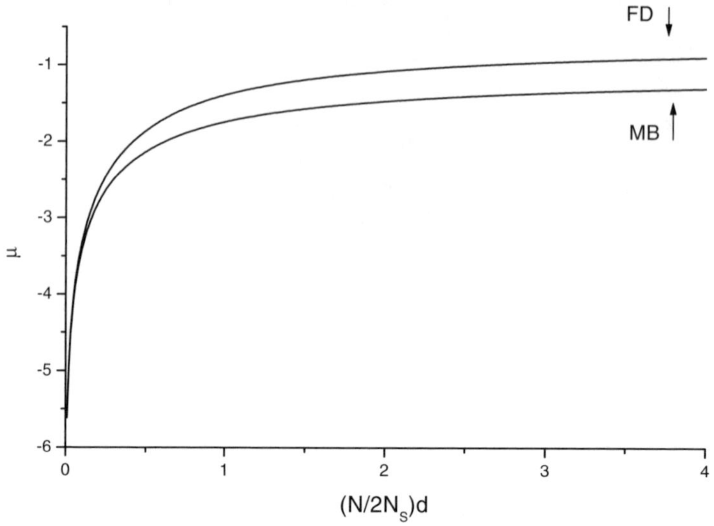

FIGURE 9.3
The trend of the chemical potential μ as a function of $N/(2N_\mathrm{S})d$ for a practical case in which the adsorption energy is $A = -1.0$ and $N_\mathrm{B}/N = 3$, both in the MB and FD approaches. Reprinted from Ref. [1] with permission from Elsevier.

MB statistics maintains the same trend, whereas in the FD statistics it changes its curvature (Figure 9.4). This behavior agrees with the model described in [4] for interacting molecules. However, according to the model of adsorption, even for very large adsorption energy the unstable loop described in [4] is absent.

For values of N_B/N relevant to the gas phase, the trends for the covering ratios and for the chemical potentials are very similar to the ones reported above.

9.2 Application: Ion adsorption in isotropic fluids

A possible application of the theory presented above concerns the ion distribution in an isotropic fluid limited by two adsorbing surfaces, as we have considered in Chapter 8. For the sake of simplicity we consider again a slab of thickness d. The liquid is supposed globally neutral, containing $N_+ = N_- = N$ ions, per unit volume, in the absence of the adsorption phenomenon. We assume that the surfaces (whose adsorption energies are $A_\pm$ for positive/negative charges) are identical and adsorb only positive ions such that $A_+ = A$ and $A_- \to \infty$. Due to the adsorption phenomenon the sample will

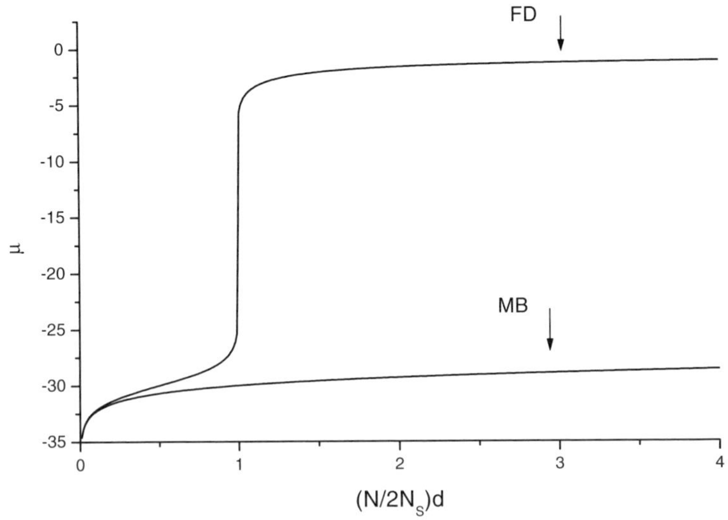

FIGURE 9.4
The same as in Figure 9.3 but for $A = -10.0$. Reprinted from Ref. [1] with permission from Elsevier.

be locally charged. Our aim is to obtain, in the framework of Fermi statistics, the Poisson's equation, the electrical potential (V) and the chemical potential (μ). Since the problem is symmetric with respect to the middle of the sample, $V(z) = V(-z)$, where the z-axis is normal to the bounding walls with the origin in the center of the cell. We indicate by $\psi(z) = qV(z)/k_{\rm B}T$ the electrostatic energy of a positive ion, of charge q, in $k_{\rm B}T$ units.

In the equilibrium at a given temperature T, in the bulk, the distribution of charged particles is given by Eq. (9.1), now rewritten as

$$n_{\rm B}^{(\pm)} = \frac{N_{\rm B}}{1 + e^{-\mu \pm \psi}}, \tag{9.12}$$

while the distribution of surface charges is given by

$$n_{\rm S} = \frac{N_{\rm S}}{1 + e^{-\mu + \psi_s + A}}. \tag{9.13}$$

In Eq. (9.12) and (9.13) $N_{\rm B}$ and $N_{\rm S}$ indicate, as before, the bulk and surface *densities* of sites, whereas $n_{\rm B}^{(\pm)}$ and $n_{\rm S}$ are the bulk and surface *densities* of particles. Notice that now the effective adsorption energy is $\psi_s + A$, i.e., it contains also the repulsive electrostatic part connected with the adsorption of positive ions.

The net charge density of the system is $\rho = q(n_{\rm B}^{(+)} - n_{\rm B}^{(-)})$ which, with the help of Eq. (9.12), can be written as

$$\rho = -2qN_{\rm B}\frac{e^{-\mu}\sinh\psi}{h(\psi,\mu)}, \tag{9.14}$$

where $h(\psi,\mu) = 1 + 2e^{-\mu}\cosh\psi + e^{-2\mu}$. In the steady state the charge distribution and the electrical potential are connected by Poisson's equation (6.4), which, by using the definition of $\psi(z)$ and ρ, can be put in the form

$$\frac{d^2\psi}{dz^2} = \frac{e^{-\mu}}{L_{\rm B}^2}\frac{\sinh\psi}{h(\psi,\mu)}, \tag{9.15}$$

where $L_{\rm B} = \sqrt{\epsilon k_{\rm B}T/2N_{\rm B}q^2}$ is the intrinsic length of the problem, defined in (8.12), but now rewritten with the subscript "B" to remember its dependence on $N_{\rm B}$. In Section 6.1.2 the Debye's screening length λ_0 is introduced by means Eq. (6.9). As follows from this definition of λ_0, it depends on the ion concentration in the absence of adsorption phenomenon. In this aspect, it is a bulk equilibrium property ($d \to \infty$). It follows that λ_0 is thickness independent. In the model we are discussing, the quantity playing the role of λ_0 is $e^{\mu/2}L_{\rm B}$. Since the chemical potential μ depends on d, the Debye's screening length depends on d too, at least for small values of d.

Equation (9.15) could be called Poisson-Fermi equation in analogy with the Poisson-Boltzmann equation. In fact, in the "classical limit" it reduces to the Poisson-Boltzmann equation considered in Section 6.1.2. Since the electrical potential is an even function of z, from Eq. (9.15) it follows that

$$\frac{d\psi}{dz} = \frac{1}{L_{\rm B}}\sqrt{\ln[h(\psi,\mu)/h(\psi_0,\mu)]}, \tag{9.16}$$

where $\psi_0 = \psi(0)$. Therefore, the potential $\psi(z)$ can be determined through the relations

$$I^{\rm FD}[\psi(z),\psi_0;\mu] = \frac{z}{L_{\rm B}} \tag{9.17}$$

and

$$I^{\rm FD}[\psi_s,\psi_0;\mu] = \frac{d}{2L_{\rm B}}, \tag{9.18}$$

where $\psi_s = \psi(\pm d/2)$ is the surface potential, and

$$I^{\rm FD}[\beta,\psi_0;\mu] = \int_{\psi_0}^{\beta}\frac{d\psi}{\sqrt{\ln[h(\psi,\mu)/h(\psi_0,\mu)]}}. \tag{9.19}$$

In this system the electric field $E(z) = -dV/dz$ is identically zero for $z > d/2$ and for $z < -d/2$, presenting a discontinuity for $z = \pm d/2$, where its value is $E(-d/2) = n_{\rm S}q/\epsilon$, which implies that

$$\left(\frac{d\psi}{dz}\right)_{z=-d/2} = -\frac{q^2N_{\rm S}}{\epsilon k_{\rm B}T}\frac{1}{1+e^{-\mu+\psi_s+A}} = -\frac{1}{L_{\rm S}}\frac{1}{1+e^{-\mu+\psi_s+A}}, \tag{9.20}$$

where $L_S = \epsilon k_B T/N_S q^2$ is the characteristic length connected with the surface and introduced in Eq. (7.6), but here rewritten in terms of N_S. By using Eq. (9.16) we can rewrite Eq. (9.20) in the form

$$e^{-\mu} = e^{-\psi_s - A}\left[\frac{L_B/L_S}{\sqrt{\ln\left[h(\psi_s,\mu)/h(\psi_0,\mu)\right]}} - 1\right]. \tag{9.21}$$

The conservation of the number of particles is expressed by

$$\int_{-d/2}^{d/2} n_B^{(+)}\, dz + 2n_S = Nd,$$

which, using Eq. (9.12) and (9.13), can be put in the form

$$\int_{-d/2}^{d/2} \frac{N_B}{1+e^{-\mu+\psi}}\, dz + \frac{2N_S}{1+e^{-\mu+\psi_s+A}} = Nd, \tag{9.22}$$

or, if use is made of Eq. (9.16), in the form

$$N_B L_B J^{FD}[\psi_s,\psi_0;\mu] + \frac{N_S}{1+e^{A-\mu+\psi_s}} = N I^{FD}[\psi_s,\psi_0;\mu] L_B, \tag{9.23}$$

where

$$J^{FD}[\psi_s,\psi_0;\mu] = \int_{\psi_0}^{\psi_s} \frac{1}{1+e^{-\mu+\psi}} \frac{d\psi}{\sqrt{\ln\left[h(\psi,\mu)/h(\psi_0,\mu)\right]}}. \tag{9.24}$$

The fundamental equations of the model are, therefore, (9.18), (9.21) and (9.23).

If the same problem is considered in the framework of the Maxwell-Boltzmann statistics, the fundamental equations of the model can be obtained by operating in the same manner as above, but starting from

$$n_B^{\pm} = N_B e^{\mu \mp \psi(z)} \tag{9.25}$$

and

$$n_S = N_S e^{\mu-\psi_s-A}, \tag{9.26}$$

instead of Eq. (9.12) and (9.13), or by taking the appropriated "classical" limit of Eq. (9.18), (9.21) and (9.23). We have for the expression connecting ψ_s and ψ_0 with the chemical potential μ

$$I^{MB}[\psi_s,\psi_0;\mu] = \int_{\psi_0}^{\psi_s} \frac{d\psi}{\sqrt{\cosh\psi - \cosh\psi_0}} = \frac{\sqrt{2}e^{\mu/2}}{L_B}\frac{d}{2}, \tag{9.27}$$

which plays the same role as Eq. (9.18) in the MB approach; we also obtain

$$e^{\mu} = 2\left(\frac{L_{\rm S}}{L_{\rm B}}\right)^2 e^{2(\psi_s+A)}(\cosh\psi_s - \cosh\psi_0), \tag{9.28}$$

as the equivalent of Eq. (9.21) in the MB approach. Finally, from the conservation of the number of particles we obtain

$$\sqrt{2}N_{\rm B}L_{\rm B}e^{\mu/2}J^{\rm MB}[\psi_s,\psi_0;\mu] + 2N_{\rm S}e^{\mu}e^{-(\psi_s+A)} = Nd, \tag{9.29}$$

where

$$J^{\rm MB}[\psi_s,\psi_0;\mu] = \int_{\psi_0}^{\psi_s} \frac{e^{-\psi}}{\sqrt{\cosh\psi - \cosh\psi_0}}d\psi, \tag{9.30}$$

which replaces Eq. (9.23). In summary, if the problem is considered in the framework of Fermi statistics, the fundamental set of equations is given by (9.18), (9.21) and (9.23) while in the MB statistics the fundamental set of equations is represented by (9.27), (9.28) and (9.29). Through these equations one can determine ψ_0, ψ_s and μ. Once these sets of coupled non-linear equations are solved, the surface charge density, $\sigma = qn_{\rm S}$ as well as the electrical potential $V(z)$ can be obtained in both approaches.

9.2.1 Limits

It is instructive to analyze the behavior of the system for $d \to \infty$ and $d \to 0$, corresponding to a very thick (semi-infinite), and very thin sample, and to compare the predictions of the two approaches.

Let us consider first the limit of a semi-infinite sample ($d \to \infty$). In this case we can consider that $\psi_0 \to 0$, whereas ψ_s tends to a constant value, independent of the thickness of the sample, as can be easily deduced from Eq. (9.18). In fact, when $\psi_0 \to 0$ one observes that the leading term in $I^{\rm FD}$ is given by

$$I[\psi_s,\psi_0 \to 0;\mu] = G(\psi_s,\mu) - G(\psi_0,\mu), \tag{9.31}$$

where

$$G(x,\mu) = 2\cosh(\mu/2)\ln x. \tag{9.32}$$

Therefore, in the limit $\psi_0 \to 0$, $I^{\rm FD}$ presents a logaritmic divergence. Let us now consider Eq. (9.23). In the limit in which $d \to \infty$, Eq. (9.24) can be written as

$$J^{\rm FD}[\psi_s,\psi_0 \to 0;\mu] = F(\psi_s,\mu) - F(\psi_0,\mu), \tag{9.33}$$

where

$$F(x,\mu) = \frac{1}{2\cosh(\mu/2)}[(1+e^{\mu})\ln x - 1], \tag{9.34}$$

which presents also a logaritmic divergence. From Eq. (9.31) and (9.33) one obtains

$$\frac{J^{\rm FD}[\psi_s,\psi_0\to 0;\mu]}{I^{\rm FD}[\psi_s,\psi_0\to 0;\mu]}=\frac{1}{2\cosh(\mu/2)}. \tag{9.35}$$

In the limit $\psi_0 \to 0$ we can rewrite Eq. (9.23) in the form

$$\frac{N_{\rm B}}{N}\frac{J^{\rm FD}[\psi_s,\psi_0\to 0;\mu]}{I^{\rm FD}[\psi_s,\psi_0\to 0;\mu)]}=1, \tag{9.36}$$

which implies that

$$e^{\mu}=\frac{N/N_{\rm B}}{1-N/N_{\rm B}}=\frac{N}{N_{\rm B}}+\mathcal{O}[(N/N_{\rm B})^2], \tag{9.37}$$

if we consider that $N \ll N_{\rm B}$, which corresponds to a weak electrolyte.

Let us establish now the value of ψ_s in this limit. We have to use the above result in Eq. (9.21), which can be written as

$$e^{-\mu+\psi_s+A}+1=\frac{L_{\rm B}/L_{\rm S}}{\sqrt{\ln[1+e^{\mu+\psi_s}+e^{2\,\mu}]}}, \tag{9.38}$$

since $e^{-\mu} \gg 1$ and $e^{\psi_s} \gg 1$. Equation (9.38) can be solved perturbatively by considering $e^{-\psi_s}$ as an expansion parameter. In this way, at the first order, one obtains

$$\psi_s=-\frac{2}{3}A+\frac{2}{3}\ln\left(\frac{L_{\rm B}}{L_{\rm S}}\sqrt{\frac{N}{N_{\rm B}}}\right) \tag{9.39}$$

In this limit, the covering ratio, according to Eq. (9.13), is

$$\theta^{\rm FD}=\frac{n_{\rm S}}{N_{\rm S}}=\frac{1}{1+(L_{\rm B}N_{\rm B}/L_{\rm S}N)^{2/3}e^{A/3}}, \tag{9.40}$$

which is thickness independent. It represents the saturation value for θ. By substituting Eq. (9.40) into Eq. (9.20) we obtain that the surface derivative of ψ tends to $\theta^{\rm FD}/L_{\rm S}$, that, for relatively large adsorption energy is of the order of $1/L_{\rm S}$, as predicted in [2].

The other limit, of very small thickness, can be obtained by taking into account that $\psi_0 \to \psi_s$, and $\psi_0 \gg 1$ and $\psi_s \gg 1$ as can be verified *a posteriori* in the same manner as we have done in Section 8.4. We consider Eq. (9.18) in the form

$$\begin{aligned} I^{\rm FD}=\frac{d}{2L_{\rm B}} &\simeq \frac{1}{\sqrt{f(\psi_0)}}\int_{\psi_0}^{\psi_s}\frac{d\psi}{\sqrt{\psi-\psi_0}}, \\ &= 2\frac{\sqrt{\psi_s-\psi_0}}{\sqrt{f(\psi_0)}}, \end{aligned} \tag{9.41}$$

where

$$f(\psi_0) = \frac{e^{\psi_0}}{e^{\mu} + e^{-\mu} + e^{\psi_0}}. \tag{9.42}$$

From Eqs. (9.41) and (9.42) it follows that

$$\psi_s - \psi_0 = \left(\frac{d}{4L_{\rm B}}\right)^2 f(\psi_0) = \mathcal{O}(d^2). \tag{9.43}$$

On the other hand, in this limit, Eq. (9.21) can be put in the form

$$\begin{aligned} e^{-\mu} &= e^{-\psi_s - A}\left[\frac{L_{\rm B}/L_{\rm S}}{\sqrt{\psi_s - \psi_0}}\frac{1}{\sqrt{f(\psi_0)}} - 1\right] \\ &= e^{-\psi_s - A}\left[\frac{2N_{\rm S}/N\,d}{f(\psi_0)} - 1\right], \end{aligned} \tag{9.44}$$

which is equivalent to

$$e^{\psi_0 - \mu} = e^{-A}\left[\frac{2N_{\rm S}}{Nd}\left(\frac{1 + e^{-2\mu}}{e^{\psi_0 - \mu}} + 1\right) - 1\right], \tag{9.45}$$

if Eq. (9.43) is considered. From Eq. (9.24) we obtain

$$J^{\rm FD} \simeq \frac{2e^{\mu}\sqrt{\psi_s - \psi_0}}{e^{\mu} + e^{\psi_0}}\frac{1}{\sqrt{f(\psi_0)}}, \tag{9.46}$$

and Eq. (9.22) can be written as

$$\frac{N_{\rm B}/N}{1 + e^{\psi_0 - \mu}} + \frac{2N_{\rm S}/Nd}{1 + e^{\psi_0 - \mu}e^{-A}} = 1. \tag{9.47}$$

The determination of the limit we are looking for can proceed in a straightforward way. With the help of Eq. (9.45) and (9.47) one determines e^{ψ_0} and e^{μ} in the first order in d. The results are

$$e^{\mu} = \frac{1}{\sqrt{N_{\rm B}/N - 1}} - \frac{N_{\rm B}/Ne^{A}}{\sqrt{N_{\rm B}/N - 1}}\frac{N}{2N_{\rm S}}d + \mathcal{O}(d^2), \tag{9.48}$$

and

$$e^{\psi_0} = \frac{N_{\rm B}/N}{2\sqrt{N_{\rm B}/N - 1}} - \frac{e^{-A}}{\sqrt{N_{\rm B}/N - 1}} + \frac{e^{-A}}{\sqrt{N_{\rm B}/N - 1}}\frac{2N_{\rm S}}{Nd} + \mathcal{O}(d). \tag{9.49}$$

In this limit it is possible to write Eq. (9.13) in the form

$$\theta^{\rm FD} \simeq \frac{Nd}{2N_{\rm S}} + \mathcal{O}(d^2), \tag{9.50}$$

which indicates that $\theta^{\rm FD} \to 0$ as $d \to 0$. In the lowest order we can write

$$e^{\psi_0} \simeq \frac{2e^{-A} N_S}{\sqrt{N N_B}} \frac{1}{d} \quad \text{and} \quad e^{\mu} \simeq \sqrt{\frac{N}{N_B}}, \tag{9.51}$$

indicating that the affinity ($= -\mu$) tends to a constant value [4].

The investigation of the limit $d \to \infty$ for the MB approach can be performed along the same lines. In this case one obtains

$$\frac{J^{\text{MB}}[\psi_s, \psi_0 \to 0; \mu]}{I^{\text{MB}}[\psi_s, \psi_0 \to 0; \mu]} = 1. \tag{9.52}$$

Therefore, from Eq. (9.27) and (9.29) it is easy to find that, when $d \to \infty$, e^{μ} is still given by Eq. (9.37). Likewise, by considering Eq. (9.28), one obtains that for $d \to \infty$, ψ_s is still given by Eq. (9.39). Notice, however, that the saturation value of the surface density, n_S is different from the one obtained in the framework of the Fermi statistics. In fact, the surface density of adsorbed charges in the MB approach is written as (9.26) which, in the present limit, gives for the covering ratio

$$\theta^{\text{MB}} = (\frac{n_S}{N_S})^{\text{MB}} = \frac{e^{-A/3}}{(L_B N_B / L_S N)^{2/3}}. \tag{9.53}$$

We observe that

$$\frac{\theta^{\text{MB}}}{\theta^{\text{FD}}} = 1 + e^{-A/3} \left(\frac{N L_S}{N_B L_B} \right)^{2/3}, \tag{9.54}$$

which, in cases of practical interest, tends to unity only when $N \ll N_B$, i.e., in the case of a very diluted system and if the adsorption energy is not very high.

In the MB approach, the chemical potential in the limit $d \to 0$ depends on the adsorption energy and presents a logarithmic divergence. In fact, by operating in the same way as before, but now using the set of equations (9.27), (9.28), and (9.29) it is easy to show that

$$e^{\mu} = \frac{1}{2} \sqrt{\frac{2N^2 d}{N_S N_B}} e^{A/2} \tag{9.55}$$

and

$$e^{\psi_0} = \sqrt{\frac{2N_S}{N_B d}} e^{-A/2}. \tag{9.56}$$

The above results lead to the following limit for the covering ratio:

$$\theta^{\text{MB}} \simeq \frac{Nd}{2N_S} + \mathcal{O}(d^2), \tag{9.57}$$

which is, obviously, coincident with the one predicted in (9.50) for the Fermi statistics.

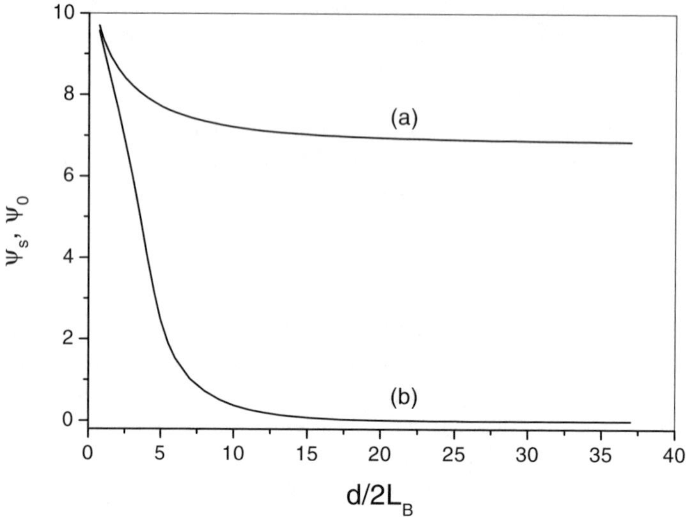

FIGURE 9.5
The surface electrical potential ψ_s (a) and the potential at the middle of the sample ψ_0 (b) versus the thickness of the sample $d/2L_B$. The curves have been depicted for $A = -6.0$ and $N_B/N = 10.0$. Reprinted from Ref. [1] with permission from Elsevier.

9.2.2 Numerical results

Let us now discuss some numerical results obtained by solving the set of simultaneous non-linear equations (9.18), (9.21) and (9.23). The calculations have been performed by considering that $N/N_B = 1/10$ and an adsorption energy $A = -6.0$ (see the estimations in Section 8.4). In Figure 9.5, the surface electrical potential ψ_s (curve a) and the potential in the middle of the sample ψ_0 (curve b) are shown as a function of the thickness of the sample $d/2L_B$. For small values of d we observe a logarithmic divergence for these quantities, as predicted analytically by (9.51). In the opposite limit, of very large thickness, ψ_s tends to a constant value, independent of the thickness of the sample, in agreement with the approximated analysis presented in Section 9.2.1. In the same limit ψ_0 tends rapidly to zero.

In Figure 9.6, the trend of the electric potential $\psi(z)$ versus $2z/(d/2L_B)$ is reported for two samples: one (curve a) of thickness $d/2L_B = 5$ and another one (curve b) of thickness $d/2L_B = 20$. The global behavior of the electrical potential is the expected one. By increasing d, $\psi_0 \to 0$; the central region of the curve becomes flatter and flatter, and the slope of ψ at the border tends to the value $1/L_S$ as discussed above.

Finally, in Figure 9.7 the covering ratio θ^{FD} is shown as a function of the thickness of the sample $d/2L_B$. Also this quantity presents a behavior which departs from the behavior found in the MB approach. For small d, θ^{FD} is

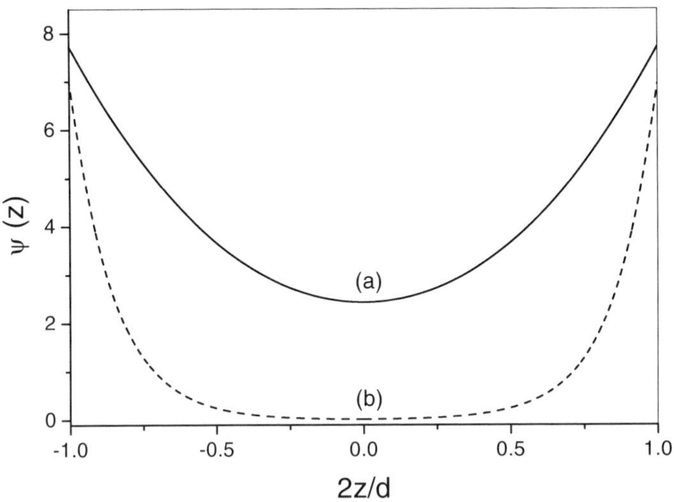

FIGURE 9.6
The electrical potential $\psi(z)$ versus $2z/(d/2L_{\rm B})$ for two samples whose thicknesses are $d/2L_{\rm B} = 5$ (curve a) and $d/2L_{\rm B} = 20$ (curve b). The values of the parameters are the same as in Figure 9.5. Reprinted from Ref. [1] with permission from Elsevier.

linear in d, as analytically predicted in (9.50); for very large d, it presents a saturation value, i.e., it does not depend on the thickness of the sample [see Eq. (9.40)].

Also in this case it is possible to introduce an intrinsic length connected with the saturation phenomenon, as we have done at the end of Section 9.1. Through Eq. (9.9), (9.53), and (9.57) we obtain

$$D^{\rm MB} = 2\frac{N_{\rm S}}{N}\left(\frac{L_{\rm B}N_{\rm B}}{L_{\rm S}N}\right)^{-2/3} e^{-A/3}, \tag{9.58}$$

for MB statistics, and by means of Eqs. (9.9), (9.40), and (9.50)

$$D^{\rm FD} = 2\frac{N_{\rm S}}{N}\left[1 + \left(\frac{L_{\rm B}N_{\rm B}}{L_{\rm S}N}\right)^{2/3} e^{A/3}\right]^{-1}, \tag{9.59}$$

for FD statistics. Also in this case $D^{\rm FD} < D^{\rm MB}$.

Conclusions

In summary, the Fermi-like description of the phenomenon of surface adsorption can be applied to charged and neutral particles in several physical systems. It has the advantage to describe, in a natural way, the phenomenon

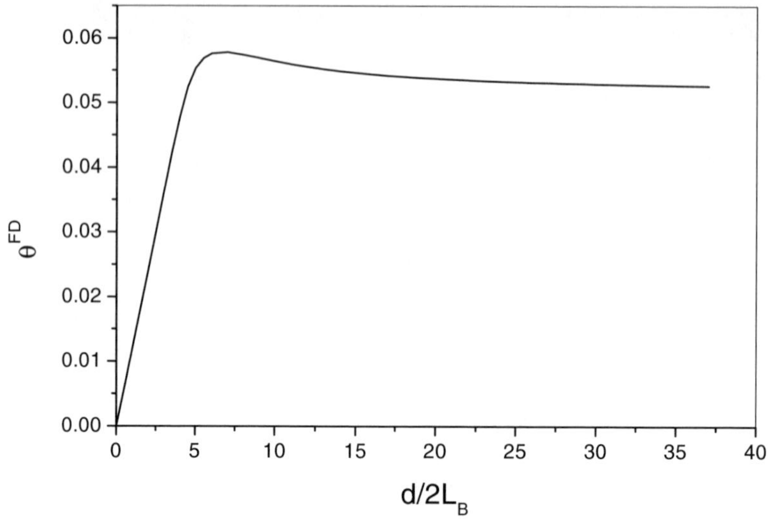

FIGURE 9.7
The covering ratio θ^{FD} versus the thickness of the sample $d/2L_{\mathrm{B}}$. The values of the parameters are the same as in Figure 9.5. Reprinted from Ref. [1] with permission from Elsevier.

of the saturation in the coverage of adsorbed particles found in many real systems. The "classical" (MB) distribution presents severe limitations when dealing with those systems for which the "bulk concentration" of potentially adsorbed particles is very high. The Fermi-Dirac distribution avoids some artificial procedures of introducing cut-offs in the interaction potential and permits us to explain the saturation phenomenon also in the case of non-interacting systems. In the form discussed in this chapter, we are still working in the hypothesis of monolayer adsorption, because it was assumed a short range adsorption energy. However, this aspect of the theory can be improved if we consider a long range adsorption energy for each site. In the case of an ideal gas there exists an intrinsic length, connected with the saturation phenomenon, that is found to be dependent on the bulk density of potentially adsorbed particles, when the MB approach predicts an intrinsic length independent of this quantity. On the other hand, the problem of the adsorption of ionic impurities in an isotropic fluid is governed by a set of three coupled non-linear equations. The problem is reduced to quadratures, and the basic equations can be numerically solved. The general trends for the electrical potential at the surface and in the bulk, and the covering ratio versus the thickness of the sample have the same trends as the ones obtained by means of the MB approach. However, the most important point of the proposed statistics is the saturation in the covering ratio also for very large adsorption energy. The main advantages of the approach are its simplicity, and the fact that the number of phenomenological parameters entering in it is reduced to the number of sites in the bulk, the number of sites at the surface and the adsorption energy which is a rather well known parameter. Furthermore, it allows one to consider the density of possible adsorbed particles as control parameter of the adsorption phenomenon.

[1] Barbero G, Batalioto F, and Evangelista LR. Fermi-like description of the adsorption phenomenon. *Physics Letters A*, **283**, 257 (2001).

[2] Nazarenko VG, Pergamenshchik VM, Koval'chuk OV, Nych AB, and Lev BI. Non-Debye screening of a surface charge and a bulk-ion-controlled anchoring transition in a nematic liquid crystal. *Physical Review E*, **60**, 5580 (1999).

[3] Adamson AW. *Physical Chemistry of Surfaces*. John Wiley & Sons, New York, 1997.

[4] Garrod C. *Statistical Mechanics and Thermodynamics*. Oxford University Press, Oxford, 1995.

10

DYNAMICAL APPROACH FOR THE ADSORPTION PHENOMENA

In the previous chapters, the analysis of the adsorption phenomenon was limited to the steady state. No time variation of the surface density of adsorbed particles or evolution of the bulk density of particles was considered. In the present chapter this aspect of the adsorption phenomenon will be treated. For this reason, a general discussion on the continuity equation, diffusion and drift currents is presented. After that, typical problems concerning the adsorption of neutral particles from the surfaces are discussed. The role of the kinetic equations at the bounding surfaces is considered in details. A simple model of a diffusion-drift phenomenon giving rise to a classical kinetic equation at the surface is proposed for neutral particles. The intrinsic times in the drift-diffusion phenomenon are investigated in the simple case where the drift field is independent of the adsorption phenomenon of neutral particles. The final part of this chapter is dedicated to presenting a dynamical analysis of the problem relevant to a liquid containing ions submitted to an external voltage. It is assumed that the external voltage is step-like. The problem is a little bit more complicated than the one considered in the first sections for the following reasons. First, the actual electric field in the sample has to be determined in a self-consistent manner by means of the Poisson equation. Second, there are two kinds of ions: the positive and the negative ones, and hence, two different continuity equations have to be solved. Third, the kinetic equations at the limiting surfaces for the two kind of ions involve different phenomenological parameters. In this general case, the equations governing the phenomenon, i.e., continuity, Poisson and equilibrium of the nematic torques, will be numerically solved.

10.1 Fundamental equations

10.1.1 Continuity equation

For a continuous medium (e.g. a pure fluid) there is a fundamental equation stating the conservation of the number of particles called *continuity equation.* Let us indicate by $\rho(\mathbf{r}, t)$ the density of particles around the point $\mathbf{r}$ at the time t. To deduce the continuity equation we consider an ideal volume τ limited by a surface $S(\tau)$. The number of particles contained in τ is given by

$$N = \int\int\int_\tau \rho \, d\tau. \tag{10.1}$$

The rate of variation of N is given by

$$\frac{dN}{dt} = \frac{d}{dt}\int\int\int_\tau \rho \, d\tau = \int\int\int_\tau \frac{\partial \rho}{\partial t} \, d\tau. \tag{10.2}$$

If we indicate by $\mathbf{j} = \rho \mathbf{v}$, where $\mathbf{v}$ is the fluid velocity, the current density we can write

$$\frac{dN}{dt} = -\oint_{S(\tau)} \mathbf{j} \cdot \nu \, dS = -\int\int\int_\tau \nabla \cdot \mathbf{j} \, d\tau, \tag{10.3}$$

where ν is the normal to $S(\tau)$ directed outward, and the divergence theorem has been applied. Equations (10.2) and (10.3) together imply

$$\int\int\int_\tau \left(\frac{\partial \rho}{\partial t} + \nabla \cdot \mathbf{j} \right) d\tau = 0. \tag{10.4}$$

Since τ is an arbitrary volume, from Eq. (10.4) it follows that

$$\frac{\partial \rho}{\partial t} + \nabla \cdot \mathbf{j} = 0, \tag{10.5}$$

which is the well known continuity equation. In this context, it is an expression for the conservation of mass.

10.1.2 Diffusion equation

In the investigation of diffusion of some substances in continuous media, like the diffusion of neutron in a nuclear reactor, diffusion of a chemical product in a solvent, or diffusion of ions in a medium, the *diffusion equation* plays a fundamental role. The major part of the diffusion phenomena obeys a linear relation known as Fick's law:

$$\mathbf{j} = -D \nabla \rho, \tag{10.6}$$

where D is the diffusion coefficient, which depends on the properties of the medium. This expression is valid in the hypothesis that $\nabla \rho$ is a small quantity. Otherwise, other terms in higher order derivatives have to be considered. If one supposes that the diffused substance is not absorbed nor emitted by the medium, then Eq. (10.5) has to be verified. By invoking Fick's law introduced above, Eq. (10.5) becomes:

$$\frac{\partial \rho}{\partial t} - D \nabla^2 \rho = 0, \tag{10.7}$$

and is known as the *diffusion equation.*

Typical idealized problems deal with densities (or concentrations) depending only on one spatial coordinate such that $\rho = \rho(z, t)$. In this case, Eq. (10.7) reduces to

$$\frac{\partial \rho}{\partial t} - D\frac{\partial^2 \rho}{\partial z^2} = 0. \tag{10.8}$$

A practical situation is to consider the diffusion process in typical slab cell of thickness d, whose limiting surfaces are placed at $z = \pm d/2$. For this geometry it is convenient to introduce the re-scaled coordinate $\zeta = z/d$, such that $-1/2 \leq \zeta \leq 1/2$. Equation (10.8) becomes

$$\frac{\partial \rho}{\partial t} - \frac{1}{\tau_{\rm D}}\frac{\partial^2 \rho}{\partial \zeta^2} = 0, \tag{10.9}$$

where $\tau_{\rm D} = d^2/D$ has dimensions of time and is known as the *diffusion time.* It is the characteristic time for a particular diffusion process.

10.1.3 No adsorption from the surfaces

To obtain the equilibrium profile of the density (or concentration) $\rho = \rho(z, t)$, for the case in which the surfaces are not adsorbing, one has to search for solutions of the Eq. (10.8).

In this case the current reduces to the *diffusion current*

$$j = -D\frac{\partial \rho}{\partial z}. \tag{10.10}$$

Before we consider the boundary and the initial conditions it is necessary to emphasize that two general requirements have to be fulfilled by the fluctuating quantity, namely:

$$\rho(z, t) = \rho(-z, t), \tag{10.11}$$

imposed by the symmetry of the problem, and

$$\int_{-d/2}^{d/2} \rho(z, t)\, dz = \rho_0\, d, \tag{10.12}$$

where $\rho_0 = \rho(z,\, t = 0)$ is the initial homogeneous density across the sample. Equation (10.12) is the conservation of the number of particles per unit area in the cell. For what concerns the boundary conditions, in the absence of adsorption, one has to consider

$$j\,(\pm d/2, t) = 0, \qquad \forall\, t. \tag{10.13}$$

The solution of Eq. (10.8) will be searched in the form:

$$\rho(z,t) = \rho_{\rm eq}(z) + \delta\rho(z,t), \tag{10.14}$$

where

$$\rho_{\rm eq}(z) = \lim_{t\to\infty} \rho(z,t) \tag{10.15}$$

is the distribution of the particles in the steady state. Hence

$$\lim_{t\to\infty} \delta\rho(z,t) = 0. \tag{10.16}$$

Substitution of Eq. (10.14) into Eq. (10.8) yields

$$\frac{d^2\rho_{\rm eq}(z)}{dz^2} = 0, \tag{10.17}$$

and

$$\frac{\partial(\delta\rho)}{\partial t} = D\frac{\partial^2(\delta\rho)}{\partial z^2}. \tag{10.18}$$

From Eq. (10.17) one has

$$\rho_{\rm eq}(z) = \rho_{\rm eq} + a\, z, \tag{10.19}$$

where, in view of Eq. (10.11), $a = 0$. Therefore,

$$\rho_{\rm eq}(z) = \rho_{\rm eq} = \text{constant}. \tag{10.20}$$

Equation (10.18) can be solved by means of separation of variables, by assuming that $\delta\rho(z,t) = Z(z)T(t)$. A general solution, by exploiting the linearity of Eq. (10.8), can be obtained in the form:

$$\delta\rho(z,t) = \sum_\beta C_\beta \cos\left(\omega_\beta\, z\right)\, e^{-\beta^2\, t}, \tag{10.21}$$

where $\omega_\beta = \beta/\sqrt{D}$. Substitution of Eqs. (10.20) and (10.21) into Eqs. (10.12) and (10.13) yields, $\forall\, t$,

$$\sum_\beta C_\beta \frac{\sin\left(\omega_\beta d/2\right)}{\omega_\beta d/2}\, e^{-\beta^2\, t} = \rho_0 - \rho_{\rm eq}, \tag{10.22}$$

and

$$\sum_\beta C_\beta\, \left(\omega_\beta d/2\right)\, \sin\left(\omega_\beta d/2\right)\, e^{-\beta^2\, t} = 0. \tag{10.23}$$

The obvious solution is $\rho_{\rm eq} = \rho_0$ and $C_\beta = 0$. This indicates that the presence of the surface does not perturb the particle distribution, if there is no adsorption from the limiting surfaces.

10.1.4 Adsorption from the surfaces

In this case the bulk equations are still Eqs. (10.8) and, for identical surfaces, Eq. (10.11). The surface density of adsorbed particles will be denoted by $\sigma = \sigma(t)$. The requirements to be fulfilled now are expressed by Eq. (5.14), namely,

$$2\sigma(t) + \int_{-d/2}^{d/2} \rho(z,t)\, dz = \rho_0\, d, \tag{10.24}$$

and

$$j\,(-d/2,t) = -D\left(\frac{\partial \rho}{\partial z}\right)_{z=-d/2} = \frac{d\sigma}{dt}. \tag{10.25}$$

To investigate the physical consequences of the phenomenon of adsorption, a kinetic equation at the limiting surfaces has to be imposed. A widely used balance equation at the boundary is Eq. (5.22), which we rewrite here as

$$\frac{d\sigma}{dt} = \kappa\, \rho(-d/2,t) - \frac{1}{\tau}\sigma(t), \tag{10.26}$$

where κ and τ are parameters describing the adsorption phenomenon, whose meaning has been discussed in Section 5.2. Equation (10.26) simply states that the time variation of the surface density of adsorbed particles depends on the bulk density of particles just in front of the adsorbing surface, and on the surface density of particles already adsorbed. In Eq. (10.26), τ has the dimension of time, whereas κ of a length/time. Consequently, if the adsorption phenomenon is present, from the kinetic equation (10.26), it follows that there is an intrinsic thickness $\kappa\tau$. This is evident if Eq. (10.26) is rewritten as

$$\tau\frac{d\sigma}{dt} = \kappa\,\tau\,\rho(-d/2,t) - \sigma.$$

$\kappa\tau$ is expected to be connected with the penetration of the surface forces responsible for the adsorption phenomenon.

To solve the problem one assumes again that $\rho(z,t)$ has the form given in Eq. (10.14), together with Eqs. (10.15) and (10.16). In analogy with (10.14), one puts

$$\sigma(t) = \sigma_{\text{eq}} + \delta\sigma(t), \tag{10.27}$$

where

$$\lim_{t\to\infty} \delta\sigma(t) = 0. \tag{10.28}$$

In the limit $t \to \infty$, by taking into account Eqs. (10.15), (10.16), (10.27) and (10.28), from Eq. (10.26) one obtains

$$\kappa\,\rho_{\rm eq} - \frac{1}{\tau}\,\sigma_{\rm eq} = 0, \tag{10.29}$$

from which one gets

$$\sigma_{\rm eq} = \kappa\,\tau\,\rho_{\rm eq}. \tag{10.30}$$

Equation (10.24), in the limit $t \to \infty$, becomes:

$$2\sigma_{\rm eq} + \rho_{\rm eq}\,d = \rho_0 d. \tag{10.31}$$

If use is made of Eq. (10.30), $\rho_{\rm eq}$ and $\sigma_{\rm eq}$ are found to be

$$\rho_{\rm eq} = \frac{\rho_0}{1 + 2\,\kappa\,\tau/d} \quad \text{and} \quad \sigma_{\rm eq} = \frac{\kappa\tau/d}{1 + 2\kappa\tau/d}\,\rho_0 d. \tag{10.32}$$

A few considerations of the adsorption parameters are still necessary and will be discussed in Section 10.6.2.

10.2 Time evolution of the bulk and surface densities

The time evolution of $\delta\rho(z,t)$ and $\delta\sigma(z,t)$ can be now determined [1]. By substituting $\rho(z,t) = \rho_{\rm eq} + \delta\rho(z,t)$ into Eq. (10.8) one obtains Eq. (10.18), whose solution can be written in the form (10.21).

Moreover, by substituting the expressions for $\rho(z,t)$ and $\sigma(t)$ into Eq. (10.26), one has

$$\frac{d\,(\delta\sigma)}{dt} = \kappa\delta\rho - \frac{1}{\tau}\,\delta\sigma, \qquad \forall\, t. \tag{10.33}$$

By Eq. (10.21) and Eq. (10.33), one obtains

$$\delta\sigma = M e^{-t/\tau} + \sum_\beta \delta\sigma_\beta\, e^{-\beta^2 t}, \tag{10.34}$$

where

$$\delta\sigma_\beta = \kappa\,\frac{C_\beta}{\tau^{-1} - \beta^2}\,\cos\left(\omega_\beta d/2\right), \tag{10.35}$$

and M has to be determined by means of the condition concerning the conservation of the number of particles.

By substituting $\rho(z,t)$ and $\sigma(t)$ into Eq. (10.24) one gets

$$2\,\delta\sigma(t) + \int_{-d/2}^{d/2} \delta\rho(z,t)dz = 0, \tag{10.36}$$

that, for Eqs. (10.21) and (10.34), can be written as

$$M\,e^{-t/\tau}+\sum_{\beta}\left[\delta\sigma_{\beta}+\frac{C_{\beta}}{\omega_{\beta}}\,\sin\left(\omega_{\beta}d/2\right)\right]\,e^{-\beta^{2}t}=0, \tag{10.37}$$

from which, by taking into account Eq. (10.35), one obtains $M = 0$ and $\tan\left(\omega_\beta\, d/2\right) = [\kappa/(\beta^2 - \tau^{-1})]\,\omega_\beta$, which determine the eigenvalues of the problem and can be rewritten as

$$\tan\,X_{\beta}=\left(\frac{\tau_{\rm D}}{4\tau_{\kappa}}\right)\,\frac{X_{\beta}}{X_{\beta}^{2}-\tau_{\rm D}/4\tau}, \tag{10.38}$$

where $X_\beta = \omega_\beta d/2$. In summary, three time scales govern the phenomenon, namely, $\tau_{\rm D} = d^2/D$, $\tau_\kappa = d/2\kappa$, and τ. It follows that when the adsorption-desorption phenomenon is present, not only an intrinsic length $\delta = \kappa\tau$ is present, but also a new intrinsic time $\tau_\kappa = d/2\kappa$ connected only with the adsorption coefficient κ and with the thickness of the sample. The eigenvalues of the problem depend on the two ratios $\tau_{\rm D}/\tau_\kappa$ and $\tau_{\rm D}/\tau$.

In a practical problem it is important to know the first eigenvalue $\beta \neq 0$ responsible for the lowest relaxation time in the phenomenon under consideration. The function on the right hand side of Eq. (10.38) has a vertical asymptote at $X_\beta = \sqrt{\tau_{\rm D}/4\tau}$. If $\tau_{\rm D} \ll \tau$, Eq. (10.38) can be approximated by $X_\beta \tan X_\beta = \tau_{\rm D}/(4\tau_\kappa)$, showing that X_β depends on $\tau_{\rm D}/\tau_\kappa$. For $\tau_{\rm D} \ll \tau_\kappa$, $X_\beta \sim \sqrt{\tau_{\rm D}/4\tau_\kappa}$. In this case ($\tau_{\rm D} \ll \tau$, $\tau_{\rm D} \ll \tau_\kappa$), one gets $\beta^2 = 1/\tau$. This means that when the diffusion process is a rapid phenomenon, the time dependence of the particle distribution is τ. In the opposite limit of $\tau_{\rm D} \gg \tau$, Eq. (10.38) gives $\tan X_\beta = -(\tau/\tau_\kappa)\,X_\beta$, whose solution is $\pi/2 < X_\beta < \pi$, and the relevant relaxation time $\tau_{\rm R}$ is in the range $\tau_{\rm D}/(4\pi^2) \leq \tau_{\rm R} \leq \tau_{\rm D}/\pi^2$. Finally, from Eq. (10.38) it follows that, for large X_β, i.e., $X_\beta \gg \sqrt{\tau_{\rm D}/4\tau}$ the eigenvalues are $X_n \approx n\,\pi$.

If the eigenvalues are known, one can calculate the coefficients C_β appearing in Eq. (10.21), by means of which one determines the time evolution of $\delta\rho(z,t)$ and $\delta\sigma(t)$. From $\rho(z,t) = \rho_{\rm eq} + \delta\rho(z,t)$, written in the limit of $t \to 0$, one has $\delta\rho(z,0) = 2\,(\tau/\tau_\kappa)/(1+\tau/\tau_\kappa)\,\rho_0 = 2\,\sigma_{\rm eq}/d$, that, by using Eq. (10.21), becomes

$$\sum_{\beta}C_{\beta}\cos\left(\omega_{\beta}\,z\right)=\frac{2\,\sigma_{\rm eq}}{d}. \tag{10.39}$$

The main problem is that the eigenvectors $u_\beta = \cos\left(\omega_\beta\, z\right)$ are not orthogonal. In this case one can orthogonalyze the set of eigenvectors by a procedure similar to the Schmidt procedure [2]. By indicating the eigenvalues with $\beta_1(\neq 0)$, β_2, β_3,...,β_n,... one writes $u_i = \cos\left(\omega_{\beta_i} z\right)$, where u_i are linearly independent. It is possible to set $v_i = \lambda_{ij} u_j$, where $\lambda_{ij} = 0$ for $i < j$, and $\lambda_{ii} = 1$. Thus, the matrix $\mathbf{\Lambda}$, of elements λ_{ij} is such that $\det\Lambda = 1$. The coefficients λ_{ij} for $i > j$ are determined by putting

$$\langle v_i | v_j \rangle = \int_{-d/2}^{d/2} v_i(z) v_j(z) dz = 0, \tag{10.40}$$

for $i \neq j$. The relation among v_i and u_i can be written in the matrix form as $\mathbf{v} = \Lambda \mathbf{u}$, from which $u_i = \left(\Lambda^{-1}\right)_{ij} v_j$. Consequently, if Eq. (10.39) is written as $C_\beta u_\beta(z) = 2\,\sigma_{\text{eq}}/d$ one gets $C_\beta \left(\Lambda^{-1}\right)_{\beta j} v_j = 2\,\sigma_{\text{eq}}/d$, from where

$$C_\beta \left(\Lambda^{-1}\right)_{\beta j} \langle v_j | v_k \rangle = \frac{2\,\sigma_{\text{eq}}}{d} \langle v_k \rangle, \tag{10.41}$$

with

$$\langle v_k \rangle = \int_{-d/2}^{d/2} v_k(z) dz. \tag{10.42}$$

Since $v_i(z)$ form a set of orthogonal functions, $\langle v_j | v_k \rangle = N_k \delta_{jk}$, where $N_k = \langle v_k | v_k \rangle$. Consequently, from Eq. (10.41), one deduces that $\left(\Lambda^{-1}\right)_{\beta k} C_\beta = (2\,\sigma_{\text{eq}}/d) \langle v_k \rangle / N_k$. In the matrix form one has for the preceding equation:

$$\mathbf{C} = \frac{2\sigma_{\text{eq}}}{d} \Lambda^{\text{T}} \mathbf{R} \tag{10.43}$$

where $\mathbf{R}$ is the vector of elements $R_k = \langle v_k \rangle / N_k$. The coefficients one is looking for are then given by

$$C_\beta = \frac{2\,\sigma_{\text{eq}}}{d} \Lambda_{\alpha\beta} R_\alpha, \tag{10.44}$$

which represents the solution of the problem. There is another way to obtain explicit formulae connecting v_q with u_q, which gives directly the elements of the matrix Λ, and, consequently, the coefficients C_β. It can be written in the form [3]:

$$v_q = \sum_{n=1}^{q} \frac{M_{nq}}{M_{qq}} u_n, \tag{10.45}$$

where M_{nq} is the minor of the element

$$d_{nq} = \int_{-d/2}^{d/2} u_n(z) u_q(z)\, dz$$

in the determinant D_q defined as

$$D_1 = d_{11}$$

$$D_2 = \begin{vmatrix} d_{11} & d_{12} \\ d_{21} & d_{22} \end{vmatrix}$$

$$D_3 = \begin{vmatrix} d_{11} & d_{12} & d_{13} \\ d_{21} & d_{22} & d_{23} \\ d_{31} & d_{32} & d_{33} \end{vmatrix}; \quad \text{etc.}$$

This alternative way is more suitable to be numerically implemented.

To study the time evolution of the densities, it is useful to rewrite the final equations governing them. The coefficient of the cosine in (10.35) can be put in the form

$$\kappa \frac{C_\beta}{\tau^{-1} - \beta^2} = \frac{d}{2} \left(\frac{\tau_D}{4\tau_\kappa} \right) \frac{C_\beta}{\left(\frac{\tau_D}{4\tau} - X_\beta^2 \right)},$$

which, by using (10.38), can be cast in the final form:

$$\kappa \frac{C_\beta}{\tau^{-1} - \beta^2} = -\frac{d}{2} \frac{\tan X_\beta}{X_\beta} C_\beta.$$

This permits us to rewrite (10.35) as

$$\delta\sigma_\beta = -\frac{d}{2} \frac{\sin X_\beta}{X_\beta} C_\beta,$$

giving for $\sigma(t)$ the re-scaled form:

$$\frac{2\,\sigma(t^*)}{d} = \rho_0 \frac{r_1/r_2}{1 + r_1/r_2} - \sum_\beta \frac{\sin X_\beta}{X_\beta} C_\beta \, e^{-X_\beta^2 t^*}, \tag{10.46}$$

where $r_1 = \tau_D/4\tau_\kappa$, $r_2 = \tau_D/4\tau$, and $t^* = 4t/\tau_D$. Likewise, by considering that $\rho_{eq} = \rho_0 - 2\sigma_{eq}/d$, and using (10.21), one obtains:

$$\rho(Z, t^*) = \rho_0 \frac{1}{1 + r_1/r_2} + \sum_\beta C_\beta \cos\left(X_\beta Z\right) \, e^{-X_\beta^2 t^*}, \tag{10.47}$$

where $-1 \leq Z = 2z/d \leq 1$.

In Figure 10.1, the behavior of $\rho(Z, t^*)/\rho_0$ versus Z, as given by Eq. (10.47), is shown for a significant set of parameters giving the ratios of the characteristic times entering in the problem. The curves show that as r_2 increases in comparison with r_1, i.e., as the importance of κ decreases when compared with τ, there is an increasing accumulation of particles near the surfaces, placed at $Z = \pm 1$. This indicates that the time characterizing the adsorption phenomenon, represented by τ_κ, becomes increasingly large, leading to an accumulation of particles (not adsorbed) near the limiting surfaces.

In Figure (10.2), the behavior of $2\sigma(t)/\rho_0 d$, as predicted by Eq. (10.46), is shown as a function of the rescaled time $t^* = 4t/\tau_D$ for three representative sets of ratios r_1 and r_2. Solid curve was depicted for $r_1 = 10$ and $r_2 = 1$, i.e., for $\tau_D = 4\tau = 40\tau_\kappa$. The curve indicates that the characteristic time governing the behavior of $\sigma(t)$ is such that $t^* = 4t/\tau_D \approx 1$. For this case, numerical calculations give the first nonzero eigenvalue as $X_1 = 1.5$, $\tau_D \approx 4$, and $\tau \approx 1$ and $\tau_\kappa \approx 0.10$. Therefore, the time behavior of $\sigma(t)$ is governed

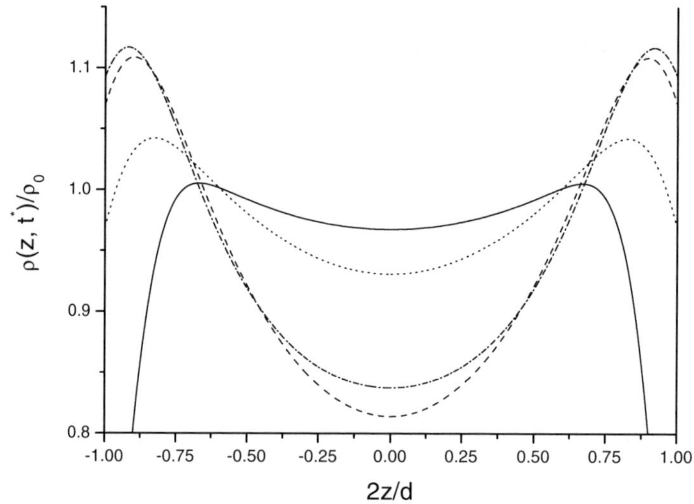

FIGURE 10.1
Behavior of $\rho(Z, t^*)/\rho_0$ versus Z, as predicted by Eq. (10.47) for $t^* = 4t/\tau_D = 0.01$. The curves are depicted for a representative set of the parameters r_1 and r_2. Solid line corresponds to $r_1 = 10.0$ and $r_2 = 1.0$, dotted line to $r_1 = 1.0$ and $r_2 = 1.0$, dashed line to $r_1 = 1.0$ and $r_2 = 5.0$, and dashed dotted line to $r_1 = 1.0$ and $r_2 = 10.0$. Reprinted with permission from Ref. [1]. Copyright (2004) by the American Physical Society.

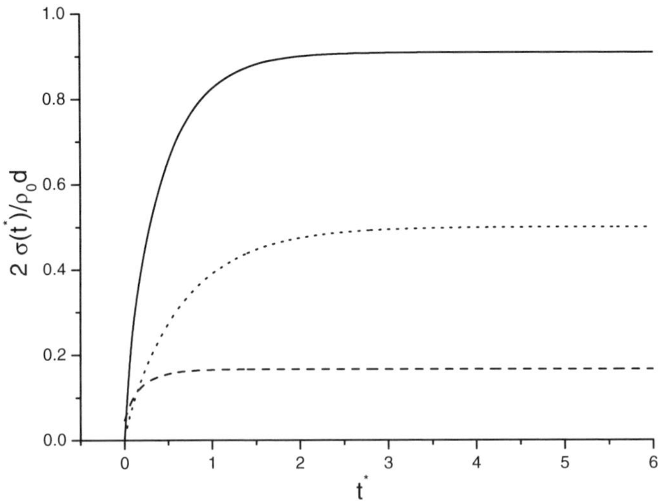

FIGURE 10.2
Behavior of $2\sigma(t*)/\rho_0 d$ versus $t^* = 4t/\tau_D$. Solid line was depicted for $r_1 = 10.0$ and $r_2 = 1.0$, dotted line for $r_1 = 1.0$ and $r_2 = 1.0$, and dashed line for $r_1 = 1.0$ and $r_2 = 5.0$. Reprinted with permission from Ref. [1]. Copyright (2004) by the American Physical Society.

by τ which is the greatest one among τ and τ_κ. A dotted curve was depicted for $r_1 = r_2 = 1.0$, i.e., for $\tau_D = 4\tau = 4\tau_\kappa$. The first non-zero eigenvalue is $X_1 = 1.21$, giving $\tau_D \approx 4$ and $\tau_\kappa \sim \tau \approx 1$. In this case, both characteristic times are important for the behavior of $\sigma(t)$. Dashed curve refers to $r_1 = 1$ and $r_2 = 5$, i.e., for $\tau_D = 20\tau = 4\tau_\kappa$. Numerical calculations give $X_1 = 2.01$, $\tau_D \approx 3.5$, and $\tau_\kappa \approx 0.9$ e $\tau \approx 0.2$. In this case, the time behavior of $\sigma(t)$ is governed by τ_κ.

The entire analysis, carried out with the help of numerical calculations, shows that, as expected on analytical grounds, the time behavior of the surface density of particles is governed by the largest one of the two characteristic times τ_κ and τ when τ_D is kept unchanged.

10.3 Drift current

In the previous section only the diffusion current originating from a non-homogeneous distribution of ρ was considered. It is given by $\mathbf{j}_D = -D\nabla\rho$.

Consider now that on ρ acts an external force $\mathbf{F} = -\nabla U(z)$, where $U(z)$ is the potential of $\mathbf{F}$. In this case $\mathbf{F}$ is responsible for a net current that, in the limit of small $|\mathbf{F}|$, is given by

$$\mathbf{j}_{\mathrm{F}} = \mu\rho\,\mathbf{F}, \tag{10.48}$$

where μ is the mobility and $\mathbf{j}_{\mathrm{F}}$ is the drift current due to $\mathbf{F}$. The total current is then

$$\mathbf{j} = \mathbf{j}_{\mathrm{D}} + \mathbf{j}_{\mathrm{F}} = -D\nabla\rho + \mu\rho\,\mathbf{F}. \tag{10.49}$$

The continuity equation (10.5) reads in this case

$$\frac{\partial \rho}{\partial t} = \nabla\cdot\left(D\nabla\rho + \mu\rho\nabla U\right), \tag{10.50}$$

which, in the one-dimensional case, is reduced to

$$\frac{\partial \rho}{\partial t} = \frac{\partial}{\partial z}\left(D\frac{\partial \rho}{\partial z} + \mu\rho\frac{\partial U}{\partial z}\right). \tag{10.51}$$

In the steady state $\rho = \rho(z)$ only and, from Eq. (10.51), one obtains

$$D\frac{d\rho}{dz} + \mu\rho\frac{dU}{dz} = \text{constant} = 0, \tag{10.52}$$

because in the steady state the net current vanishes. From Eq. (10.52) it follows that

$$\frac{1}{\rho}\frac{d\rho}{dz} = -\frac{\mu}{D}\frac{dU}{dz}. \tag{10.53}$$

Differential equation (10.53) can be easily integrated to give

$$\rho(z) = A\exp\left(-\frac{\mu U(z)}{D}\right), \tag{10.54}$$

where A is an integration constant to be determined by imposing the conservation of the number of particles. However, according to Boltzmann statistical mechanics the distribution $\rho(z)$ is given by

$$\rho(z) = A\exp\left(-\frac{U(z)}{k_{\mathrm{B}}T}\right). \tag{10.55}$$

By comparing Eq. (10.54), obtained by putting $j_{\mathrm{D}} + j_{\mathrm{F}} = 0$, and Eq. (10.55), obtained by means of the statistical mechanics, one derives

$$\frac{\mu}{D} = \frac{1}{k_{\mathrm{B}}T} \tag{10.56}$$

known as Einstein-Smoluchowski equation. As examples of the influence of the drift on the evolution of the density of particles in a slab, two meaningful cases will be discussed.

10.3.1 Drift in the presence of a constant field in the absence of adsorption

An ideal case relevant to a solution of particles in an external field will be considered. If the particles are neutral, as the dyes in a liquid crystal, the homogeneous external field can be identified with the one due to the gravity. If the particles are electrically charged, the external field is due to an external power supply. However, since in the analysis the external field is assumed constant, the density of ions is supposed so small that the actual electric field across the sample coincides with the external one. In other words, the actual electric field is assumed, in a first approximation, as independent of the adsorption phenomenon. It will be shown below that in the real case of ionic impurities, the actual electric field has to be deduced in a self-consistent manner. However, for the sake of simplicity, this feedback is neglected here.

Consider the simple case in which

$$\frac{dU}{dz} = h \tag{10.57}$$

is a constant. It is assumed that there is no adsorption from the limiting surfaces. In this framework Eq. (10.51) has to be solved by imposing the conditions (10.12) and (10.13).

By putting $\rho(z,t) = \rho_{\rm eq}(z) + \delta\rho(z,t)$, where $\rho(z,0) = \rho_0$, and $\rho_{\rm eq}(z)$ is given by (10.15), and hence $\lim_{t\to\infty} \delta\rho(z,t) = 0$, one obtains

$$\rho_{\rm eq}(z) = \rho_{\rm eq}(0) e^{-2\Omega\, z}, \tag{10.58}$$

where $\Omega = \mu h/(2D)$. The integration constant $\rho_{\rm eq}$ is obtained by imposing Eq. (10.12). One gets

$$\rho_{\rm eq}(0) = \rho_0 \frac{\Omega\, d}{\sinh(\Omega\, d)}. \tag{10.59}$$

From Eq. (10.59) it follows that for $\Omega = 0$, i.e., $h = 0$, $\rho_{\rm eq}(0) = \rho_0$, and, hence, $\rho_{\rm eq}(z) = \rho_0$. In this case, $\delta\rho(z,t) = 0$, for all $-d/2 \le z \le d/2$ and $0 \le t < \infty$. In other words, in the absence of adsorption phenomena at the limiting surfaces, without the external field, the diffusion phenomenon is obviously absent.

10.4 Relaxation time

The function $\delta\rho(z,t)$ is the solution of the linear partial differential equation

$$\frac{\partial(\delta\rho)}{\partial t} = D \frac{\partial^2(\delta\rho)}{\partial z^2} + \mu h \frac{\partial(\delta\rho)}{\partial z}. \tag{10.60}$$

A solution to Eq. (10.60) can have the form

$$\delta\rho(z,t) = e^{-\beta^2 t}\phi_\beta(z), \tag{10.61}$$

where $\beta \neq 0$. By substituting Eq. (10.61) into Eq. (10.60) one concludes that $\phi_\beta(z)$ is the solution of the ordinary differential equation with constant coefficients

$$\frac{d^2\phi_\beta(z)}{dz^2} + 2\Omega\frac{d\phi_\beta(z)}{dz} + \frac{\beta^2}{D}\phi_\beta(z) = 0. \tag{10.62}$$

The characteristic exponents relevant to $\phi_\beta(z)$ are then $m_\beta = -\Omega \pm i\omega_\beta$, where

$$\omega_\beta = \sqrt{\frac{\beta^2}{D} - \Omega^2}. \tag{10.63}$$

It follows that

$$\phi_\beta(z) = e^{-\Omega\, z}\left[a_\beta \cos\left(\omega_\beta z\right) + b_\beta \sin\left(\omega_\beta z\right)\right], \tag{10.64}$$

and, making use of Eq. (10.61), one has for $\delta\rho(z,t)$ the expression

$$\delta\rho(z,t) = e^{-\beta^2 t - \Omega\, z}\left[a_\beta \cos\left(\omega_\beta z\right) + b_\beta \sin\left(\omega_\beta z\right)\right], \tag{10.65}$$

whose functional dependence on t and z is typical of a drift in presence of diffusion. The total current density is, in the present case, given by

$$j = -D\frac{\partial\rho}{\partial z} - \mu h\rho. \tag{10.66}$$

By taking into account that $\rho(z,t) = \rho_{\text{eq}}(z) + \delta\rho(z,t)$, where $\rho_{\text{eq}}(z)$ is given by Eq. (10.58), one obtains

$$j = -D\frac{\partial(\delta\rho)}{\partial z} - \mu h(\delta\rho), \tag{10.67}$$

that, for Eq. (10.65), can be rewritten as

$$\begin{aligned} j = -D\, e^{-(\beta^2 t + \Omega\, z)} \times \{&\omega_\beta\left[-a_\beta \sin\left(\omega_\beta z\right) + b_\beta \cos\left(\omega_\beta z\right)\right] \\ &+ \Omega\left[a_\beta \cos\left(\omega_\beta z\right) + b_\beta \sin\left(\omega_\beta z\right)\right]\}. \end{aligned} \tag{10.68}$$

The boundary conditions $j(\pm d/2, t) = 0$ give the homogeneous system

$$\begin{aligned} a_\beta\left(\Omega\cos X_\beta - \omega_\beta \sin X_\beta\right) + b_\beta\left(\Omega\sin X_\beta + \omega_\beta\cos X_\beta\right) &= 0 \\ a_\beta\left(\Omega\cos X_\beta + \omega_\beta \sin X_\beta\right) - b_\beta\left(\Omega\sin X_\beta - \omega_\beta\cos X_\beta\right) &= 0, \end{aligned} \tag{10.69}$$

where $X_\beta = \omega_\beta d/2$. A non-trivial solution for a_β and b_β is possible when the determinant of the coefficients of the system (10.69) vanishes. This condition gives $[\Omega^2+\omega_\beta^2]\,\sin(2X_\beta)=0$, from which it follows that $X_n=\omega_n d/2=n\pi/2$, where $n=1,2,\dots$. Hence $\omega_n=n\,\pi/d$ and $\beta_n^2=D\left[\Omega^2+n^2\pi^2/d^2\right]$. The characteristic times $\tau_n=1/\beta_n^2$ are then given by [4]

$$\tau_n=\frac{1}{\beta_n^2}=\frac{1}{1/t_{\rm h}+n^2/t_{\rm D}}, \tag{10.70}$$

where $t_{\rm h}=1/(\Omega^2\,D)$ is an intrinsic time connected with the presence of the external field, and $t_{\rm D}=d^2/(\pi^2\,D)$ is the diffusion time in the present case. The longest characteristic time, for $n=1$, is then

$$\tau_1=\frac{1}{1/t_{\rm h}+1/t_{\rm D}}=\frac{t_{\rm D}\,t_{\rm h}}{t_{\rm h}+t_{\rm D}}. \tag{10.71}$$

For $t_{\rm D}\ll t_{\rm h}$, i.e., $\pi/d\gg\Omega$, from Eq. (10.71), it follows that $\tau_1\sim t_{\rm D}$. In the opposite case where $\pi/d\ll\Omega$ one finds $\tau_1\sim t_{\rm h}$.

If the diffusing particles are ions contained in an isotropic liquid, and the drift is due to an external electric field, by taking into account that $h=qE$, where q is the electric charge of the ion, and $\mu/D=1/k_{\rm B}T$, the relation $\tau_1\sim t_{\rm h}$ reads

$$\tau_1\sim t_{\rm E}=4\left(\frac{d^2}{\mu\,q\,V_0}\right)\frac{k_{\rm B}T}{qV_0}, \tag{10.72}$$

where V_0 is the applied voltage. Equation (10.72) holds in the hypothesis that the concentration of ions is so low that the effective electric field in the sample practically coincides with the external one. This characteristic time is usually determined in elementary textbooks as $\tau_{\rm d}=d/v_{\rm d}=d/(\mu\,q\,E)=d^2/(\mu\,q\,V_0)$. From the expression of $\tau_{\rm d}$ we derive that it depends on the temperature by means of the ion mobility μ. Consequently, the relative variation of $\tau_{\rm d}$ occurring for a variation of the temperature T of δT is

$$\frac{\delta\tau_{\rm d}}{\tau_{\rm d}}=\frac{1}{\mu}\frac{d\mu}{dT}\,\delta T. \tag{10.73}$$

The relation reported above for τ_1 shows that $\tau_1=a\tau_{\rm d}$, where the parameter $a=4\,k_{\rm B}T/(qV_0)$ represents the importance of the thermal agitation energy with respect to the potential energy responsible for the drift. It follows that when T changes of δT, the relative variation of the relaxation time is

$$\frac{\delta\tau_1}{\tau_1}=\left\{\frac{1}{\mu}\frac{d\mu}{dT}+\frac{1}{T}\right\}\,\delta T. \tag{10.74}$$

To evaluate $d\mu/dT$ we assume for the temperature dependence of the diffusion coefficient of the ions in the nematic material the one predicted by Stokes-Einstein law's [5], $D=k_{\rm B}T/(6\pi\eta R_0)$, where $\eta(T)$ is the viscosity of the ion in the nematic liquid, and R_0 an average dimension of the ion. By taking into account the Einstein-Smoluchowski relation $\mu/D = 1/(k_B T)$, we get for the ion mobility the expression $\mu(T) = 1/[6\pi\eta(T)R_0]$, from which we obtain

$$\frac{1}{\mu}\frac{d\mu}{dT} = -\frac{1}{\eta}\frac{d\eta}{dT}. \tag{10.75}$$

If we identify $\eta(T)$ with the viscosity of the nematic liquid crystal, then it follows that $(1/\eta)(d\eta/dT) \sim 10^{-2}$ [6]. In this case $\delta\tau_1/\tau_1$ differs from $\delta\tau_d/\tau_d$ for the term $\delta T/T$, which is not negligible with respect to $(1/\eta)(d\eta/dT)$.

Note that τ_1, given by (10.72), represents the time necessary for the system to reach the equilibrium, where the drift current is balanced by the diffusion current. On the contrary, τ_d represents the time required by an ion to travel across the entire sample. Usually $\tau_1 < \tau_d$ because not all the ions have to be moved to reach the equilibrium.

10.5 Time evolution of the density of ions

Since Eq. (10.60) is linear, the general solution for the problem can be written in the form

$$\delta\rho(z,t) = \sum_n e^{-\beta_n^2 t}\phi_n(z), \tag{10.76}$$

where

$$\phi_n(z) = e^{-\Omega z}\left[a_n \cos(\omega_n z) + b_n \sin(\omega_n z)\right]. \tag{10.77}$$

The quantities

$$p_n = \frac{a_n}{b_n} = \frac{\omega_n \cos X_n + \Omega \sin X_n}{\omega_n \sin X_n - \Omega \cos X_n}, \tag{10.78}$$

when the eigenvalues have been determined, are known quantities. Consequently, it is possible to rewrite Eq. (10.76) in the form

$$\delta\rho(z,t) = e^{-\Omega z}\sum_n b_n e^{-\beta^2 t}\,\psi_n(z), \tag{10.79}$$

where

$$\psi_n(z) = \sin(\omega_n z) + p_n \cos(\omega_n z). \tag{10.80}$$

From (10.80) it follows that

$$\langle\psi_n|\psi_m\rangle = \int_{-d/2}^{d/2} \psi_n(z)\psi_m(z)dz = 0, \tag{10.81}$$

for $n \neq m$, and

$$\langle e^{-\Omega z} | \psi_m \rangle = \int_{-d/2}^{d/2} e^{-\Omega z} \, \psi_n(z) dz = 0. \tag{10.82}$$

Notice that

$$\int_{-d/2}^{d/2} \delta\rho(z,t) dz = \sum_n b_n e^{-\beta^2 t} \int_{-d/2}^{d/2} e^{-\Omega z} \psi_n(z) dz = 0, \tag{10.83}$$

for Eq. (10.82). Consequently, Eq. (10.12) is verified for all t, as required.

It is now possible to determine the coefficients b_n appearing in Eq. (10.79). From the expression $\rho(z,t) = \rho_{\text{eq}}(z) + \delta\rho(z,t)$ one has, in the limit $t = 0$,

$$\delta\rho(z,0) = \rho_0 - \rho_{\text{eq}}(z). \tag{10.84}$$

By taking into account Eq. (10.79), one can rewrite Eq. (10.84) in the form

$$\sum_n b_n \psi_n(z) = f(z), \tag{10.85}$$

where, for Eq. (10.59),

$$f(z) = e^{\Omega z} \, \delta\rho(z,0) = \rho_0 \left[e^{\Omega z} - \frac{\Omega d}{\sinh(\Omega d)} \, e^{-\Omega z} \right]. \tag{10.86}$$

From Eq. (10.85), for Eq. (10.81), one obtains

$$b_m = \frac{1}{N_m} \int_{-d/2}^{d/2} f(z) \psi_m(z) dz, \tag{10.87}$$

where $N_m = \langle \psi_m | \psi_m \rangle$.

In Figure 10.3, we show the time evolution of $\rho(z,t) = \rho_{\text{eq}} + \delta\rho(z,t)$ for three typical values of the external field h, such that $\tau_1 \sim t_{\text{D}}$, $\tau_1 \sim t_{\text{D}} t_{\text{h}}/(t_{\text{D}} + t_{\text{h}})$, and $t_1 \sim t_{\text{h}}$. As is evident from Figure 10.3c, in the limit of a large external field, that in practical units means $V_0 \gg \pi(k_{\text{B}}\, T/q) \approx 0.075\,\text{V}$, the equilibrium distribution is reached after a time $t_{\text{eq}} \approx 0.4 t_{\text{D}}$.

10.6 Drift-diffusion phenomenon in nematic liquid crystals

The analysis reported above is, strictly speaking, valid only for isotropic liquids. The application of our results to NLC requires some supplementary hypothesis. As is known, nematic liquid crystals are anisotropic liquids, whose optical axes coincide with the average molecular orientation, called nematic

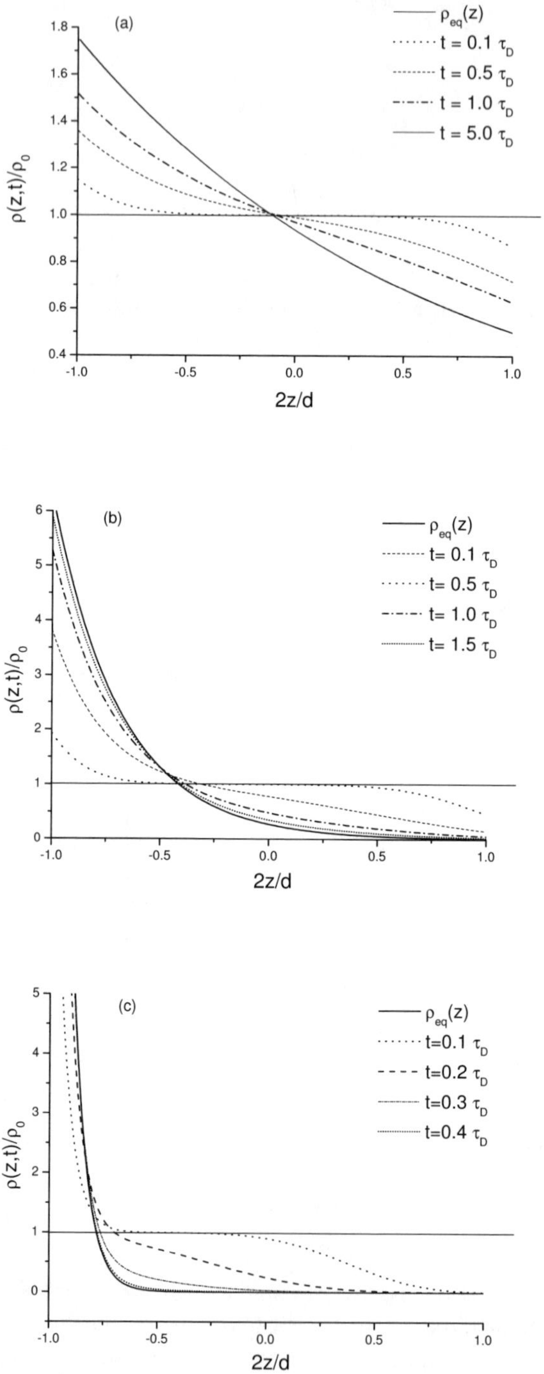

FIGURE 10.3
Time evolution of $\rho(z,t)$ for three typical values of the external voltage $V_0 = \alpha\pi(k_B T/q)$ for (a) $\alpha = 1/5$, (b) $\alpha = 1$ and (c) $\alpha = 5$. Reprinted from Ref. [4].

director and indicated by **n**. The physical parameters of nematic materials are described by symmetric tensors of second order [7]. In particular, the dielectric constant, diffusion coefficient, and the ion mobility of the nematic medium have different values along and perpendicular to the director. Their representing tensors are of the kind

$$\begin{aligned} \varepsilon_{ij} &= \varepsilon_{\perp}\delta_{ij} + \varepsilon_a n_i n_j, \\ D_{ij} &= D_{\perp}\delta_{ij} + D_a n_i n_j, \\ \mu_{ij} &= \mu_{\perp}\delta_{ij} + \mu_a n_i n_j, \end{aligned} \tag{10.88}$$

where $\varepsilon_a = \varepsilon_{\|} - \varepsilon_{\perp}$, and $\|$ and $\perp$ refers to **n**. Similar considerations hold for the tensors of diffusion and ion mobility. This circumstance implies that the effective dielectric constant, diffusion and ion mobility depend on the nematic orientation described by **n**.

Since we are considering a nematic sample submitted to an external field, other limitations on the values of the field have to be imposed. The nematic materials present an electric polarization connected with the nematic distortion that we have not considered in the analysis. Furthermore, if the electric field is strong enough it can induce a distortion. It follows that our results are valid also for NLC only if they are uniformly oriented. This implies that the external field has to be smaller than the critical field for the Fréedericksz transition. In this case our results can be directly applied to analyze the typical relaxation time in a drift-diffusion phenomenon. However, since the values of the parameters along and normal to the director are of the same order of magnitude [8], the analysis presented above gives the correct order of magnitude of the relaxation time for the ions in an NLC even in the case in which it is distorted.

Therefore, we can apply our result to typical experimental situations concerning NLC. To have an estimation we assume as experimental values $d \sim 8$ μm, $D \sim 10^{-11}\,\mathrm{m^2/s}$ [9] we have $t_{\mathrm{D}} = d^2/(\pi^2 D) \sim 0.65\,\mathrm{s}$, and hence $t_{\mathrm{eq}} \approx 0.26\,\mathrm{s}$. This means that the ions follow the external field variations, if the external field is changing with a characteristic time larger than 0.3 s. However, in the case in which the ions are macroparticles coming from the deterioration of the aligning layers [9], the scale times can be completely different. In this case for $D \sim 10^{-12}\,\mathrm{m^2/s}$, which corresponds to a radius of the ion of the order of 20 nm [9], we have $t_{\mathrm{D}} \sim 6.5\,\mathrm{s}$ and hence $t_{\mathrm{eq}} \sim 2.6\,\mathrm{s}$. In this situation, if the external field is changing with a period of the order of 1 s, the macroparticles coming from the surfaces do not participate in the phenomenon. Of course, the ions dissolved in the liquid crystal, for which $D \sim 10^{-11}\,\mathrm{m^2/s}$, on the contrary, participate. This conclusion can be of some importance to the experimental determination of physical parameters of NLC when external electric fields are applied [10]–[12].

10.6.1 Drift in the presence of a non-homogeneous field in the absence of adsorption

In a real problem one has a species of particles interacting with the limiting surfaces via van der Waals forces. In a first approximation, the corresponding potential responsible for the drift can be written as

$$U(z) = -U_0 \frac{\cosh(z/\lambda)}{\cosh(d/2\lambda)}, \tag{10.89}$$

where λ is of the order of the penetration of the van der Waals forces. Since λ is in the mesoscopic range, whereas d is a macroscopic quantity, $U(z)$ is strongly localized close to the bounding surfaces. In this situation, when the surfaces are absent, the density of particles is homogeneous across the sample. However, when the surfaces are switched-on, i.e., the interaction between the particles and the substrate is present, a current of drift takes place close to the boundary until the diffusion current balances the drift current. This problem can be solved by means of the same approach used in the previous section. The aim is now to show that the drift-diffusion phenomenon, in a system characterized by a potential localized on mesoscopic lengths close to the boundary can be used to interpret the kinetic equation at the adsorbing surface.

It is supposed that $U(z)$ can be approximated by

$$U(z) = \begin{cases} U_1(z) = h\,(z + z^*), & -d/2 \leq z \leq z_1 \longrightarrow \text{Region 1} \\ U_{\rm B}(z) = 0, & z_1 \leq z \leq z_2 \longrightarrow \text{Region B} \\ U_2(z) = -h\,(z - z^*), & z_2 \leq z \leq d/2 \longrightarrow \text{Region 2,} \end{cases} \tag{10.90}$$

where, as before, $h = dU/dz = U_0/\lambda$, $z_1 = -z^*$, and $z_2 = z^*$, with $z^* = d/2 - \lambda$, schematically represented in Figure 10.4.

The problem under consideration concerns two surface layers where the potential is such that $dU(z)/dz =$ constant, and a bulk region where $U(z) = 0$, and the current reduces to the diffusion current. The continuity equations in the three regions are

$$\frac{\partial \rho_\alpha}{\partial t} = -\frac{\partial j_\alpha}{\partial z}, \tag{10.91}$$

in the respective $z-$regions, where $\alpha = 1, \mathrm{B}, 2$ and

$$j_\alpha = -\left(D \frac{\partial \rho_\alpha}{\partial z} + \varepsilon_\alpha \mu h \rho_\alpha \right) \tag{10.92}$$

with $\varepsilon_1 = 1$, $\varepsilon_{\rm B} = 0$ and $\varepsilon_2 = -1$. Equations (10.91) have to be solved with the boundary conditions

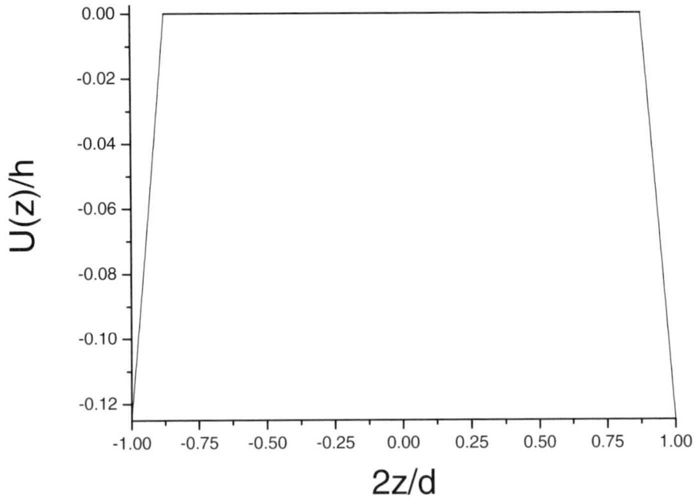

FIGURE 10.4
Profile of the piecewise continuous potential responsible for the adsorption phenomenon.

$$\rho_1(z_1,t) = \rho_{\rm B}(z_1,t) \quad \text{and} \quad \rho_{\rm B}(z_2,t) = \rho_2(z_2,t), \tag{10.93}$$

for what concerns the densities, and

$$\begin{aligned} j_1(-d/2,t) = 0, \quad j_1(z_1,t) = j_{\rm B}(z_1,t), \\ j_{\rm B}(z_2,t) = j_2(z_2,t), \quad j_2(d/2,t) = 0 \end{aligned} \tag{10.94}$$

for what concerns the currents. By putting again $\rho_\alpha(z,t) = \rho_{{\rm eq}\,\alpha}(z) + \delta\rho_\alpha(z,t)$, one obtains, by taking into account (10.93),

$$\rho_{{\rm eq}\,\alpha}(z) = p\,\exp\left[-\mu\, U_\alpha(z)/D\right]. \tag{10.95}$$

The constant p is deduced by imposing

$$\int_{-d/2}^{d/2} \rho_{\rm eq}(z)\,dz = \rho_0\,d. \tag{10.96}$$

By using Eq. (10.95) and the Einstein-Smoluchowski relation (10.56), we have

$$p = \rho_0 \left\{ 1 + \frac{2\lambda}{d} \left[\frac{e^{U_0/k_{\rm B}T} - 1}{U_0/k_{\rm B}T} - 1 \right] \right\}^{-1}. \tag{10.97}$$

The functions $\delta\rho_\alpha(z,t)$ are solutions of the differential equations

$$\frac{\partial(\delta\rho_\alpha)}{\partial t} = D\frac{\partial^2(\delta\rho_\alpha)}{\partial z^2} - \varepsilon_\alpha \mu h \frac{\partial \rho_\alpha}{\partial z}, \tag{10.98}$$

which are the continuity equations for the three layers. The solutions of Eqs. (10.98) are searched in the form

$$\delta\rho_\alpha(z,t) = e^{-\beta_\alpha^2\, t}\,\phi_\alpha(z). \tag{10.99}$$

Since (10.93) must be verified for all t, it follows that $\beta_1^2 = \beta_{\rm B}^2 = \beta_2^2 = \beta^2$, and Eq. (10.99) can be rewritten as

$$\delta\rho_\alpha(z,t) = e^{-\beta^2\, t}\,\phi_\alpha(z), \tag{10.100}$$

where the functions $\phi_i(z)$ are such that

$$\phi_1(z_1) = \phi_{\rm B}(z_1) \qquad \text{and} \qquad \phi_{\rm B}(z) = \phi_2(z_2) \tag{10.101}$$

for Eq. (10.93). By substituting Eq. (10.100) into Eq. (10.98) one gets

$$\frac{d^2\phi_\alpha}{dz^2} + \varepsilon_\alpha \frac{\mu}{D} h \frac{d\phi_\alpha}{dz} + \frac{\beta^2}{D}\phi_\alpha = 0. \tag{10.102}$$

The characteristic exponents of the ϕ_α–functions are then $m_1 = -\Omega \pm i\omega_{\rm S}$, $m_{\rm B} = \pm i\omega_{\rm B}$, and $m_2 = \Omega \pm i\omega_{\rm S}$, where

$$\omega_{\rm S} = \sqrt{\frac{\beta^2}{D} - \Omega^2} \qquad \text{and} \qquad \omega_{\rm B} = \frac{\beta}{\sqrt{D}}. \tag{10.103}$$

It follows that

$$\phi_\alpha(z) = e^{-\varepsilon_\alpha\,\Omega\, z}\left[a_\alpha \cos(\omega_\alpha z) + b_\alpha \sin\left(\omega_\alpha\, z\right)\right]. \tag{10.104}$$

By imposing the boundary conditions (10.93) and (10.94), one obtains a homogeneous system in the coefficients a_α and b_α. By putting

$$a_1 = C_1, \quad b_1 = C_2, \quad a_{\rm B} = C_3, \quad b_{\rm B} = C_4, \quad a_2 = C_5, \quad \text{and} \quad b_2 = C_6, \tag{10.105}$$

one can write the boundary conditions (10.93) and (10.94) in the matricial form

$$M_{ij}C_j = 0, \tag{10.106}$$

where the elements of the matrix M are

$$
\begin{aligned}
M_{11} &= \cos\left(\omega_{\mathrm{S}}\, z^*\right), M_{12} = -\sin\left(\omega_{\mathrm{S}}\, z^*\right), \\
M_{13} &= -\cos\left(\omega_{\mathrm{B}}\, z^*\right), \quad M_{14} = \sin\left(\omega_{\mathrm{B}}\, z^*\right), \\
M_{15} &= M_{16} = 0, M_{21} = M_{22} = 0, \\
M_{23} &= -\cos\left(\omega_{\mathrm{B}}\, z^*\right), M_{24} = -\sin\left(\omega_{\mathrm{B}}\, z^*\right), \\
M_{25} &= \cos\left(\omega_{\mathrm{S}}\, z^*\right), M_{26} = \sin\left(\omega_{\mathrm{S}}\, z^*\right), \\
M_{31} &= \omega_{\mathrm{S}} \sin\left(\omega_{\mathrm{S}}\, d/2\right) + \Omega \cos\left(\omega_{\mathrm{S}}\, d/2\right), \\
M_{32} &= \omega_{\mathrm{S}} \cos\left(\omega_{\mathrm{S}}\, d/2\right) - \Omega \sin\left(\omega_{\mathrm{S}}\, d/2\right) \\
M_{33} &= M_{34} = M_{35} = M_{36} = 0, \\
M_{41} &= \omega_{\mathrm{S}} \sin\left(\omega_{\mathrm{S}}\, z^*\right) + \Omega \cos\left(\omega_{\mathrm{S}}\, z^*\right), \\
M_{42} &= \omega_{\mathrm{S}} \cos\left(\omega_{\mathrm{S}} z^*\right) - \Omega \sin\left(\omega_{\mathrm{S}} z^*\right), \\
M_{43} &= -\omega_{\mathrm{B}} \sin\left(\omega_{\mathrm{B}}\, z^*\right), M_{44} = -\omega_{\mathrm{B}} \cos\left(\omega_{\mathrm{B}}\, z^*\right), \\
M_{45} &= M_{46} = 0, M_{51} = M_{52} = 0, \\
M_{53} &= \omega_{\mathrm{B}} \sin\left(\omega_{\mathrm{S}}\, z^*\right), M_{54} = -\omega_{\mathrm{B}} \cos\left(\omega_{\mathrm{B}}\, z^*\right), \\
M_{55} &= -\omega_{\mathrm{S}} \sin\left(\omega_{\mathrm{S}}\, z^*\right) - \Omega \cos\left(\omega_{\mathrm{S}} z^*\right), \\
M_{56} &= \omega_{\mathrm{S}} \cos\left(\omega_{\mathrm{S}}\, z^*\right) - \Omega \sin\left(\omega_{\mathrm{S}} z^*\right), \\
M_{61} &= M_{62} = M_{63} = M_{64} = 0, \\
M_{65} &= -\omega_{\mathrm{S}} \sin\left(\omega_{\mathrm{S}}\, d/2\right) - \Omega \cos\left(\omega_{\mathrm{S}}\, d/2\right), \\
M_{66} &= \omega_{\mathrm{S}} \cos\left(\omega_{\mathrm{S}}\, d/2\right) - \Omega \sin\left(\omega_{\mathrm{S}}\, d/2\right).
\end{aligned}
\tag{10.107}
$$

By imposing the condition $\det M = 0$ one obtains the equation for the eigenvalues $\beta \neq 0$ of the problem. Simple calculations give

$$
\begin{aligned}
4\Omega\omega_{\mathrm{B}} \cos(\delta\, \omega_{\mathrm{B}}) + 4\cos(2\lambda\sqrt{\omega_{\mathrm{B}}^2 - \Omega^2})\, h_\Omega \\
- 4\sqrt{\omega_{\mathrm{B}}^2 - \Omega^2}\, \sin(2\lambda\sqrt{\omega_{\mathrm{B}}^2 - \Omega^2})\, h_{\mathrm{B}} = 0,
\end{aligned}
\tag{10.108}
$$

with

$$
\begin{aligned}
h_\Omega &= \left[-\Omega\, \omega_{\mathrm{B}} \cos(\delta\, \omega_{\mathrm{B}}) + (\Omega^2 - \omega_{\mathrm{B}}^2) \sin(\delta\, \omega_{\mathrm{B}})\right] \quad \text{and} \\
h_{\mathrm{B}} &= \left[\omega_{\mathrm{B}} \cos(\delta\, \omega_{\mathrm{B}}) - \Omega \sin(\delta\, \omega_{\mathrm{B}})\right],
\end{aligned}
$$

where $\delta = d - 2\lambda$ is the thickness of the bulk. When the eigenvalues are determined, one has

$$
\delta\rho_\alpha(z,t) = e^{-\varepsilon_\alpha\, \Omega\, z} \sum_{n=1}^{\infty} e^{-\beta_n^2\, t} \left[a_{\alpha n} \cos\left(\omega_{Sn} z\right) + b_{\alpha n} \sin\left(\omega_{\alpha n} z\right)\right], \tag{10.109}
$$

where $\omega_{1n} = \omega_{2n} = \omega_{Sn} = \sqrt{\beta_n^2/D - \Omega^2}$ and $\omega_{\mathrm{B}} = \omega_n = \beta_n/D$. The analysis from now on is standard.

10.6.2 Validity of the phenomenological model

For simplicity, in the analysis reported above, it is assumed that there is no adsorption on the limiting surfaces. However, since at the end, in the steady state, the density of the particles close to $z = \pm d/2$ will be larger than the one in the bulk, the considered system is, actually, a model for the adsorption. By comparing the results of the present analysis with the analysis of the adsorption from the surfaces, it is possible to connect the phenomenological parameters κ and τ appearing in Eq. (10.26) with the microscopic parameters U_0 and λ. According to Eq. (10.95), in the steady state the bulk density of particles is $\rho_{\rm eq} = \rho_{\rm eqB} = \rho_0/(1+2R)$, where, for Eq. (10.97),

$$R = \frac{\lambda}{d}\left[\frac{e^{U_0/k_{\rm B}T}-1}{U_0/k_{\rm B}T} - 1\right]. \tag{10.110}$$

Close to the bounding walls, for Eq. (10.95), the profiles of the particle densities are $\rho_{{\rm eq}\,\alpha} = p\exp[-U_\alpha(z)/k_{\rm B}T]$. Since $\lambda \ll d$ it is possible to introduce a surface density of particles by means of the relation

$$\sigma_{{\rm eq}\,2} = \int_{z_2}^{d/2} [\rho_{{\rm eq}\,2}(z) - \rho_{{\rm eq\,B}}(z)]\, dz, \tag{10.111}$$

and a similar relation for $\sigma_{{\rm eq}\,1}$. Simple calculations give $\sigma_{{\rm eq}\,1} = \sigma_{{\rm eq}\,2} = \sigma_{\rm eq}$, where $\sigma_{\rm eq} = [R/(1+2R)]\,\rho_0 d$.

By comparing $\rho_{\rm eq}$ and $\sigma_{\rm eq}$ obtained above with the ones given by (10.32) one obtains

$$\kappa\tau = \lambda\left[\frac{e^{U_0/k_{\rm B}T}-1}{U_0/k_{\rm B}T} - 1\right]. \tag{10.112}$$

As stated above $\kappa\tau$ is proportional to the penetration length of the surface forces, and the constant of proportionality depends on the adsorption energy $A = U_0/k_{\rm B}T$.

In order to show that the system under consideration implies a kinetic equation at the surface of the kind (10.26), we define an effective surface density $\sigma(t)$ by means of the relation

$$\sigma = \int_{z_2}^{d/2} [\rho_2(z,t) - \rho_{\rm B}(z,t)]\; dz \approx \int_{z_2}^{d/2} \rho_2(z,t)\, dz, \tag{10.113}$$

because λ is a mesoscopic length. By assuming that in the surface layer the bulk density is well approximated by

$$\rho_2(z,t) = \rho_2(z_2,t) + \left(\frac{\partial \rho_2}{\partial z}\right)_{z_2}(z - z_2), \tag{10.114}$$

and using Eq. (10.91) and the boundary conditions, Eqs. (10.93) and (10.94), one obtains

$$\frac{d\sigma_2(t)}{dt} = \left(\mu\, h + 2\frac{D}{\lambda}\right) \rho_{\rm B}(z_2, t) - \frac{2\,D}{\lambda^2}\,\sigma_2(t), \tag{10.115}$$

which coincides with the kinetic equation written phenomenologically at the adsorbing surface, i.e., Eq. (10.26). In (10.115) the coefficient of $\rho_{\rm B}(z_2, t)$ contains two terms. The first represents the drift velocity due to the surface field in the surface layer, and it is responsible for the adsorption phenomenon. A simple analysis shows that when the adsorption energy U_0 is large with respect to the thermal agitation energy, $\mu h + 2D/\lambda \sim \mu h$. In fact, by taking into account that $\mu/D = 1/(k_{\rm B}T)$ and that $h = U_0/\lambda$, we get $\mu h + 2D/\lambda = (D/\lambda)[(U_0/k_{\rm B}T) + 2]$, that in the limit of $U_0 \gg k_{\rm B}T$ reduces to the first term, as stated above. Hence, by comparing Eq. (10.115) with Eq. (10.26) we deduce that the effective $\kappa = \mu\, h$ and $1/\tau = 2D/\lambda^2$. Therefore, $\kappa\,\tau/\lambda = U_0/(2\,k_{\rm B}T)$, which coincides with the power expansion of the Eq. (10.112) in the limit of $U_0 \ll k_{\rm B}T$. With this type of analysis it is possible to connect, separately, κ and τ with the microscopic parameters of the model U_0 and λ. In particular, τ is of the order of the diffusion time of the particles in the surface layer [13].

The problem considered above on the adsorption of neutral particles from the limiting surfaces is of some importance in the physics of liquid crystals. In fact, one of the more delicate points in the display realization is to align in a reproducible manner the nematic sample. Usual techniques are based on mechanical treatment of the limiting surfaces covered by appropriate surfactants, as discussed in Chapter 2. Recently, an alternative technique has been proposed. According to this new procedure, the nematic liquid crystal is doped with a dye. After some time, the dye molecules are adsorbed by the limiting surfaces. This occurs because the interaction energy of the dye molecule with the substrate is larger than the one of the nematic molecules. Since the dye molecules are similar to the nematic molecules during the adsorption phenomenon they are oriented along the nematic field imposed by a reference surface. If, in this situation, the polarized light illuminates the surface covered by dye molecules, a structural transformation is induced on them. The resulting effect is a change of the easy axis, controlled by the light [14]–[16].

10.7 Molecular reorientation dynamics in a nematic cell

The adsorption of neutral particles is relatively important because it can describe the dye adsorption in nematic samples. In that case, the simple microscopical model proposed for the adsorption invokes the existence of a surface potential responsible for a drift current in the bounding layers that can be interpreted in terms of a kinetic equation at the surface. In the analysis, the surface field has been considered independent of the density of particles in the surface layer, as well as the phenomenological parameters entering in the kinetic equation at the surface. These approximations work reasonably well in the case of dye molecules, which are neutral from the electrical point of view. However, as discussed in the preceding chapters, devoted to the static influence of the ions on the nematic properties, when the adsorbed particles are electrically charged the analysis of the problem has to be faced in a different manner. In fact, since the ions are charged, during the adsorption phenomenon the effective electric field in the sample is changing, until the adsorption phenomenon takes place. When the steady state is reached, the actual electric field in the sample is the sum of the external one and of the one due to the ion separation induced by the field itself. In this case, as discussed in Chapter 7, the actual field can be completely different from the external one, and the dynamical solution presented in the section devoted to the analysis of the drift in the presence of a constant field does not work at all. As already underlined, that type of analysis works well only if the density of ions is so small that the actual field practically coincides with the external one. But in a commercial liquid crystal this is not usually the case.

In this section, the evolution of the system towards the equilibrium state is investigated by solving numerically the continuity equation for the electric charge, taking into account the currents of drift and of diffusion. The goal is the description of the dynamical behavior of a nematic cell submitted to a step-like voltage [17]. This kind of analysis is fundamental for practical applications of ferroelectric liquid crystals, as discussed in Ref. [18].

10.7.1 Basic equations for the electrical variables

A nematic sample in the shape of a slab of thickness d is considered once more. The z-axis of the Cartesian reference frame is normal to the bounding surfaces at $z = \pm d/2$. The electrodes are assumed as being covered by a layer of thickness λ of dielectric material avoiding transfer of charge from the nematic sample to the circuit, as shown in Figure 10.5. Consequently, the electrodes can be considered as blocking. An external power supply is connected to the sample to fix the electric potential of the electrodes at $z = \mp(d/2 + \lambda)$ at $\mp V_0/2$, respectively. The external charge densities sent by the power supply to the electrodes to fix this difference of potential are indicated by $\pm q\Sigma$, where q is the unit charge and Σ the surface density of charged particles.

All physical quantities entering in the problem are assumed as depending only on the z-coordinate. In thermodynamical equilibrium, in the absence of an external electric field, and by neglecting selective ion adsorption, the liquid crystal contains a density $n_{0+}(z, 0) = n_{0-}(z, 0) = n_0$ of ions, uniformly distributed. Ionic recombination [19] is not considered, even if this phenomenon can be easily taken into account. However, according to the investigations performed by Murakami and Naito on the low frequency dielectric properties of nematic samples, this assumption seems to be supported experimen-

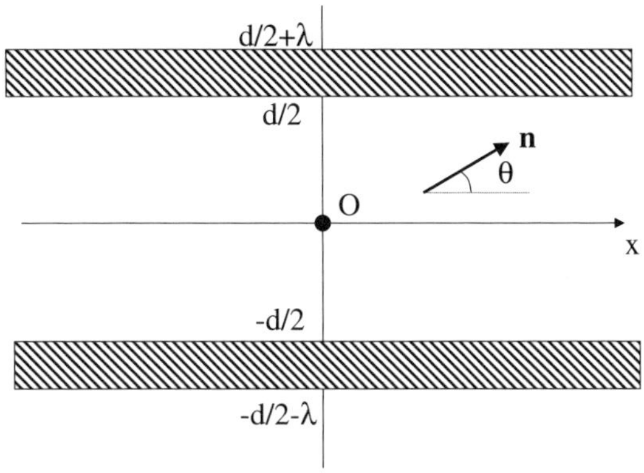

FIGURE 10.5
Nematic sample considered in the analysis developed in this section. λ is the thickness of the dielectric material, of dielectric constant ϵ_S, deposited on the electrodes to avoid charge injection. The dielectric constant of the liquid crystal is $\epsilon_\perp$. In the absence of the external field the nematic orientation is homogeneous across the sample at an angle θ_0, due to the surface treatments. The surface anchoring energy is assumed strong. Reprinted with permission from Ref. [20]. Copyright (2004) by the American Physical Society.

tally [21]. In fact, the experimental data reported in Ref. [21] show that the ionic concentration is practically temperature independent, in the limit of low applied voltage. This behavior is not consistent with a chemical reaction of association-dissociation.

The presence of the ions modifies deeply the electrical properties of the liquid crystal when an external electric field is applied to the sample, because the electric potential is no longer an harmonic function. This aspect of the problem has been already discussed in the chapters devoted to the static analysis of the influence of the ions on the liquid crystal properties. Furthermore, also the concept of electrical chemical potential no longer has a clear meaning since the ionic charge distribution depends on the applied voltage. In fact, in the presence of an external electric field, the ions move under the effect of the electric force until an equilibrium distribution is reached. In this situation the bulk density of positive ions, $n_+(z,t)$, differs from the bulk density of negative ions, $n_-(z,t)$. Some of the ions can also be chemically adsorbed by the dielectric layers deposited on the electrodes. In this case the conservation of the number of the ions requires that

$$\sigma_{1+}(t) + \sigma_{2+}(t) + \int_{-d/2}^{d/2} n_+(z,t)\,dz = n_0\,d + \sigma_{1+}(0) + \sigma_{2+}(0) \quad \text{and}$$

$$\sigma_{1-}(t) + \sigma_{2-}(t) + \int_{-d/2}^{d/2} n_-(z,t)\,dz = n_0\,d + \sigma_{1-}(0) + \sigma_{2-}(0),$$

(10.116)

where $\sigma_{1\pm}$ and $\sigma_{2\pm}$ are the surface densities of positive and negative ions adsorbed at the surfaces at $z = \mp d/2$ respectively, which generalizes (10.24).

The current densities of positive and negative charges are, respectively,

$$J_+ = -\mu_+ q \left\{ n_+ \frac{\partial V}{\partial z} + \frac{k_B T}{q} \frac{\partial n_+}{\partial z} \right\} \quad \text{and}$$
$$J_- = -\mu_- q \left\{ n_- \frac{\partial V}{\partial z} - \frac{k_B T}{q} \frac{\partial n_+}{\partial z} \right\}, \tag{10.117}$$

where $J_\pm = J_\pm(z,t)$, $n_\pm = n_\pm(z,t)$, and $V = V(z,t)$. In Eq. (10.117) the first terms represent the current densities due to the drift connected with the electric field, whereas the remaining terms are the diffusion current densities. The parameters $\mu_\pm$ are the mobilities of the positive and negative ions, and $D_\pm = (k_B T/q)\,\mu_\pm$ the diffusion coefficients of the two types of ions [19]. In Eq. (10.116), according to the usual electrochemical definition, the drift current is written as $\mathbf{j}_E = \mu q n \mathbf{E}$. On the contrary, in Section 10.3, the drift current has been written in the form $\mathbf{j}_F = \mu^* \rho \mathbf{F}$. In the present case, $\rho = q\,n$ and $\mathbf{F} = q\mathbf{E}$, and, hence, $\mathbf{j}_F = \mu^* q^2 n \mathbf{E}$. Comparing the two expressions for the drift current one deduces that $\mu^* q = \mu$; the Einstein-Smoluchowski relation $\mu^*/D = 1/k_B T$ in terms of μ reads $\mu/D = q/k_B T$ [22].

The charge conservation equations are

$$\frac{\partial n_+}{\partial t} - \mu_+ \frac{\partial}{\partial z} \left\{ n_+ \frac{\partial V}{\partial z} + \frac{k_B T}{q} \frac{\partial n_+}{\partial z} \right\} = 0 \quad \text{and}$$
$$\frac{\partial n_-}{\partial t} + \mu_- \frac{\partial}{\partial z} \left\{ n_- \frac{\partial V}{\partial z} - \frac{k_B T}{q} \frac{\partial n_-}{\partial z} \right\} = 0. \tag{10.118}$$

From Eq. (10.118), it follows the conservation of the net charge $\rho = q(n_+ - n_-)$ and net current density $J = J_+ + J_-$, expressed by the equation

$$q\,\frac{\partial n(z,t)}{\partial t} - \frac{\partial J(z,t)}{\partial z} = 0, \tag{10.119}$$

where $n(z,t) = n_+(z,t) - n_-(z,t)$.

Equations (10.118) have to be solved with the Maxwell equation

$$\operatorname{div} \mathbf{D}(z,t) = \rho(z,t),$$

where $\mathbf{D}(z,t)$ is the dielectric displacement and $\rho(z,t)$ the net electric charge density at the point z and time t. In the two dielectric layers, of thickness λ and dielectric constant ε_S, deposited over the electrodes to avoid charge injection, the net charge is zero. Consequently,

$$D_S(z,t) = D_S(t) = -q\Sigma(t),$$

for $-(d/2+\lambda) \leq z \leq -d/2$ and $d/2 \leq z \leq d/2+\lambda$. The electric field in the surface layer is then $E_S(t) = D_S(t)/\varepsilon_S = -q\Sigma(t)/\varepsilon_S$. In the liquid crystal, where the ionic charge is present, the electric field satisfies the Poisson's equation

$$\frac{d\{\varepsilon_{zz}E_B(z,t)\}}{dz} = q\,n(z,t). \qquad (10.120)$$

In Eq. (10.120), $q\,n(z,t)$ is the net bulk ionic charge, ε_{zz} is the z,z component of the dielectric tensor of the nematic liquid crystal, and $E_B(z,t)$ is the z-component of the electric field, which is the only one different from zero in the case under consideration.

In the simple case where the nematic distortion is contained in a plane, coinciding with the (x,z)-plane, the director is fully described by the tilt angle θ formed by $\mathbf{n}$ with the x-axis. In this case, to which we limit our investigation, $\varepsilon_{zz} = \varepsilon_{\parallel}\sin^2\theta + \varepsilon_{\perp}\cos^2\theta = \varepsilon_{\perp} + \varepsilon_a \sin^2\theta$. In a first approximation, for simplicity, we assume that the nematic liquid crystal can be considered as an isotropic liquid. This is equivalent to considering small deformations (i.e., small θ) or nematic media having small dielectric anisotropy. Thus, the electrical equations can be solved independently from the mechanical equation describing the nematic deformation induced by the field. The extension of the calculation to the general dielectric anisotropy can be, however, easily done.

By means of this simplifying hypotheses $\varepsilon_{zz} = \varepsilon_{\perp}$, and from Eq. (10.120) one gets that the electric field in the liquid crystal is given by

$$E_B(z,t) = \frac{q}{\varepsilon_{\perp}} \int_{-d/2}^{z} n(z',t)\,dz' + \mathcal{L}(t). \qquad (10.121)$$

The function $\mathcal{L}(t)$ is fixed by the boundary condition on the electric displacements at the interface in $z=-d/2$:

$$D_B(-d/2,t) - D_S(t) = q\sigma_1(t),$$

where $q\sigma_1(t) = q[\sigma_{1+}(t) - \sigma_{1-}(t)]$ is the total charge density adsorbed at the surface at $z=-d/2$. Consequently,

$$\mathcal{L}(t) = -q[\Sigma(t) - \sigma_1(t)]/\varepsilon_{\perp}.$$

The value of Σ is determined by imposing that the difference of potential on the sample is V_0. By means of the condition

$$V_0 = -\int_{-(d/2+\lambda)}^{d/2+\lambda} E(z,t)\,dz = -2E_S(t)\,\lambda - \int_{-d/2}^{d/2} E_B(z,t)\,dz, \qquad (10.122)$$

we obtain

$$\Sigma(t) = \gamma \left\{ \varepsilon_\perp \frac{V_0}{qd} + \sigma_1(t) + \frac{1}{d} \int_{-d/2}^{d/2} \left(\int_{-d/2}^{z} n(z',t)\, dz' \right) dz \right\}, \qquad (10.123)$$

where $\gamma^{-1} = 1 + 2(\lambda \varepsilon_\perp / d\varepsilon_S)$, is a physical parameter which depends on the dielectric properties of the surface layers. By substituting $\Sigma(t)$ given by (10.123) into (10.121) we get

$$E_B(z,t) = \frac{q}{\varepsilon_\perp} \left\{ \int_{-d/2}^{z} n(z',t)\, dz' - \frac{\gamma}{d} \int_{-d/2}^{d/2} \left(\int_{-d/2}^{z'} n(z'',t)\, dz'' \right) dz' \right\}$$
$$- \gamma \frac{V_0}{d} - \frac{\gamma - 1}{\varepsilon_\perp} \sigma_1 q. \qquad (10.124)$$

Equation (10.124) gives the electric field $E_B(z,t)$ in terms of the distribution of the ionic charge and of the applied voltage V_0.

To determine the evolution of the ionic charge with time when the external voltage V_0 is applied, it is necessary to solve Eq. (10.118), where $-\partial V/\partial z = E_B$ is given by (10.124). The boundary conditions for Eq. (10.118) are $n_+(z,0) = n_-(z,0) = n_0$, i.e., before the application of the external field the ionic distribution is homogeneous across the sample. The other boundary conditions depend on the adsorption on the limiting surfaces. If the ionic charge cannot leave the liquid crystal, and hence there are not surface adsorbed charges, $\sigma_1 = \sigma_2 = 0$, and the current densities have to vanish for $z = \pm d/2$. In this special case, the relevant boundary conditions for Eq. (10.118) are

$$J_+ = -\mu_+ q \left\{ n_+ \frac{\partial V}{\partial z} + \frac{k_B T}{q} \frac{\partial n_+}{\partial z} \right\} = 0, \quad \text{for} \quad z = \pm d/2 \quad \text{and}$$
$$J_- = -\mu_- q \left\{ n_- \frac{\partial V}{\partial z} - \frac{k_B T}{q} \frac{\partial n_+}{\partial z} \right\} = 0, \quad \text{for} \quad z = \pm d/2. \qquad (10.125)$$

In the opposite case, where there is adsorption of ions from the limiting surfaces, it is necessary to formulate the ionic exchange at the surfaces. One can assume that [23]

$$J_\pm = \pm q \frac{d\sigma_{1\pm}}{dt} = \pm q \left(k_{t\pm}\, n_\pm - \frac{1}{\tau_{t\pm}} \sigma_{1\pm} \right), \quad \text{for} \quad z = -d/2 \quad \text{and}$$
$$J_\pm = \pm q \frac{d\sigma_{2\pm}}{dt} = \pm q \left(k_{t\pm}\, n_\pm - \frac{1}{\tau_{t\pm}} \sigma_{2\pm} \right), \quad \text{for} \quad z = d/2, \qquad (10.126)$$

where $k_{t\pm}$ and $\tau_{t\pm}$ are phenomenological parameters describing the adsorption phenomenon of positive and negative ions, whose physical meaning is evident.

From Eq. (10.126) it follows that in the steady state, where $d\sigma_\pm/dt = 0$, for $t \to \infty$, the surface densities of adsorbed ions are

$$\sigma_{1\,\pm}(t \to \infty) = \tau_{\mathrm{t}\,\pm} k_{\mathrm{t}\,\pm} n_\pm(-d/2, t \to \infty) \quad \text{and}$$
$$\sigma_{2\,\pm}(t \to \infty) = \tau_{\mathrm{t}\,\pm} k_{\mathrm{t}\,\pm} n_\pm(d/2, t \to \infty). \tag{10.127}$$

In the present analysis it is supposed that there is adsorption at the limiting surfaces, and hence the boundary conditions for Eq. (10.118) are Eq. (10.126).

It is possible to rewrite Eq. (10.118) in a form giving information on the time evolution of the ion distribution in the bulk. By introducing the reduced quantities $u_\pm = n_\pm/n_0$, $\zeta = z/d$, $v = V/V_0$, $\tau_{\mathrm{d}\pm} = d^2/(\mu_\pm V_0)$, and $r = k_{\mathrm{B}}T/(qV_0)$, we get

$$\frac{\partial u_+}{\partial t} - \frac{1}{\tau_{\mathrm{d}+}} \frac{\partial}{\partial \zeta}\left\{ u_+ \frac{\partial v}{\partial \zeta} + r \frac{\partial u_+}{\partial \zeta} \right\} = 0 \quad \text{and}$$
$$\frac{\partial u_-}{\partial t} - \frac{1}{\tau_{\mathrm{d}-}} \frac{\partial}{\partial \zeta}\left\{ u_- \frac{\partial v}{\partial \zeta} - r \frac{\partial u_-}{\partial \zeta} \right\} = 0. \tag{10.128}$$

It follows that the characteristic times connected with the evolution of $n_\pm$ are of the order of $\tau_{\mathrm{d}\pm} = d^2/(\mu_\pm V_0)$ or of $\tau/\tau_{\mathrm{d}\pm}$, according to the value of r and of $\partial v/\partial \zeta$. Furthermore, from Eq. (10.126), the characteristic times connected with the evolution of $\sigma_{1\pm}$ and $\sigma_{2\pm}$ depend on $\tau_{\mathrm{t}\pm}$ and $\tau_{\mathrm{d}\pm}$.

10.7.2 Numerical solution of the electrical equations

Equations (10.118) have been numerically solved in Ref. [20] by discretizing both time and space in a usual finite difference forward scheme to define the derivatives [24]. Convergence of the numerical solution has been carefully verified by choosing fine discretization grids and testing the independence of the solution from the choice of the discretization time and space steps [25, 26]. As expected, convergence is extremely sensitive to the choice of the latter [27]. Finally, the boundary conditions Eq. (10.125) have been discretized through the usual formalism for adsorption [28].

The ions dissolved in the nematic sample are supposed to be identical in all the aspects. It follows that $\mu_+ = \mu_- = \mu$, $k_{\mathrm{t}+} = k_{\mathrm{t}-} = k_{\mathrm{t}}$, and $\tau_{\mathrm{t}+} = \tau_{\mathrm{t}-} = \tau_{\mathrm{t}}$, and, therefore, that $\tau_{\mathrm{d}+} = \tau_{\mathrm{d}-} = \tau_{\mathrm{d}} = d^2/(\mu V_0)$, and the characteristic time connected with the evolution of the adsorbed charge is the largest between τ_{d} and τ_{t}. The steady state values of the net surface adsorbed charge are $\sigma_{1,2}(t \to \infty) = \tau_{\mathrm{t}} k_{\mathrm{t}} n(\mp d/2, t \to \infty)$.

The values of the physical parameters used for the numerical calculations are reported in Table 10.1. A step-like applied voltage was simulated. The order of magnitude of the contact resistance is $R \sim 6$ kΩ [29]. The effective capacitance of the sample is $C = \gamma \epsilon_\perp (S/d)$, where γ takes into account the two layers deposited over the electrodes and $S \sim 2$ cm^2 is the surface of the

TABLE 10.1
Values of the physical parameters used in the simulations in SI units. q=1.6·10^{-19}C and $\varepsilon_0 = 8.85 \times 10^{-12}$F/m.

Variable	Symbol	Value	Ref.
Initial ion density	n_0	$5 \cdot 10^{19}$	[19]
Ion mobility	μ	$0.8, 2.1, 10 \times 10^{-11}$	[18, 30]
Specimen length	d	30×10^{-6}	-
Dielectric layer thickness	λ	0.02×10^{-6}	-
External potential	V_0	$1, 1.1, 1.2, 1.3$	-
Temperature	T	$290°K$	-
Dielectric constant of the layer	ε_S	$2\varepsilon_0$	-
Dielectric constant of NLC	$\varepsilon_\perp$	$6.7\varepsilon_0$	[31]
Dielectric anisotropy	ε_a	$13\varepsilon_0$	[31]
Absorption coefficient	k_t	10^{-6}	[23]
Desorption coefficient	τ_t	0.01	[23]
Initial orientation	θ_0	0.05rad	–
Frank's elastic constant	K	6.4×10^{-12}	[31]
Rotational viscosity	η	0.094	[31]
Ordinary refraction index	n_{ord}	1.53	[31]
Extraordinary refraction index	n_{ext}	1.71	[31]
Total flexoelectric coefficient	e	$5{\times}10^{-11}$	[32]

limiting surface. Using the values of Table 10.1, one gets C ~60 nF. The time constant of the electric circuit is then $\tau = RC$ ~3.6 μs. Since τ is negligibly small with respect to τ_d and τ_t one can suppose that the applied voltage passes immediately from 0 to V_0.

In the present analysis, where the applied voltage is step-like, the adsorption phenomenon does not play an important role. In fact, the bulk distribution of the electric field is, practically, independent of the chemical adsorption of the ions from the limiting surfaces. Only the value of the surface electric field depends on it, but its influence on the nematic orientation is negligible.

Figure 10.6 shows the time dependence of the positive (Figure 10.6a), σ_{1+}, and negative (Figure 10.6b), σ_{1-}, adsorbed densities of ions, and the net surface charge density (Figure 10.6c), $\sigma_1 = \sigma_{1+} - \sigma_{1-}$, at the surface at $z = -d/2$, for three representative values of the ionic mobility [30, 33]. σ_{1+} is a monotonic increasing function, whereas σ_{1-} is a monotonic decreasing function of t. Since the electrode at $z = -(d/2 + \lambda)$ is at $-V_0/2$, $\sigma_{1+} \gg \sigma_{1-}$ and hence $\sigma_1 \sim \sigma_{1+}$.

In Figure 10.7, the behavior of the ion density is analyzed. In the upper row, the profile of $n(-d/2,t) = n_+(-d/2,t) - n_-(-d/2,t)$ is shown several times for the three representative values of μ reported in Table 10.1 . As expected, the charge distribution is localized close to the limiting surface, in a region whose thickness is slightly smaller than Debye's screening length $\lambda_0 = \sqrt{\varepsilon_\perp k_\mathrm{B}T/(2q^2 n_0)}$(~*0.3* μ*m)*. Note also that the distribution at $t = 12.5$ s and

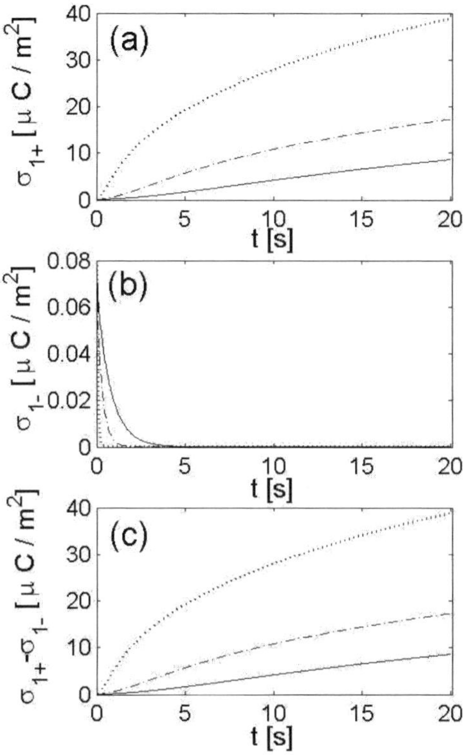

FIGURE 10.6
Surface densities of adsorbed positive (a) and negative (b) ions. The net surface charge density adsorbed on the dielectric layer deposited on the electrodes is $q(\sigma_{1+} - \sigma_{1-})$ (c). The applied voltage is V_0=1.3 V. Solid $\mu = 0.8 \times 10^{-11}\text{m}^2\text{s}^{-1}\text{V}^{-1}$, dash-dot $\mu = 2.1 \times 10^{-11}\text{m}^2\text{s}^{-1}\text{V}^{-1}$, dot $\mu = 10 \times 10^{-11}\text{m}\ ^2\text{s}^{-1}\text{V}^{-1}$. Reprinted with permission from Ref. [20]. Copyright (2004) by the American Physical Society.

$t = 17.5\,\mathrm{s}$ almost coincide (more for the largest mobility value). Furthermore, in the bulk $N_+ - n_- \approx 0$, as follows from the first row of Figure 10.7. This result justifies the approximated analysis presented in the previous section, where we assumed that $n = n_0 - \sigma/d$. Of course in the previous calculations the meaning of σ is

$$\sigma = \int_{-d/2}^{0} (n_+ - n_-)dz \approx \int_{-d/2}^{-d/2+b} (n_+ - n_-)dz,$$

where b is the thickness over which the rapid variations of n_+ and n_- are limited.

In the second row of Figure 10.7, the time dependencies of $n(-d/2, t)$, for the same values of μ, are reported. With the values of the physical parameters reported in Table 10.1 , $n_+(-d/2, t)$ is a monotonic increasing function, whereas $n_-(-d/2, t)$ is a monotonic decreasing function of t (not reported). With the representative values of the ionic mobility chosen in the numerical calculations the drift time falls in the range 9 s $\leq \tau_\mathrm{d} \leq$ 112 s, whereas the trapping time is τ_t=0.01 s. In this case $\tau_\mathrm{d} \gg \tau_\mathrm{t}$. Consequently, the ions pushed by the electric force close to the border of the sample, are rapidly adsorbed, and the phenomenon of accumulation of ions close to the limiting surfaces is absent. Of course, in the opposite case where $\tau_\mathrm{t} \gg \tau_\mathrm{d}$, there are accumulations of ions just in front of the surface. These ions are subsequently adsorbed by the surface with the characteristic time τ_t. In this situation, $n_+(-d/2, t)$ presents a maximum for a time of the order of τ_d. After that it relaxes to the equilibrium value on a time of the order of τ_t. A similar behavior is expected for the surface electric field. The numerical calculations confirm these trends for $n_+(-d/2, t)$ and $E_\mathrm{B}(-d/2, t)$.

In the bottom row of Figure 10.7 is shown the time dependence of $n(0, t)$. As expected, both $n_+(0, t)$ and $n_-(0, t)$ decrease uniformly with t, whereas $n(0, t)$ remains zero, indicating that the center of the sample remains locally neutral.

In Figure 10.8a, the charge density, $q\Sigma$, sent by the power supply on the electrodes to maintain the difference of potential at the value V_0 is reported versus time, for $\mu = 2.1 \times 10^{-11}$ m^{2}s^{-1} V^{-1} and several voltages. In the present case $q\Sigma$ is connected only with the ionic charge moving inside the liquid crystal, because it is assumed that the nematic can be considered isotropic from the dielectric point of view. In the more general case, where also the dielectric anisotropy is taken into account, the external charge is due also to time variation of the z, z-dielectric constant connected with the time dependence of the tilt angle, $\theta(t)$. The electric current density flowing in the circuit is given by $J_\mathrm{ext}(t) = q\, d\Sigma(t)/dt$. By taking into account Eq. (10.123) we get

$$J_\mathrm{ext}(t) = q\,\frac{d\Sigma}{dt} = q\gamma \left\{ \frac{d\sigma_1}{dt} + \frac{1}{d} \int_{-d/2}^{d/2} \left(\int_{-d/2}^{z} \frac{\partial n(z', t)}{\partial t}\, dz' \right) dz \right\}. \quad (10.129)$$

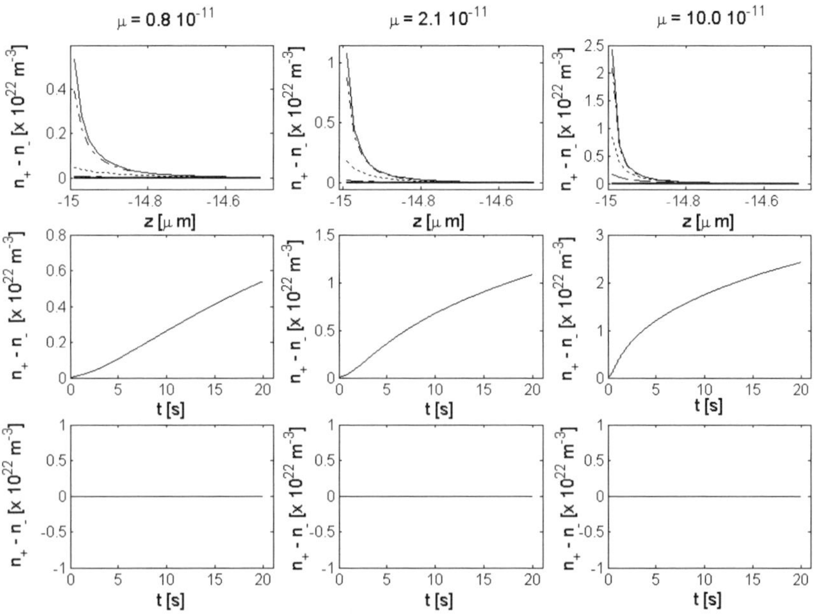

FIGURE 10.7
Net density of ions $n(z,t) = n_+(z,t) - n_-(z,t)$ in the nematic sample, for $\mu = 0.8 \times 10^{-11}\mathrm{m}^2\mathrm{s}^{-1}\mathrm{V}^{-1}$, first column, $\mu = 2.1 \times 10^{-11}\mathrm{m}^2\mathrm{s}^{-1}\mathrm{V}^{-1}$, second column, $\mu = 10 \times 10^{-11}\mathrm{m}^2\mathrm{s}^{-1}\mathrm{V}^{-1}$, third column. First row: profiles at several times after the application of the external voltage, solid-thick $t = 0.0\,\mathrm{s}$, dash $t = 0.5\,\mathrm{s}$, dot $t = 2.5\,\mathrm{s}$, dash-dot $t = 7.5$ s, solid-thin $t = 17.5\,\mathrm{s}$. Second row: net density of ions just in front to the limiting surface $n(-d/2, t)$. Third row: net density of ions in the middle of the sample $n(0,t)$. Reprinted with permission from Ref. [20]. Copyright (2004) by the American Physical Society.

By using Eq. (10.119), $J_{\rm ext}(t)$ can be rewritten in the form

$$J_{\rm ext}(t) = \gamma \left\{ q\,\frac{d\sigma_1}{dt} - J(-d/2) + \frac{1}{d}\int_{-d/2}^{d/2} J(z,t)\,dz \right\}, \qquad (10.130)$$

that, making use of (10.126), is equivalent to

$$J_{\rm ext}(t) = \frac{\gamma}{d}\int_{-d/2}^{d/2} J(z,t)\,dz. \qquad (10.131)$$

By means of (10.117), one gets for $J_{\rm ext}(t)$ the final expression

$$J_{\rm ext}(t) = -\,q\,\mu\,\frac{\gamma}{d}\int_{-d/2}^{d/2} \left\{ [n_+(z,t) + n_-(z,t)]\,\frac{\partial V}{\partial z} + \frac{k_{\rm B}T}{q}\left[\frac{\partial n_+(z,t)}{\partial z} - \frac{\partial n_-(z,t)}{\partial z}\right]\right\} dz, \qquad (10.132)$$

in the present case, where $\mu_+ = \mu_- = \mu$. Figure 10.8b shows $J_{\rm ext}(t)$ versus t. As is clear from Figure 10.8c, the time dependence of $J_{\rm ext}(t)$ is not simply exponential. This result can be easily understood because the transit time of the ions in the nematic cell depends on the ions already present close to the limiting surfaces. In other words, the transit time depends on the effective profile of the electrical potential, that, in turns, depends on the distribution of ions. This feedback is responsible for the deviation of $J_{\rm ext}(t)$ from a simple exponential law. It is also to be noted that the current decays very slowly with time, due to the rather large value of the characteristic time τ_d for the given value of mobility and specimen length. In Figure 10.8d, the actual difference of potential across the nematic sample, given by

$$V_{\rm B}(t) = V(d/2,t) - V(-d/2,t) = -\int_{-d/2}^{d/2} E_{\rm B}(z,t)\,dz, \qquad (10.133)$$

is reported. As is evident from this figure, the actual $V_{\rm B}$ tends to a saturation value, which is well below V_0. This reduction is due to the back electric field connected with the separation of ions induced by the external electric field due to the power supply.

The profiles of the electric field $E_{\rm B}(z,t)$ and of the electric potential $V(z,t)$ in the nematic sample for several times are shown in Figure 10.9a and Figure 10.9b, respectively, for $\mu = 2.1 \times 10^{-11}$ m^{2}s^{-1} V^{-1} . The electric field is more and more localized close to the nematic surface with increasing time, while at long times, the bulk of the liquid crystal is almost equipotential (zero electric field). The time dependencies of the electric field at the surface, $E_{\rm B}(-d/2,t)$, and in the middle of the sample, $E_{\rm B}(0,t)$, are shown in Figure 10.9c and Figure 10.9d for the representative values of μ reported in Table 10.1. As is evident from Figure 10.9c the surface electric field is, for the chosen values of μ, $k_{\rm t}$ and $\tau_{\rm t}$, a monotonic increasing function of t. On the contrary, the

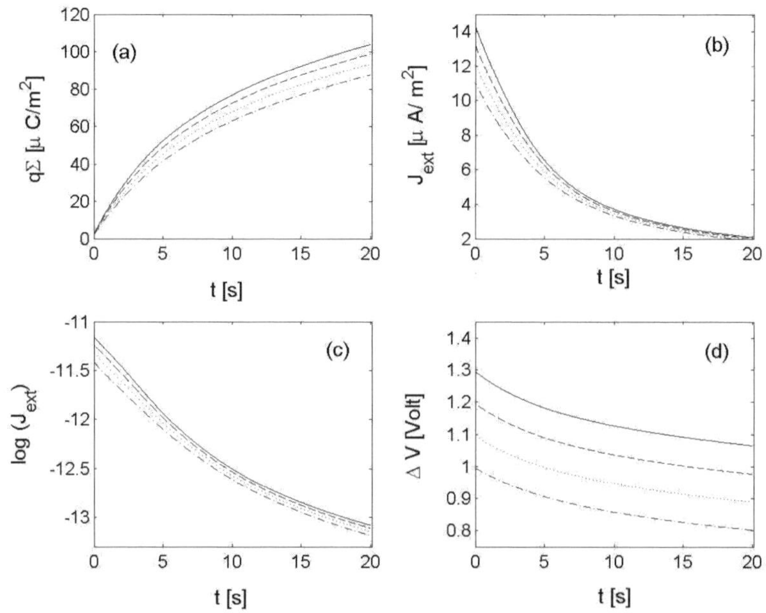

FIGURE 10.8
Electric charge $q\Sigma(t)$ versus t sent by the power supply on the electrodes to maintain the difference of potential on the sample at V_0(a). Electric density current $J_{\text{ext}}(t)$ flowing in the circuit versus t (b), and $\log J_{\text{ext}}$ versus t (c). The time dependence of $J_{\text{ext}}(t)$ is not simply exponential. Actual difference of potential on the nematic sample versus t (d). The value to which tends $V_{\text{B}}(t)$ is well below V_0, for the screening effect due to the presence of the ionic charge. The ionic mobility is $\mu = 2.1 \times 10^{-11}\text{m}^2\text{s}^{-1}\text{V}^{-1}$. Solid $V_0 = 1.3\,\text{V}$, dash $V_0 = 1.2\,\text{V}$, dot $V_0 = 1.1\,\text{V}$, dash-dot $V_0 = 1.0\,\text{V}$. Reprinted with permission from Ref. [20]. Copyright (2004) by the American Physical Society.

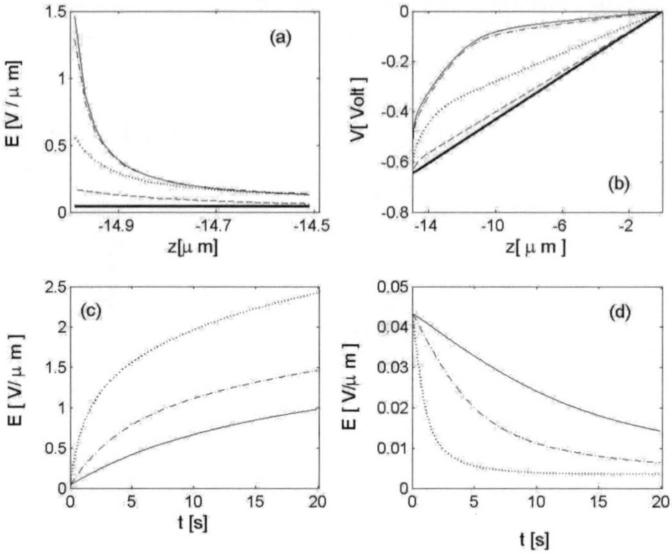

FIGURE 10.9
Profile of the electric field $E_{\rm B}(z,t)$ close to the limiting surface (a), and of the electrical potential $V(z,t)$ across the sample (b), for different times t after the application of the external voltage $V_0 = 1.3$ V. In (a) and (b) the ionic mobility is $\mu = 2.1 \times 10^{-11}\mathrm{m^2 s^{-1} V^{-1}}$, solid-thick $t = 0.0\,$s, dash $t = 0.5\,$s, dot $t = 2.5\,$s, dash-dot $t = 7.5\,$s, solid-thin $t = 17.5\,$s. Time dependence of the surface electric field $E_{\rm B}(-d/2,t)$, (c), and of the electric field in the middle of the sample $E_{\rm B}(0,t)$, (d). In (c) and (d) solid $\mu = 0.8 \times 10^{-11}\mathrm{m^2 s^{-1} V^{-1}}$, dash-dot $\mu = 2.1 \times 10^{-11}\mathrm{m^2 s^{-1} V^{-1}}$, dot $\mu = 10 \times 10^{-11}\mathrm{m^2 s^{-1} V^{-1}}$. Reprinted with permission from Ref. [20]. Copyright (2004) by the American Physical Society.

electric field in the bulk decreases monotonically from the initial value to an equilibrium value equal to the external field diminished of the back electric field created by the separation of ions induced by the external field itself.

By means of the profile $n(z,t) = n_+(z,t) - n_-(z,t)$ it is possible to evaluate the contribution of ionic origin to the surface polarization. It is given by

$$\mathcal{P}(t) = \int_{-d/2}^{0} q\, z\, n(z,t)\, dz + q\,\sigma_1(t)\,\frac{d}{2}, \qquad (10.134)$$

where the latter contribution is connected with the charge adsorbed on the limiting surface. The time dependence of $\mathcal{P}(t)$ is shown in Figure 10.10a. The order of magnitude of $\mathcal{P}$ is the same as the one experimentally determined, and due, according to the authors of [34], to molecular properties of the liquid crystal. On the contrary, the present analysis shows that it is simply due to the ions dissolved in the liquid crystal. As follows from Figure 10.9a the electric field across the sample is not homogeneous. Its spatial variation is localized in a surface layer close to the limiting surfaces. In this case, as discussed in Chapter 8, the electrostatic energy connected with the interaction of the electric field localized in the surface layer with the nematic liquid crystal can be considered as a surface energy. Following the analysis reported in Section 7.3, the contribution connected with the dielectric anisotropy is

$$W_D(t) = -\frac{1}{2}\epsilon_a \int_{-d/2}^{0} [E_{\rm B}^2(z,t) - E_{\rm B}^2(0,t)]\, dz, \qquad (10.135)$$

whereas the contribution associated to the flexoelectric polarization is

$$W_Q(t) = -e[E_{\rm B}(-d/2,t) - E_{\rm B}(0,t)], \qquad (10.136)$$

where $e = e_{11} + e_{33}$ is the total flexoelectric coefficient. The anchoring energy strengths $W_D(t)$ and $W_Q(t)$ versus t are shown in Figure 10.10b and Figure 10.10c, respectively.

10.7.3 Deformations induced by the external field

The nematic sample is a slab of thickness d. In the absence of the external field the nematic director $\mathbf{n}$ is oriented, by means of the surface treatment, at an angle θ_0 very small with respect to the x-axis, homogeneous across the sample. The parameter θ_0 could be measured by means of the birefringence of the sample in the absence of the external field. However, in our case it will be considered as a free parameter, and determined by the fit of the curve of the birefringence versus time, for different values of the applied voltage. We consider the case of $\epsilon_a = \epsilon_\parallel - \epsilon_\perp > 0$. If the sample is submitted to an electric field $\mathbf{E} = E(t)\mathbf{k}$, where $\mathbf{k}$ is the unit vector parallel to the z-axis, the electric torque can destabilize the initial orientation. We are in the situation to observe a kind of Fréedericksz transition [35]. In this section we investigate the dynamics of the orientation induced by the external field in the case of strong anchoring on the surfaces. In this framework $\theta(\pm d/2) = \theta_0$ for any external field.

The dynamical evolution of the system is governed, in the one-constant approximation, by the equation[8]

$$K\frac{\partial^2\theta}{\partial z^2} + \frac{1}{2}\epsilon_a E^2(z,t)\sin(2\theta) - \eta\frac{\partial\theta}{\partial t} = 0, \qquad (10.137)$$

where K is the Frank's elastic constant, and η the rotational viscosity coefficient of the liquid crystal. We note that in the strong anchoring hypothesis the flexoelectric effect does not enter in the dynamical equilibrium equation. In the following we write Eq. (10.137) in the form

$$\frac{\partial^2\theta}{\partial\zeta^2} + \frac{1}{2}\pi^2 h^2(\zeta,t)\sin(2\theta) - \tau_{\rm v}\frac{\partial\theta}{\partial t} = 0, \qquad (10.138)$$

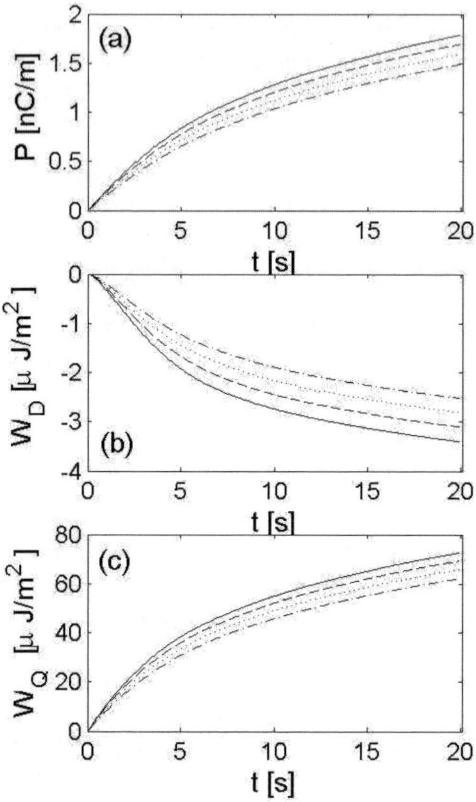

FIGURE 10.10
Surface polarization $\mathcal{P}$ due to the ion separation connected with the external field versus t, (a). Surface anchoring energy $W_D(t)$ connected with the dielectric anisotropy of the nematic liquid crystal due to the surface field of ionic origin versus t, (b), and surface anchoring energy of flexoelectric origin $W_Q(t)$ versus t, (c). Solid $V_0 = 1.3\,\mathrm{V}$, dash $V_0 = 1.2\,\mathrm{V}$, dot $V_0 = 1.1\,\mathrm{V}$, dash-dot $V_0 = 1.0\,\mathrm{V}$. The ionic mobility is $\mu = 2.1 \times 10^{-11}\mathrm{m^2 s^{-1} V^{-1}}$. Reprinted with permission from Ref. [20]. Copyright (2004) by the American Physical Society.

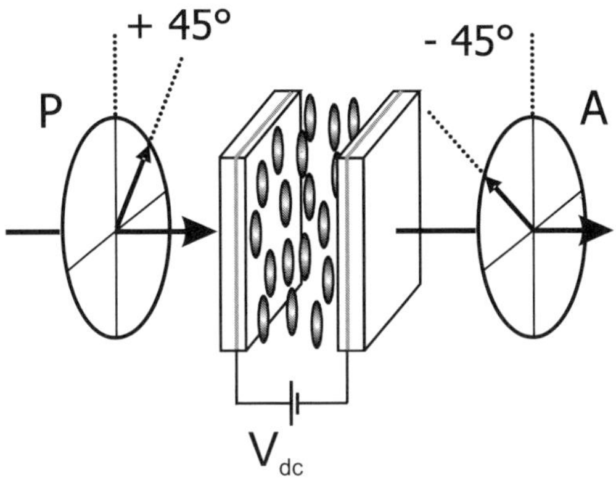

FIGURE 10.11
Typical experimental setup for optical phase retardation measurement. The planar cell is placed between crossed polarizers (P and A), with its optical axis oriented at 45° with respect to both the polarizer (P) and the analyzer (A) axes. A low power He-Ne laser beam ($\lambda_{\rm op} = 632.8\,\rm nm$) is used to probe the NLC cell birefringence, while a stepwise DC voltage is applied. Reprinted from Ref. [17] with permission from American Institute of Physics.

where $\zeta = z/d$, $h(\zeta, t) = E(\zeta, t)/E_{\rm th}$ is the electric field expressed in units of the threshold field ($E_{\rm th} = (\pi/d)\sqrt{K/\epsilon_a}$), and $\tau_{\rm v} = (\eta/K)d^2$ the viscous relaxation time of the sample under consideration. In the limit of small θ, i.e., for E of the order of the critical field $E_{\rm th}$, Eq. (10.137) at the first order in θ can be written as

$$\frac{\partial^2\theta}{\partial\zeta^2} + \pi^2 h^2(\zeta, t)\,\theta - \tau_{\rm v}\,\frac{\partial\theta}{\partial t} = 0. \tag{10.139}$$

From Eq. (10.139) it follows that the time evolution of the tilt angle is the largest between the characteristic times $\tau_{\rm d}$ and $\tau_{\rm v}$.

A typical experimental setup to measure the change in the effective birefringence is shown in Figure 10.11. For an NLC with positive dielectric anisotropy an electric field normal to the glass walls tends to orient the nematic director and hence the optical axis along the field direction, i.e., homeotropically, thus reducing the cell's average birefringence. The change in the effective birefringence can be detected with a linearly polarized He-Ne probe beam by measuring the intensity transmitted through a crossed analyzer.

The present analysis has been carried out by means of a forward finite difference scheme, taking care of stability conditions for the choice of the ratio between space and time steps [24]. Furthermore, provided the space

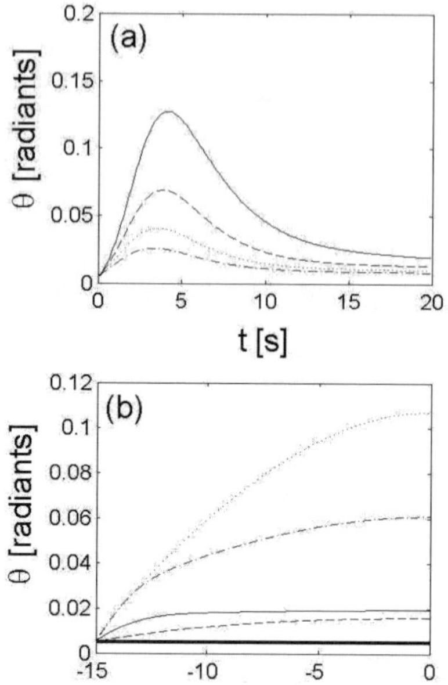

FIGURE 10.12
Tilt angle in the middle of the sample $\theta(0,t)$ versus t, solid $V_0 = 1.3\,\mathrm{V}$, dash $V_0 = 1.2\,\mathrm{V}$, dot $V_0 = 1.1\,\mathrm{V}$, dash-dot $V_0 = 1.0$ V, (a). Tilt angle profiles $\theta(z,t)$ for several t, after the application of the external voltage $V_0 = 1.3\,\mathrm{V}$, solid-thick $t = 0.0\,\mathrm{s}$, dash $t = 0.5\,\mathrm{s}$, dot $t = 2.5\,\mathrm{s}$, dash-dot $t = 7.5\,\mathrm{s}$, solid-thin $t = 17.5\,\mathrm{s}$, (b). The ionic mobility is $\mu = 2.1 \times 10^{-11}\mathrm{m}^2\mathrm{s}^{-1}\mathrm{V}^{-1}$. Reprinted with permission from Ref. [20]. Copyright (2004) by the American Physical Society.

discretization is very fine, different implementations for the strong anchoring boundary conditions have been proven to be equivalent (with differences of the results of the order of a few per thousand). Hence, convergence of the numerical solutions has been ensured. In Figure 10.12a, the tilt angle in the middle of the sample versus t is shown. In Figure 10.12b, the profiles of the tilt angle at different times are presented. As follows from this figure, the time dependence of the tilt angle at a given point is a non-monotonic function of t. At low times the bulk deformation is determined by the appearance of the electric field, while, as soon as the field in the bulk decays to low values (see Figure 10.9a) the tilt angle is dragged by the one close to the limiting surface, where the electric field is stronger than the one in the bulk.

The validity of the model has been experimentally verified by means of the measurement of the phase retardation of a nematic cell submitted to an external voltage. By indicating with n_{ext} and n_{ord}, respectively, the extraordinary and the ordinary indices of refraction of the nematic liquid crystal, the phase shift between the two components of a linearly polarized beam impinging perpendicularly on the sample is given by

$$\Delta\phi(t) = \frac{2\pi}{\lambda_{\mathrm{op}}} \int_{-d/2}^{d/2} \left[n_{\mathrm{eff}}(z,t) - n_{\mathrm{ord}}\right] dz = \\ = -\frac{2d\pi n_{\mathrm{ord}}}{\lambda_{\mathrm{op}}} + \frac{2d}{\lambda_{\mathrm{op}}} \int_{-1/2}^{1/2} n_{\mathrm{eff}}(\zeta,t)\, d\zeta, \tag{10.140}$$

where

$$n_{\mathrm{eff}}(\zeta,t) = \frac{n_{\mathrm{ext}} n_{\mathrm{ord}}}{\sqrt{n_{\mathrm{ext}}^2 \sin^2\theta(\zeta,t) + n_{\mathrm{ord}}^2 \cos^2\theta(\zeta,t)}}, \tag{10.141}$$

and λ_{op} is the wavelength of light.

The measurements of $\Delta\phi$ were performed in the standard manner, for different applied voltages. The results are reported in Figure 10.13. Here, where the mobility of positive and negative ions is assumed the same, a good agreement between the theoretical prediction and the experimental data is obtained for an average mobility $\mu = 2.1 \times 10^{-11} \mathrm{m^2 s^{-1} V^{-1}}$. This value of $\mu \sim (\mu_- + \mu_+)/2$ is comparable to the one reported in Ref. [23], which is rather small with respect to the one reported by [19].

The agreement is reasonably good, taking into account the simplifying hypotheses performed in the presented analysis. In particular, we have assumed that in the liquid crystal there is only one kind of positive and negative ion, with the same mobility, and probably that is not realistic. Furthermore, as discussed by Sawada *et al.* [30, 33], different kinds of ions, with different mobility, can be present simultaneously in the nematic cell. Some of these ions can also be originated by the film deposited on the electrode to align the liquid crystal [21, 36]. Consequently, the model has to be extended to consider a population of different types of ions.

The main results of the preceding analysis can be summarized as follows. In the simple case in which there is only one type of negative and positive ions with the same mobility, $\mu_- = \mu_+ = \mu$, the calculations show that for μ in the range $10^{-9} - 10^{-10}$ $\mathrm{m^2 s^{-1} V^{-1}}$ the field and ionic distribution across the sample is a rapid phenomenon. For a thickness of the order of 30 μm, the characteristic time for the ion redistribution, τ_{d}, is in the range $0.9\,\mathrm{s} \leq \tau_{\mathrm{d}} \leq 9\,\mathrm{s}$. The characteristic time relevant to the optical response, using the values of Table 10.1, is of the order of $\tau_{\mathrm{v}} \sim 13.2$ s. The same characteristic times for a sample of thickness $d = 8\,\mu\mathrm{m}$ are $0.056\,\mathrm{s} \leq \tau_{\mathrm{d}} \leq 0.56\,\mathrm{s}$, and $\tau_{\mathrm{v}} = 0.8\,\mathrm{s}$. In this case the ions play an important role on the anchoring energy and flexoelectric coefficients that cannot be neglected [34, 37, 38].

In a real situation, the positive and negative ions have different mobility, and furthermore, there are several types of ions with different dimensions. In

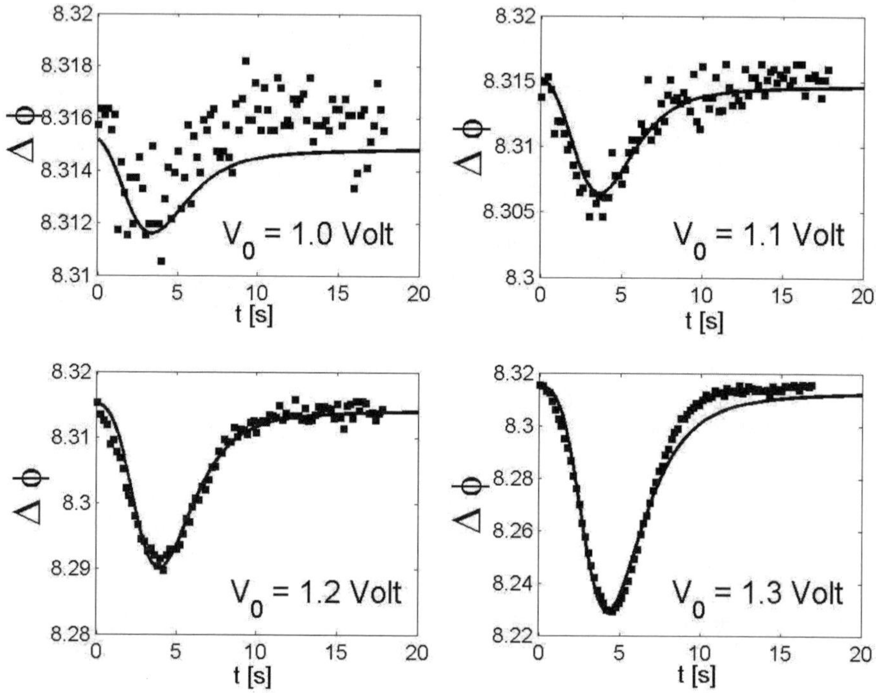

FIGURE 10.13
Phase shift $\Delta\phi(t)$ versus t theoretically evaluated by using $\mu = 2.1 \times 10^{-11}\mathrm{m}^2\mathrm{s}^{-1}\mathrm{V}^{-1}$ and the physical parameter values reported in Table 10.1, for different applied voltages. The dots represent the corresponding experimental data. Values of the physical parameters used in the simulations in SI units: $q = 1.6\times10^{-19}$C and $\varepsilon_0 = 8.85\times10^{-12}$F/m. Reprinted with permission from Ref. [20]. Copyright (2004) by the American Physical Society.

this case only a careful analysis of the different mobility of the several ions can give information on the influence of the ionic charge on the physical properties of the nematic liquid crystal. However, since small ions with large mobility are always present, before any statements on the meaning of flexoelectric properties and anchoring energy of nematic cells submitted to dc voltage can be made, it is necessary to know the characteristics of the ions dissolved in the nematic liquid crystal.

[1] Barbero G and Evangelista LR. Adsorption phenomenon of neutral particles and a kinetic equation at the interface. *Physical Review E*, **70**, 031605 (2004).

[2] Butkov E. *Mathematical Physics.* Addison-Wesley Publishing Co., Reading, MA, 1973.

[3] Morse PM and Feshbach H. *Methods of Theoretical Physics.* McGraw-Hill, New York, 1953.

[4] Evangelista LR and Barbero G. Intrinsic characteristic times in the drift-diffusion problem. *Liquid Crystals*, **31**, 1399 (2004).

[5] Kittel C and Kroemer H. *Thermal Physics.* WH Freeman, New York, 1980 (2nd edition).

[6] Tsykalo AL. *Thermophysical Properties of Liquid Crystals.* Gordon and Breach, Philadelphia, 1991.

[7] Barbero G and Evangelista LR. *An Elementary Course on the Continuum Theory for Nematic Liquid Crystals.* World Scientific, Singapore, 2001.

[8] Thurston RN, Cheng J, Meyer RB, and Boyd GD. Physical mechanisms of DC switching in a liquid crystal bistable boundary layer display. *Journal of Applied Physics*, **56**, 263 (1984).

[9] Murakami S and Naito H. Electrode and interface polarizations in nematic liquid crystal cells. *Japanese Journal of Applied Physics*, **36**, 2222 (1997).

[10] Meister R and Jerome B. Influence of a surface electric field on the anchoring characteristics of nematic phases at rubbed polyimides. *Journal of Applied Physics*, **86**, 2473 (1999).

[11] Nazarenko VG and Lavrentovich OD. Anchoring transition in a nematic liquid crystal composed of centrosymetric molecules. *Physical Review E*, **49**, R990 (1994).

[12] Fazio VSU and Komitov L. Alignment transition in a nematic liquid crystal due to field-induced breaking of anchoring. *Europhysics Letters*, **46**, 38 (1999).

[13] Helseth LE, Wen HZ, Fische TM, and Johansen TH. Adsorption and diffusion in a one-dimensional potential well. *Physical Review E*, **68**, 011402 (2003).

[14] Jánossy I. Molecular interpretation of the absorption-induced optical reorientation of nematic liquid crystals. *Physical Review E*, **49**, 2957 (1994).

[15] Francescangeli O, Slussarenko S, Simoni F, Andrienko D, Reshetnyak V, and Reznikov Y. Light-induced surface sliding of the nematic director in liquid crystals. *Physical Review Letters*, **82**, 1855 (1999).

[16] Ouskova E, Reznikov Y, Shiyanovskii SV, Su L, West JL, Kuksenok OV, Francescangeli O, and Simoni F. Photo-orientation of liquid crystals due to light-induced desorption and adsorption of dye molecules on an aligning surface. *Physical Review E*, **64**, 051709 (2001).

[17] Pagliusi P, Zappone B, Cipparrone G, and Barbero G. Molecular reorientation dynamics due to direct current voltage-induced ion redistribution in undoped nematic planar cell. *Journal of Applied Physics*, **96**, 218 (2004).

[18] Zhang H and D'Have K. Surface trapping of ions and symmetric addressing scheme for FLCDs. *Molecular Crystals and Liquid Crystals*, **351**, 27 (2000).

[19] Thurston RN. Equilibrium distributions of electric field in a cell with adsorbed charge at the surfaces. *Journal of Applied Physics*, **55**, 4154 (1984).

[20] Scalerandi M, Pagliusi P, Cipparrone G, and Barbero G. Influence of the ions on the dynamical response of a nematic cell submitted to a dc voltage. *Physical Review E* , **9**, 051708 (2004).

[21] Murakami S and Naito H. Electrode and interface polarizations in nematic liquid crystal cells. *Japanese Journal of Applied Physics*, **36**, 2222 (1997).

[22] Bockris JO, Reddy AKN, and Gamboa-Aldeco M. *Modern Electrochemistry: Ionics.* Plenum Press, New York, 1998 (2nd edition).

[23] Maximus B, de Ley E, de Meyere A, and H. Pauwels H. Ion transport in SSFLCD's. *Ferroelectrics*, **121**, 103 (1991).

[24] Strickwerda J. *Finite Difference Schemes and Partial Differential Equations.* Wadsworth-Brooks, London, 1989.

[25] Vliegenthart AC. Finite-difference methods for Korteweg-de Vries equation. *Journal of Engineering Mathematics*, **5**, 137 (1971).

[26] Scalerandi M, Romano A, and Condat CA. Korteweg-de Vries solitons under additive stochastic perturbations. *Physical Review E*, **58,** 4166 (1998).

[27] Scalerandi M, Giordano M, Delsanto PP, and Condat CA. A stable finite-difference scheme for the Boussinesq equation. *Nuovo Cimento D*, **20**, 1 (1998).

[28] Kaniadakis G, Delsanto PP, Condat CA. A local interaction simulation approach to the solution of diffusion problems. *Mathematical and Computer Modelling*, **17**, 31 (1993).

[29] Sugimura A, Matsui N, Takahashi Y, Sonomura H, Naito H, and Okuda M. Transient currents in nematic liquid crystals. *Physical Review B* , **43**, 8272 (1991).

[30] Sawada A, Tarumi K, and Naemura S. Novel characterization method of ions in liquid crystal materials by complex dielectric constant measurements. *Japanese Journal of Applied Physics*, **38**, 1423 (1999).

[31] Blinov LM and Chigrinov VG. *Electrooptic Effects in Liquid Crystal Materials.* Springer-Verlag, New York, 1993.

[32] Evangelista LR. Weak anchoring or pretilt? *Physics Letters A*, **205**, 203 (1995).

[33] Sawada A, Tarumi K, and Naemura S. Effects of electric double layer and space charge polarization by plural kinds of ions on complex dielectric constant of liquid crystal materials. *Japanese Journal of Applied Physics*, **38**, 1418 (1999).

[34] Mazzulla A, Ciuchi F, and Sambles JR. Optical determination of flexoelectric coefficients and surface polarization in a hybrid aligned nematic cell. *Physical Review E*, **64**, 021708 (2001).

[35] Barbero G, Evangelista LR, and Madhusudana NV. Effect of surface electric field on the anchoring of nematic liquid crystals. *European Physical Journal B*, **1**, 337 (1998).

[36] Zhang H, Pauwels H, Parghi DD, and Heppke G. Observations on ionic contamination in AFLC mixtures. *Molecular Crystals and Liquid Crystals*, **368**, 145 (2001).

[37] Mazzulla A, Ciuchi F, and Sambles JR. Reply to "Comment on 'Optical determination of flexoelectric coefficients and surface polarization in a hybrid aligned nematic cell'". *Physical Review E*, **68**, 023702 (2003).

[38] Barbero G and Evangelista LR. Comment on "Optical determination of flexoelectric coefficients and surface polarization in a hybrid aligned nematic cell". *Physical Review E*, **68**, 023701 (2003).

11

IMPEDANCE SPECTROSCOPY OF A CELL: THE ROLE OF THE IONS

In this chapter, the role of the diffuse layer of the ionic cloud on the impedance spectroscopy measurements of a cell of a liquid is discussed. The analysis is performed, firstly, by assuming that the ions have the same mobility, the electrodes are perfectly blocking and the adsorption phenomenon can be neglected. It is shown that the dielectric permittivity, in the limit of large frequency ω, tends to the dielectric permittivity of the pure liquid as $\omega^{-3/2}$. The connection between the detected equivalent permittivity and chemical potential of the cell with the real and imaginary parts of the complex dielectric constant is discussed. It is shown also that the presence of the ions is responsible for a distribution of relaxation times. Subsequently, a generalization of the obtained results to the case in which the mobility of the positive ions is different from that of the negative ones is presented. It is shown that the difference between the mobilities of the positive and negative ions gives rise, in the low frequency region, to a new plateau of the real part of the electric impedance of the cell. Furthermore, it is responsible for a deviation from the ideal Cole-Cole diagram relevant to the real and imaginary parts of the complex dielectric constant. Finally, the influence of the adsorption phenomenon on the impedance spectroscopy measurements is considered. The analysis is performed by assuming that the ions have the same mobility and the electrodes are perfectly blocking. We find that in the low frequency range the presence of the adsorption phenomenon is responsible for an increasing of the real part of the impedance of the cell, similar to the one usually described by means of the impedance of the metal-electrolyte interface.

11.1 Introduction

The impedance spectroscopy technique is used to characterize liquid materials from the electrical point of view [1]. According to this technique a condenser having the shape of a slab is filled with the material to be investigated. The condenser is then submitted to an ac voltage, and the impedance, $\mathcal{Z}$, of the sample is measured versus the frequency of the applied voltage. The analysis is performed in the limit of small amplitude of the applied voltage, in such a manner that the response of the sample to the external signal is linear.

From $\mathcal{Z}$ one can obtain $\mathcal{R} = \Re\mathcal{Z}$ and $\mathcal{X} = \Im\mathcal{Z}$, which are the quantities experimentally detectable. The phenomenological parameters characterizing the physical properties of the cell are the equivalent chemical potential, σ_Z, and the equivalent dielectric constant, ε_Z, defined by

$$\sigma_Z = \frac{1}{\mathcal{R}}\frac{d}{S} \quad \text{and} \quad \varepsilon_Z = -\frac{1}{\omega\mathcal{X}}\frac{d}{S}. \tag{11.1}$$

Usually, the dielectric properties of a cell are described by introducing a complex dielectric constant $\epsilon = \epsilon' - i\epsilon''$. The real part, ϵ', is connected with the usual dielectric properties of the medium, whereas the imaginary part, ϵ'', is related to the relative dielectric loss factor. The relation between σ_Z, ε_Z and ϵ', ϵ'' can be easily determined. In fact, if the dielectric permittivity is a complex quantity, the impedance of the cell is

$$\mathcal{Z} = -i\frac{1}{\omega(\epsilon' - i\epsilon'')}\frac{d}{S}. \tag{11.2}$$

From (11.2) it follows that the real, $\mathcal{R}$, and imaginary, $\mathcal{X}$, parts of the impedance $\mathcal{Z}$ are, respectively,

$$\mathcal{R} = \frac{\epsilon''}{\omega(\epsilon'^2 + \epsilon''^2)}\frac{d}{S} \quad \text{and} \quad \mathcal{X} = -\frac{\epsilon'}{\omega(\epsilon'^2 + \epsilon''^2)}\frac{d}{S}. \tag{11.3}$$

From (11.3) and (11.1) we get

$$\sigma_Z = \omega\,\frac{\epsilon'^2 + \epsilon''^2}{\epsilon''} \quad \text{and} \quad \varepsilon_Z = \frac{\epsilon'^2 + \epsilon''^2}{\epsilon'}. \tag{11.4}$$

Consequently, ϵ' and ϵ'' can be expressed, in terms of σ_Z and ε_Z, respectively as follows

$$\epsilon' = \frac{\varepsilon_Z}{1 + \omega^2\left(\varepsilon_Z/\sigma_Z\right)^2} \quad \text{and} \quad \epsilon'' = \frac{\omega\left(\varepsilon_Z/\sigma_Z\right)}{1 + \omega^2\left(\varepsilon_Z/\sigma_Z\right)^2}\,\varepsilon_Z. \tag{11.5}$$

Equations (11.5) recall Debye's equations for relaxation phenomena in dielectrics [2]. According to this model, the frequency dependence of the complex dielectric constant is given by

$$\epsilon(\omega) = \varepsilon_\infty + \frac{\varepsilon_s - \varepsilon_\infty}{1 + i\omega\tau_\epsilon}, \tag{11.6}$$

where $\varepsilon_\infty = \lim_{\omega\to\infty}\varepsilon_Z$, $\varepsilon_s = \lim_{\omega\to 0}\varepsilon_Z$, and τ_ϵ is the relaxation time. Using (11.6) the real and imaginary parts of ϵ are found to be

$$\epsilon'(\omega) = \varepsilon_\infty + \frac{\varepsilon_s - \varepsilon_\infty}{1 + (\omega\tau_\epsilon)^2} \quad \text{and} \quad \epsilon''(\omega) = (\varepsilon_s - \varepsilon_\infty)\,\frac{\omega\tau_\epsilon}{1 + (\omega\tau_\epsilon)^2}. \tag{11.7}$$

From (11.7) it follows that

$$\left(\epsilon'^2 + \frac{\varepsilon_s + \varepsilon_\infty}{2}\right)^2 + \epsilon''^2 = \left(\frac{\varepsilon_s - \varepsilon_\infty}{2}\right)^2. \tag{11.8}$$

The parametric representation of ϵ' and ϵ'' versus ω is called the Cole-Cole diagram [1]. From (11.8) we can conclude that if Debye's model works well, the Cole-Cole diagram is a circle of radius $(\varepsilon_s - \varepsilon_\infty)/2$, centered in $[(\varepsilon_s + \varepsilon_\infty)/2, 0]$.

These considerations are quite general and concern the impedance spectroscopy technique. Our aim is to analyze the influence of the ions on the impedance spectroscopy measurements. In the limit of low frequency of the signal, the ions present in the liquid contribute to the electrical current, and so to the detected impedance. Several models have been proposed to take into account the effect of the ions on the electric response of a liquid [3]. In this chapter, we present a simple model to describe the influence of the ions on the impedance spectroscopy. We assume that the ions, monovalent of charge q, are dimensionless and dispersed in a homogeneous medium of dielectric constant ε, have the same mobility and that they are not selectively adsorbed by the electrodes. According to this last hypothesis, the ζ-potential vanishes, and the analysis of the impedance spectroscopy is greatly simplified [4, 5]. The electrodes are assumed to be perfectly blocking, in such a manner that there is no charge injection into the liquid. First we find the distribution of ions, when the external voltage depends on the time in sinusoidal way. After that the ionic contribution to the current in the external circuit is evaluated and finally, the equivalent impedance of the cell is deduced.

11.2 The physical system and basic hypotheses

Let us consider a slab of thickness d filled by an isotropic liquid. The z-axis of the Cartesian reference frame used in the description is normal to the bounding surfaces at $z = \pm d/2$, as in the preceding chapters. We assume that in thermodynamical equilibrium the liquid contains a density N of ions of positive and negative sign, uniformly distributed. In this situation, in the absence of selective adsorption, the liquid is globally and locally neutral. The presence of an external electric voltage produces a perturbation of the distribution of the ions in the liquid, in the sense that it remains globally neutral, but now it is locally charged. In the following, we suppose that the sample is submitted to an external sinusoidal voltage of amplitude V_0 and frequency $f = \omega/(2\pi)$. By indicating with n_p and n_m, respectively the density of positive and negative ions, we have $n_p(z,t) = n_m(z,t) = N$, for $V_0 = 0$, and $n_p(z,t) \neq n_m(z,t)$, for $V_0 \neq 0$. The conservation of the number of particles implies that

$$\int_{-d/2}^{d/2} n_p(z,t)\,dz = \int_{-d/2}^{d/2} n_m(z,t)\,dz = N\,d, \tag{11.9}$$

under the assumption that there is no recombination and the electrodes are perfectly blocking, as we are supposing. We assume that the amplitude of the external voltage, V_0, is such that the actual densities of ions only slightly differ from N [6]. By putting

$$n_\alpha = N + \delta n_\alpha(z,t), \tag{11.10}$$

where $\alpha = p, m$, the previous hypothesis implies that $\delta n_\alpha(z,t) \ll N$. We suppose furthermore that $V(\pm d/2, t) = \pm(V_0/2)\exp(i\omega t)$. Equation (11.9), taking into account Eq. (11.10), implies that

$$\int_{-d/2}^{d/2} \delta n_p(z,t)\,dz = \int_{-d/2}^{d/2} \delta n_m(z,t)\,dz = 0, \tag{11.11}$$

stating the global neutrality.

The fundamental equations of the problem are the equation of continuity governing the density of positive and negative ions

$$\frac{\partial n_\alpha}{\partial t} = -\frac{\partial j_\alpha}{\partial z}, \tag{11.12}$$

and the Poisson's equation

$$\frac{\partial^2 V}{\partial z^2} = -\frac{q}{\varepsilon}(n_p - n_m), \tag{11.13}$$

where q is the electrical charge of the ions, and j_α the density of currents of positive and negative ions given by

$$j_\alpha = -D_\alpha \left(\frac{\partial n_\alpha}{\partial z} \pm \frac{q}{k_B T} n_\alpha \frac{\partial V}{\partial z} \right), \tag{11.14}$$

where $+$ for $\alpha = p$, and $-$ for $\alpha = m$. Since the electrodes are supposed perfectly blocking we have the following boundary conditions on j_α

$$j_\alpha(\pm d/2, t) = 0. \tag{11.15}$$

The other boundary conditions of the problem are connected with the imposed difference of potential $V(\pm d/2, t) = \pm(V_0/2)\exp(i\omega t)$. To find the influence of the diffuse layers of ions on the impedance spectroscopy we have to evaluate first the total current in the external circuit, taking into account the presence of the ions. After that, it is necessary to evaluate the electrical impedance of the cell under investigation.

As stated above, we limit the analysis to the case in which the applied voltage V_0 is such that $\delta n_\alpha(z,t) \ll N$. In this case, as it will be shown in the next section, $\delta n_\alpha(z,t) = \eta_\alpha(z)\exp(i\omega t)$ and $V(z,t) = \phi(z)\exp(i\omega t)$, where $\eta_\alpha(z)$ and $\phi(z)$ can be easily determined. Once the mathematical problem to determine $\eta_\alpha(z)$ and $\phi(z)$ has been solved, the impedance of the cell can be determined as follows. Since $V(z,t) = V_0\,\exp(i\omega t)$ the electric field in the cell is

$$E(z,t) = -\frac{\partial V(z,t)}{\partial z} = -\phi'(z)\,\exp(i\omega t), \tag{11.16}$$

where the prime indicates derivative with respect to the z-coordinate. From the theorem of Coulomb, $E(d/2,t) = -\Sigma(t)/\epsilon$, where $\Sigma(t)$ is the density of electric charge at $z = d/2$. Hence, $\Sigma(t) = \varepsilon\phi'(d/2)\,\exp(i\omega t)$, and the total electric charge at $z = d/2$ is

$$Q(t) = \Sigma(t)\,S = \varepsilon\phi'(d/2)S\,\exp(i\omega t), \tag{11.17}$$

where S is the surface of the electrode. It follows that the electric current in the external circuit is

$$I(t) = \frac{dQ(t)}{dt} = i\omega\varepsilon\phi'(d/2)\,S\,\exp(i\omega t), \tag{11.18}$$

and the impedance of the cell we are looking for is given by

$$\mathcal{Z} = \frac{V_0\,\exp(i\omega t)}{I(t)} = \frac{V_0}{i\omega\varepsilon\phi'(d/2)S}. \tag{11.19}$$

This relation will be used in the following to determine the impedance of a cell containing ions in different situations.

11.3 Simple case in which the ions have the same mobility

In this framework, since $\mu_p = \mu_m = \mu$, we have $n_p(z,t) = n_m(-z,t)$. From Eq. (11.14), by taking into account that $\delta n_\alpha(z,t) \ll N$ and $D_p = D_m = D$, we get

$$j_\alpha = -D\left\{\frac{\partial(\delta n_\alpha)}{\partial z} \pm \frac{Nq}{k_BT}\frac{\partial V}{\partial z}\right\}. \tag{11.20}$$

Substitution of (11.20) into Eq. (11.12) yields

$$\frac{\partial(\delta n_\alpha)}{\partial t} = D\left\{\frac{\partial^2(\delta n_\alpha)}{\partial z^2} \pm \frac{Nq}{k_BT}\frac{\partial^2 V}{\partial z^2}\right\}. \tag{11.21}$$

Furthermore, by substituting (11.10) in Eq. (11.13) we have

$$\frac{\partial^2 V}{\partial z^2} = -\frac{q}{\varepsilon}\,(\delta n_p - \delta n_m). \tag{11.22}$$

Equations (11.21) and (11.22) show that if $V(\pm d/2, t) = \pm(V_0/2)\exp(i\omega t)$, in the steady state $\delta n_\alpha(z,t) = \eta_\alpha(z)\exp(i\omega t)$ and $V(z,t) = \phi(z)\exp(i\omega t)$, where, in particular,

$$\phi(\pm d/2) = \pm V_0/2, \tag{11.23}$$

for the boundary conditions imposed on the applied potential. It follows that in the steady state Eq. (11.22) can be rewritten as

$$\phi''(z) = -(q/\varepsilon)[\eta_p(z) - \eta_m(z)]. \tag{11.24}$$

The functions $\eta_\alpha(z)$ are solutions of the differential equations

$$\eta''_{p,m}(z) - \frac{1}{\ell^2}\,\eta_{p,m}(z) + \frac{1}{2\lambda_0^2}\,\eta_{m,p}(z) = 0, \tag{11.25}$$

obtained by Eq. (11.21), where $\lambda_0 = \sqrt{\varepsilon k_B T/(2Nq^2)}$ is the Debye's screening length, defined in (6.9), and $\ell^2 = 2\lambda_0^2/[1 + 2i(\omega/D)\lambda_0^2]$. We look for solutions of Eqs. (14) of the type $\eta_\alpha(z) = C_\alpha \exp(\nu z)$. By substituting this expressions into (11.25) we get

$$\left(\nu^2 - \frac{1}{\ell^2}\right) C_p + \frac{1}{2\lambda_0^2} C_m = 0,$$
$$\left(\nu^2 - \frac{1}{\ell^2}\right) C_m + \frac{1}{2\lambda_0^2} C_p = 0. \tag{11.26}$$

Equations (11.26) admit a solution different from $C_p = C_m = 0$ only if

$$\left(\nu^2 - \frac{1}{\ell^2}\right) = \frac{1}{4\lambda_0^4}. \tag{11.27}$$

From (11.27) we get that the characteristic exponents of the problem are $\nu_{1,2} = \pm\beta$ and $\nu_{3,4} = \pm\gamma$, where

$$\beta = \frac{1}{\lambda_0}\sqrt{1 + i\,\frac{\omega}{D}\lambda_0^2} \quad \text{and} \quad \gamma = \sqrt{i\,\frac{\omega}{D}}. \tag{11.28}$$

From Eq. (11.26) we obtain also that

$$\frac{C_m}{C_p} = -2\lambda_0^2\left(\nu^2 - \frac{1}{\ell^2}\right), \tag{11.29}$$

that, by taking into account (11.28), give

$$\frac{C_m^1}{C_p^1} = \frac{C_m^2}{C_p^2} = -1 \quad \text{and} \quad \frac{C_m^3}{C_p^3} = \frac{C_m^4}{C_p^4} = 1. \tag{11.30}$$

It follows that the functions $\eta_\alpha(z)$ are of the type

$$\begin{aligned}\eta_p(z) &= C_p^1 \exp(\beta z) + C_p^2 \exp(-\beta z) + C_p^3 \exp(\gamma z) + C_p^4 \exp(-\gamma z),\\ \eta_m(z) &= -C_p^1 \exp(\beta z) - C_p^2 \exp(-\beta z) + C_p^3 \exp(\gamma z) + C_p^4 \exp(-\gamma z).\end{aligned} \tag{11.31}$$

From the condition $n_p(z) = n_m(-z)$, connected with the hypothesis that the positive and negative ions have the same mobility, taking into account (11.10), it follows that $\eta_p(z) = \eta_m(-z)$. From (11.14), by imposing this condition, we get $C_p^1 + C_p^2 = 0$ and $C_p^3 - C_p^4 = 0$. Consequently $C_p^1 = -C_p^2 = p_0/2$ and $C_p^3 = C_p^4 = m_0/2$. The solutions of Eqs. (11.25) for the problem under investigation are then

$$\eta_\alpha(z) = m_0 \cosh(\gamma z) \pm p_0 \sinh(\beta z). \tag{11.32}$$

The conservation of the number of particles is contained in Eq. (11.11), that can be rewritten in the form

$$\int_{-d/2}^{d/2} \eta_\alpha(z)\, dz = 0. \tag{11.33}$$

Condition (11.33), taking into account (11.32), implies $m_0 = 0$. Hence $\eta_\alpha(z) = \pm p_0 \sinh(\beta z)$, where p_0 is an integration constant to be determined by means of the boundary conditions (11.15) and (11.23).

The profile of the electric potential is given by Eq. (11.24), that in the present case reads

$$\phi''(z) = -2(q/\varepsilon)\, p_0 \sinh(\beta z), \tag{11.34}$$

from which, by taking into account that in our framework $\phi(z) = -\phi(-z)$, we get

$$\phi(z) = -2(q/\varepsilon\beta^2)\, p_0 \sinh(\beta z) + cz. \tag{11.35}$$

The integration constant c is determined by the boundary conditions (11.15) and (11.23).

The density of currents are, according to (11.20), given by

$$j_\alpha = -D[\eta_\alpha' \pm (qN/k_BT)\phi'(z)] \exp(i\omega t), \tag{11.36}$$

that for the results reported above can be rewritten in the form

$$j_\alpha = \mp D\left[i(\omega/D\beta)\, p_0 \cosh(\beta z) + (Nq/k_BT)\, c\right] \exp(i\omega t). \tag{11.37}$$

By means of (11.35) and (11.37) the boundary conditions of the problem become

$$\begin{aligned}-2(q/\varepsilon\beta^2)\, p_0 \sinh(\beta d/2) + cd/2 &= V_0/2,\\ i(\omega/D\beta)\, p_0 \cosh(\beta d/2) + (Nq/k_BT)c &= 0.\end{aligned} \tag{11.38}$$

By solving (11.38) with respect to p_0 and c we get

$$p_0 = -\frac{Nq\beta}{2k_BT} \frac{1}{(1/\lambda_0^2\beta)\sinh(\beta d/2) + i(\omega d/2D)\cosh(\beta d/2)} V_0,$$
$$c = i\frac{\omega}{2D} \frac{\cosh(\beta d/2)}{(1/\lambda_0^2\beta)\sinh(\beta d/2) + i(\omega d/2D)\cosh(\beta d/2)} V_0. \tag{11.39}$$

The electrical problem is then solved.

The impedance of the cell is given by Eq. (11.19). In the present case, taking into account (11.39), we have

$$\mathcal{Z} = -i\frac{2}{\omega\varepsilon\beta^2 S}\left\{\frac{1}{\lambda_0^2\beta}\tanh\left(\beta\frac{d}{2}\right) + i\frac{\omega d}{2D}\right\}. \tag{11.40}$$

If a true dielectric is considered, $N = 0$, and hence $\lambda_0 \to \infty$. In this case (11.40) gives

$$\mathcal{Z} = \frac{1}{i\omega\varepsilon S/d}, \tag{11.41}$$

as expected.

Before entering into the detailed analysis of the impedance of the cell, given by (11.40), we consider the special cases of $\omega \to 0$ and $\omega \to \infty$, limiting our investigations to the case in which $\lambda_0 \ll d$, i.e., $u = d/\lambda_0 \gg 1$. Consequently, $\tanh u \sim 1$ and $\cosh u \sim \sinh u \sim e^u/2$. In the first case, from (11.40) we obtain $\mathcal{Z}(\omega \to 0) = \mathcal{R}(\omega \to 0) + i\mathcal{X}(\omega \to 0)$, where

$$\mathcal{R}(\omega \to 0) = \frac{\lambda_0^2 d}{D\varepsilon S}\left(1 - \frac{\lambda_0^4}{D^2}\omega^2\right),$$
$$\mathcal{X}(\omega \to 0) = -2\frac{\lambda_0}{\omega\varepsilon S}\left(1 + \frac{\lambda_0^3 d}{2D^2}\omega^2\right). \tag{11.42}$$

From Eqs. (11.42), taking into account Eq. (11.1), we get for $\sigma_Z(\omega \to 0)$ and $\varepsilon_Z(\omega \to 0)$ the expressions

$$\sigma_Z(\omega \to 0) = \frac{\varepsilon D}{\lambda_0^2}\left(1 + \frac{\lambda_0^4}{D^2}\omega^2\right),$$
$$\varepsilon_Z(\omega \to 0) = \frac{1}{2}\varepsilon\frac{d}{\lambda_0}\left(1 - \frac{\lambda_0^3 d}{2D^2}\omega^2\right). \tag{11.43}$$

In particular, in the dc limit, $\sigma_Z(0) = \varepsilon D/\lambda_0^2 = 2Nq\mu$, as expected, because we have two identical carriers of charge. Furthermore, $\varepsilon_Z(0) = (1/2)\varepsilon\,(d/\lambda_0)$, since in this limit the equivalent capacity of the cell is the one of two equal condensers of thickness λ_0 in series. Note that $\sigma_Z(0)$ is d-independent, whereas $\varepsilon_Z(0)$ is proportional to d [7].

In the limit of high frequency ($\omega \to \infty$) from (11.40) we obtain $\mathcal{Z}(\omega \to \infty) = \mathcal{R}(\omega \to \infty) + i\mathcal{X}(\omega \to \infty)$, where

$$\mathcal{R}(\omega \to \infty) = \frac{Dd}{\omega^2 \lambda_0^2 \varepsilon S}\left(1 - \frac{1}{d}\sqrt{\frac{2D}{\omega}}\right),$$

$$\mathcal{X}(\omega \to \infty) = -\frac{d}{\omega \varepsilon S}\left(1 - \frac{D}{\lambda_0^2 d\omega}\sqrt{\frac{2D}{\omega}}\right). \tag{11.44}$$

In this case $\sigma_Z(\omega \to \infty)$ and $\varepsilon_Z(\omega \to \infty)$ are found to be

$$\sigma_Z(\omega \to \infty) = \frac{\varepsilon \lambda_0^2}{D}\omega^2\left(1 + \frac{1}{d}\sqrt{\frac{2D}{\omega}}\right),$$

$$\varepsilon_Z(\omega \to \infty) = \varepsilon\left(1 + \frac{D}{\lambda_0^2 d\omega}\sqrt{\frac{2D}{\omega}}\right). \tag{11.45}$$

From (11.45), it follows that for $\omega \to \infty$, the equivalent chemical potential is $\sigma_Z = (\varepsilon\lambda_0^2/D)\,\omega^2 = \sigma_Z(0)\,(\lambda_0^2\omega/D)^2$. The equivalent capacitance of the cell at high frequency is $C_{\rm eq} = \varepsilon_Z\,(S/d)$, showing that $\lim_{\omega\to\infty}\varepsilon_Z = \varepsilon$, as expected because in this limit the ions do not give any contribution to the electrical response of the cell. The leading terms for $\sigma_Z(\omega \to \infty)$ and $\varepsilon_Z(\omega \to \infty)$ are independent of the thickness of the cell. We note, finally, that in this limit $\varepsilon_Z(\omega \to \infty)$ tends to ε as $\omega^{-3/2}$, as experimentally observed [8].

We can now investigate the frequency dependence of the real and imaginary parts of the impedance of the cell. In order to obtain an explicit expression for $\mathcal{R}$ and $\mathcal{X}$ we write $\beta = \beta_r + i\beta_i$. Simple calculations give

$$\beta_r = (1/\lambda_0)\sqrt{(M+1)/2} \quad \text{and} \quad \beta_i = (1/\lambda_0)\sqrt{(M-1)/2},$$

where $M = \sqrt{1 + (\omega\lambda_0^2/D)^2}$. We put now $A = \tanh(\beta_r d/2)$, $B = \tan(\beta_i d/2)$, $m = A(1+B^2)/[1+(AB)^2]$, and $n = B(1-A^2)/[1+(AB)^2]$. With these positions the explicit expressions for the real and imaginary parts of impedance of the cell are found to be

$$\mathcal{R} = \frac{2\lambda_0^2}{\omega\epsilon M^2 S}\left\{\left(\frac{n\beta_r - m\beta_i}{M} + \frac{\omega d}{2D}\right) - \frac{\omega\lambda_0^2}{D}\,\frac{m\beta_r + n\beta_i}{M}\right\},$$

$$\mathcal{X} = -\frac{2\lambda_0^2}{\omega\epsilon M^2 S}\left\{\frac{m\beta_r + n\beta_i}{M} + \frac{\omega\lambda_0^2}{D}\left(\frac{n\beta_r - m\beta_i}{M} + \frac{\omega d}{2D}\right)\right\}. \tag{11.46}$$

The quantity M introduced above is such that for $\omega \ll \omega_r = D/\lambda_0^2$, $M \to 1$, whereas for $\omega \gg \omega_r$, $M \to \omega/\omega_r \gg 1$. It follows that a change of the frequency dependence of $\mathcal{R}$ and $\mathcal{X}$ is expected around $\omega \sim \omega_r$.

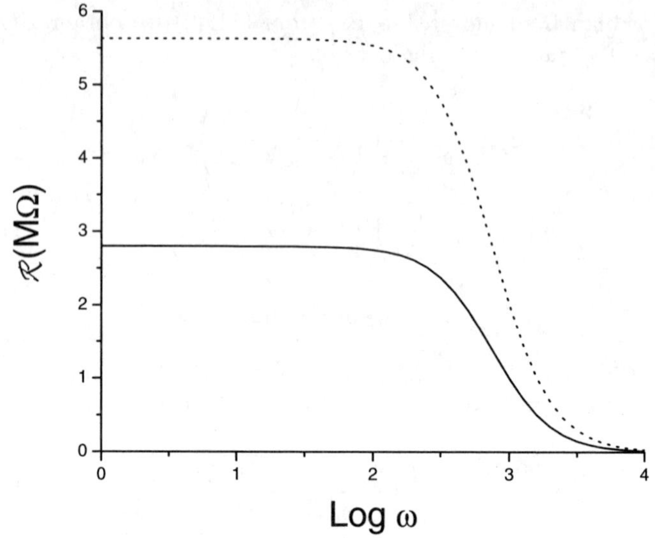

FIGURE 11.1
Real part of the impedance of the cell, $\mathcal{R}$, versus ω. Dotted line $d_1 = 50\mu$m, solid line $d_2 = 25\mu$m.

We suppose that the cell is a slab of thickness $d_1 = 25\mu$m or $d_2 = 50\mu$m, filled with the nematic liquid crystal 5CB (4-cyano-4'-n-pentylbiphenyl), planarly oriented by the surface treatment, as the one considered by Murakami *et al.* [7]. Since the applied voltage is very small with respect to the threshold voltage for the Fréedericksz transition, there is no reorientation of the nematic liquid crystal induced by the external voltage. In this case $\varepsilon = \varepsilon_\perp = 6.7\varepsilon_0$, where ε_0 is the dielectric permittivity of the free space. We assume furthermore that the density of ions is $N \sim 4.2 \times 10^{20}$ m^{-3} and $D \sim 8.2 \times 10^{-12}$ m^2/s [9, 10]. With these values $\lambda_0 \sim 10^{-7}$ m and $\omega_r \sim 740$ rad/s.

In Fig. 11.1, we show $\mathcal{R}(\omega)$, where from now on $\text{Log}\,\omega = \log_{10}\omega$. It tends to a constant value for $\omega \to 0$, and to zero for $\omega \to \infty$. It presents a large plateau until ω_r. In Figure 11.2 is reported $\mathcal{X}(\omega)$. It tends to $-\infty$ for $\omega \to 0$, and to 0, from the negative side, for $\omega \to \infty$. In between it has first a maximum and then a minimum for $\omega \sim \omega_r$. The equivalent chemical potential of the cell, $\sigma_Z(\omega)$, is shown in Figure 11.3. As discussed above, $\lim_{\omega\to 0}\sigma_Z(\omega) = \varepsilon D/\lambda_0^2$, and $\sigma_Z(\omega \to \infty) = (\varepsilon\lambda_0^2/D)\omega^2$. From Figure 11.3 it follows that σ_Z is practically independent of the thickness of the sample. Finally, in Figure 11.4 we report $\varepsilon_Z(\omega)$. In this case $\lim_{\omega\to 0}\varepsilon(\omega) = \varepsilon(d/2\lambda_0)$, and $\lim_{\omega\to\infty}\varepsilon_Z(\omega) = \varepsilon$.

In the analysis presented above we have investigated the influence of the ions on the dielectric spectroscopy of a nematic cell. According to our scheme this influence is described by the equivalent chemical potential, σ_Z, and equivalent dielectric permittivity, ε_Z, of the cell. When these quantities are known, by

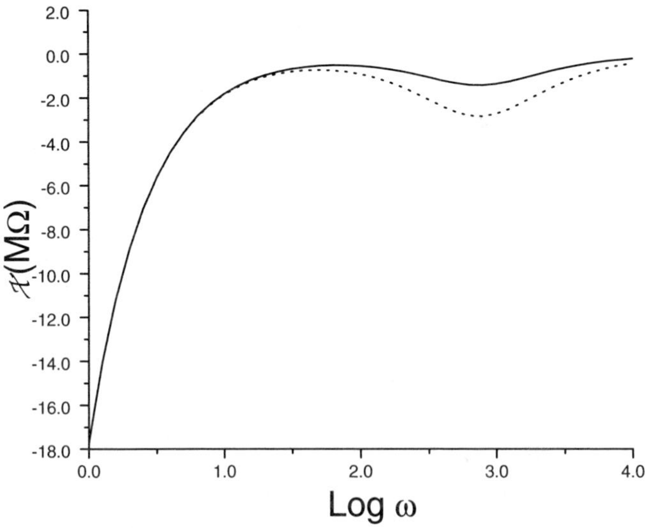

FIGURE 11.2
Imaginary part of the impedance of the cell, $\mathcal{X}$, versus ω. Dotted line $d_1 = 50\mu$m, solid line $d_2 = 25\mu$m.

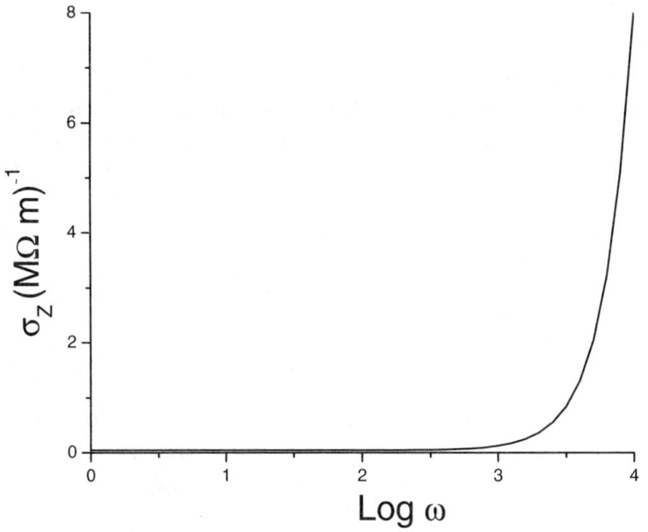

FIGURE 11.3
Equivalent chemical potential of the cell, σ_Z, versus ω. Dotted line $d_1 = 50\mu$m, solid line $d_2 = 25\mu$m. Note that σ_Z is practically independent of the thickness of the sample.

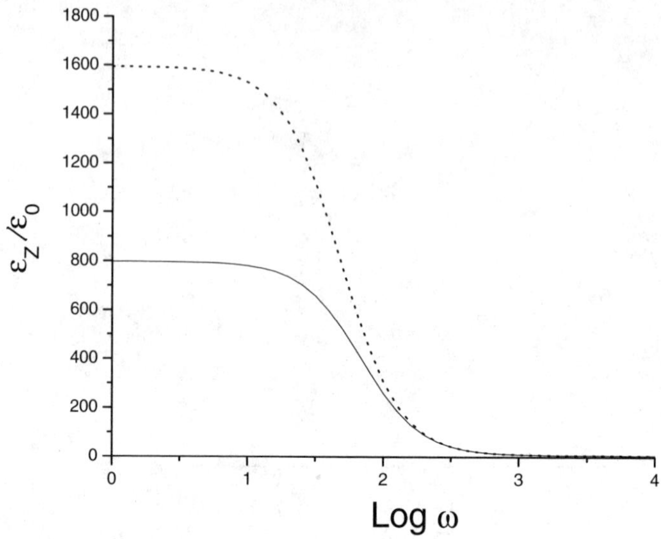

FIGURE 11.4
Equivalent relative dielectric permittivity of the cell, ε_Z, versus ω. Dotted line $d_1 = 50\mu$m, solid line $d_2 = 25\mu$m.

means of Eq. (11.5) we can easily determine the real and imaginary parts of the complex dielectric constant and the relaxation time.

In the case considered by us, the relaxation time is given by $\tau_Z = \varepsilon_Z/\sigma_Z$, where $\sigma_Z = \sigma_Z(\omega)$ and $\varepsilon_Z = \varepsilon_Z(\omega)$, due to the presence of the ions. It follows that when the ions are present, they give rise to a distribution of relaxation times.

By means of (11.5) and (11.43) we get in the limit of $\omega \to 0$, when $d \gg \lambda_0$,

$$\epsilon' = \frac{1}{2}\,\varepsilon\,\frac{d}{\lambda_0}\left(1 - \frac{1}{4}\,\frac{d^2\lambda_0^2}{D^2}\,\omega^2\right),$$

$$\epsilon'' = \frac{1}{4}\,\varepsilon\,\frac{d^2}{D}\,\omega\left(1 - \frac{1}{4}\,\frac{d^2\lambda_0^2}{D^2}\,\omega^2\right). \tag{11.47}$$

The relation for ϵ'', in the limit of $\omega \to 0$, shows that in the low frequency region ϵ'' presents a maximum for $\omega_M \sim D/(\lambda_0 d)$. Furthermore, in the dc limit, the relaxation time tends to $\tau_Z(0) = d\lambda_0/(2D)$.

In the opposite limit of high frequency ($\omega \to \infty$), by means of (11.4) and (11.45), by assuming again $d \gg \lambda_0$, we obtain

$$\epsilon' = \varepsilon\left(1 + \frac{D}{\lambda_0^2 d\omega}\sqrt{\frac{2D}{\omega}}\right) \quad \text{and} \quad \epsilon'' = \frac{D}{\omega\lambda_0^2}\,\varepsilon. \tag{11.48}$$

In this limit the relaxation time tends to zero as $\tau_Z = D/(\omega^2\lambda_0^2)$.

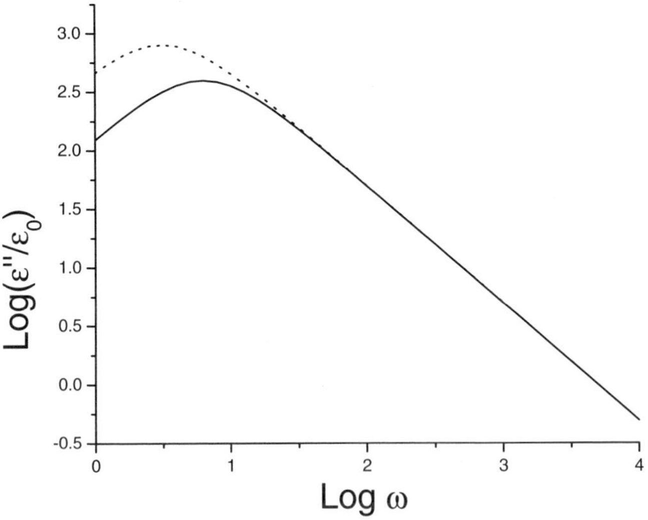

FIGURE 11.5
Imaginary part of the relative complex dielectric permittivity of the cell, ϵ''/ε_0, versus ω. Dotted line $d_1 = 50\mu$m, solid line $d_2 = 25\mu$m. Note that ϵ'/ε_0 is practically independent of the thickness of the sample for a large frequency range.

In Figure 11.5 is reported Log $(\epsilon''/\varepsilon_0)$ versus Log ω. In Figure 11.6, we show Log $(\epsilon'/\varepsilon_0)$ versus Log ω, which has the usual trend. In Figure 11.7, we show Log $(\epsilon'/\varepsilon_0)$ and Log $(\epsilon''/\varepsilon_0)$ versus Log ω. Figure 11.8 shows the frequency dependence of the relaxation time of ionic origin. Finally, by taking into account that according to our calculations $\varepsilon_s = (1/2)\varepsilon(d/\lambda_0)$ and $\varepsilon_\infty = \varepsilon$, in Figure 11.9 we show the Cole-Cole diagram for our cell containing ions.

A final remark on the amplitude of the applied voltage is necessary. As stated at the beginning of this section, V_0 has to be such that $\delta n_\alpha(z,t) \ll N$, because only in this case the system behaves in a linear manner. Since $\delta n_\alpha(z,t) = \eta_\alpha(z)\exp(i\omega t)$, the previous condition can be rewritten in the form $2|\eta_\alpha(z)| \ll N$. By taking into account that $\eta_\alpha(z) = \pm p_0 \sinh(\beta z)$, where p_0 is given by Eq. (11.39), we find that the system can be considered as a linear system for

$$V_0 \ll U = V_T \left| \frac{1}{\lambda_0^2\beta^2} + i\frac{\omega d}{2D\beta}\coth\left(\beta\frac{d}{2}\right)\right|, \tag{11.49}$$

where we have used again the thermal potential V_T introduced in Section 7.1. In Figure 11.10, we show the frequency dependence U. As is evident from this figure, in the limit of $\omega \to 0$, U tends to V_T, as expected. In fact, in this region the profile of the electric potential is given by the Poisson-Boltzmann

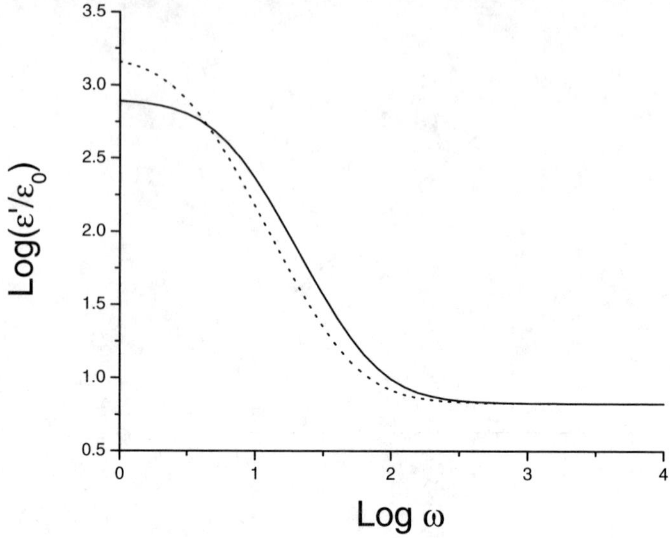

FIGURE 11.6
Real part of the relative complex dielectric permittivity of the cell ϵ'/ε_0 versus ω. Dotted line $d_1 = 50\mu$m, solid line $d_2 = 25\mu$m.

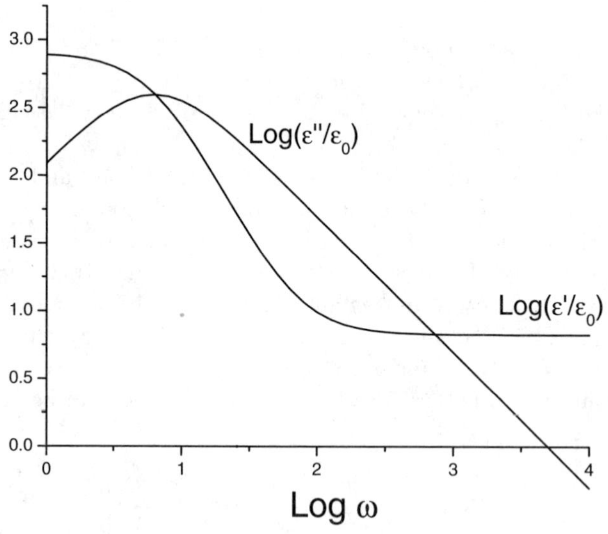

FIGURE 11.7
Real ϵ'/ε_0 and imaginary ϵ''/ε_0 parts of the relative complex dielectric permittivity of the cell relevant to $d = 25\mu$m.

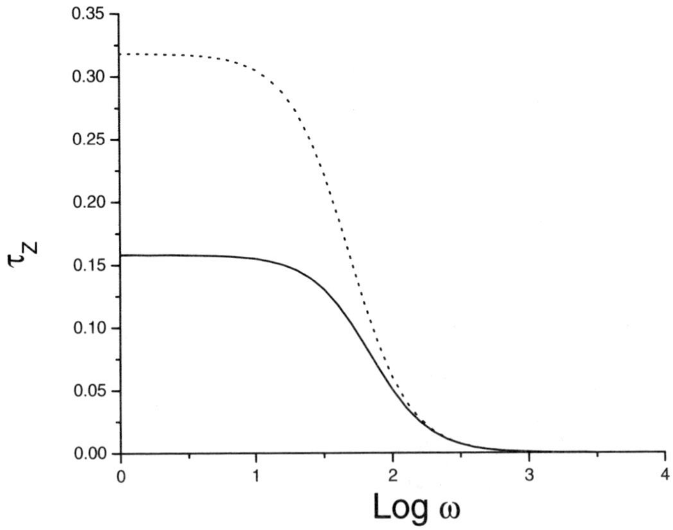

FIGURE 11.8
Relaxation time τ_Z versus ω. Dotted line $d_1 = 50\mu$m, solid line $d_2 = 25\mu$m.

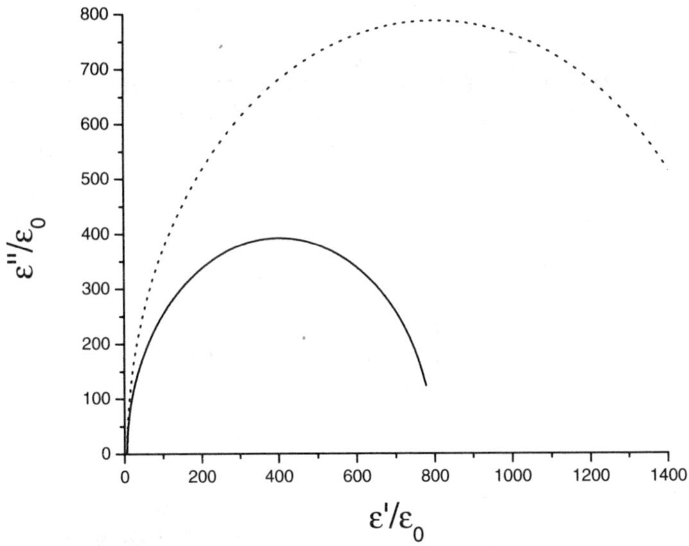

FIGURE 11.9
Cole-Cole diagram for the electrolytic cell according to the proposed model.

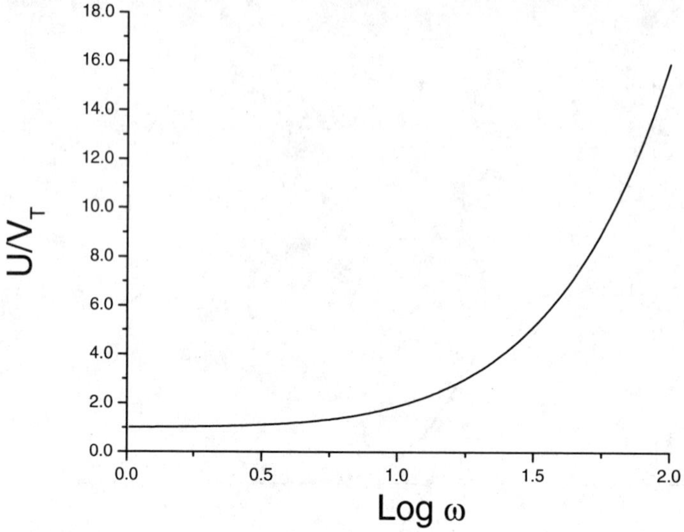

FIGURE 11.10
Frequency dependence of the critical voltage U for an electrolyte cell in which the ions have the same mobility. In the low frequency region $U(\omega \to 0) = V_T[1 + (1/2)(\lambda_0 d\omega/2D)^2]$. On the contrary, in the high frequency range $U(\omega \to \infty) = (V_T/2)\sqrt{\omega d^2/D}$.

equation, where the thermal energy $k_B T$ is compared with the electrostatic energy qV. On the contrary, for large ω, U diverges as $\omega^{1/2}$. By means of (11.49) a simple calculation gives, in the limit of $d \gg \lambda_0$, the asymptotic expressions

$$U(\omega \to 0) = V_T \left\{ 1 + \frac{1}{2}\left(\frac{\lambda_0 d}{2D}\right)^2 \omega^2 \right\},$$

$$U(\omega \to \infty) = \frac{V_T}{2}\frac{d}{\sqrt{D}}\sqrt{\omega}. \tag{11.50}$$

From the discussion reported above it follows that if the investigations of the dielectric properties of the material have to be performed in the very low frequency region, the amplitude of the signal applied to the sample has to be small with respect to V_T. On the contrary, for large ω this condition is substituted by a new one that depends on the frequency [11].

11.4 Effect of different anionic and cationic mobilities on the impedance spectroscopy

The theoretical analysis of the impedance of an electrolytic cell limited by blocking electrodes by assuming that the positive and negative ions have the same mobility has been presented in the previous section. The aim of this section is to investigate the case in which the mobility of the positive ions, D_p, is different from that of the negative ions, D_m [12]. We assume that there is only one group of ions. However, the extension to the case in which several groups are present is straightforward.

The current densities, containing the diffusion and drift components, are

$$j_\alpha(z,t) = -D_\alpha \left\{ \frac{\partial n_\alpha(z,t)}{\partial z} \pm \frac{q}{k_B T}\, n_\alpha(z,t)\, \frac{\partial V(z,t)}{\partial z} \right\}, \tag{11.51}$$

where $+$ for $\alpha = p$ and $-$ for $\alpha = m$. As in the previous section, in the limit of $\delta n_\alpha(z,t) \ll N$, the system behaves as a linear system. In this situation, when $V(\pm d/2, t) = V_0 \exp(i\omega t)$, an analysis of the same type presented above shows that $\delta n_\alpha(z,t) = \eta_\alpha(z) \exp(i\omega t)$ and $V(z,t) = \phi(z) \exp(i\omega t)$. The function $\phi(z)$ is still the solution of Eq. (11.24), whereas $\eta_\alpha(z)$ represent solutions of the equations

$$\eta''_{p,m}(z) - \frac{1}{\ell^2_{p,m}}\, \eta_{p,m}(z) + \frac{1}{2\lambda_0^2}\, \eta_{m,p}(z) = 0, \tag{11.52}$$

where $\ell^2_\alpha = 2\lambda_0^2/[1 + 2i(\omega/D_\alpha)\lambda_0^2]$ are intrinsic parameters of the problem. Equations (11.52) generalize Eqs. (11.25) to the case in which $D_p \neq D_m$.

We look for a solution of Eqs. (11.52) of the type $\eta_\alpha(z) = C_\alpha \exp(\nu z)$. By substituting this ansatz into Eqs. (11.52) we obtain

$$\begin{aligned} \left(\nu^2 - \frac{1}{\ell_p^2}\right) C_p + \frac{1}{2\lambda_0^2}\, C_m &= 0, \\ \left(\nu^2 - \frac{1}{\ell_m^2}\right) C_m + \frac{1}{2\lambda_0^2}\, C_p &= 0. \end{aligned} \tag{11.53}$$

Equations (11.53) form a linear and homogeneous system in C_α. A solution different from the trivial one is possible only if

$$\left(\nu^2 - \frac{1}{\ell_p^2}\right)\left(\nu^2 - \frac{1}{\ell_m^2}\right) - \frac{1}{4\lambda_0^4} = 0. \tag{11.54}$$

Equation (11.54) determines the characteristic exponents of the problem. From (11.54) we obtain $\nu_{1,2} = \pm\gamma_1$, and $\nu_{3,4} = \pm\gamma_2$, where

$$\gamma_{1,2} = \sqrt{\frac{1}{2}\left(\frac{1}{\ell_p^2} + \frac{1}{\ell_m^2}\right) \pm \sqrt{\left[\frac{1}{2}\left(\frac{1}{\ell_p^2} - \frac{1}{\ell_m^2}\right)\right]^2 + \frac{1}{4\lambda_0^4}}}. \tag{11.55}$$

The solutions of Eqs. (11.52) are then

$$\eta_\alpha(z) = C_\alpha^1 \exp(\gamma_1 z) + C_\alpha^2 \exp(-\gamma_1 z) + C_\alpha^3 \exp(\gamma_2 z) + C_\alpha^4 \exp(-\gamma_2 z). \tag{11.56}$$

From Eqs. (11.53) it follows also that

$$\frac{C_m}{C_p} = -2\lambda_0^2\left(\nu^2 - \frac{1}{\ell_p^2}\right). \tag{11.57}$$

Consequently

$$\begin{aligned} \frac{C_m^1}{C_p^1} &= \frac{C_m^2}{C_p^2} = -2\lambda_0^2\left(\gamma_1^2 - \frac{1}{\ell_p^2}\right) = k_1, \\ \frac{C_m^3}{C_p^3} &= \frac{C_m^4}{C_p^4} = -2\lambda_0^2\left(\gamma_2^2 - \frac{1}{\ell_p^2}\right) = k_2, \end{aligned} \tag{11.58}$$

and (11.56) can be rewritten as

$$\begin{aligned} \eta_p(z) &= C_p^1 \exp(\gamma_1 z) + C_p^2 \exp(-\gamma_1 z) + C_p^3 \exp(\gamma_2 z) + C_p^4 \exp(-\gamma_2 z), \\ \eta_m(z) &= k_1[C_p^1 \exp(\gamma_1 z) + C_p^2 \exp(-\gamma_1 z)] + k_2[C_p^3 \exp(\gamma_2 z) + C_p^4 \exp(-\gamma_2 z)]. \end{aligned} \tag{11.59}$$

The integration constants C_p^a, $a = 1, 2, 3, 4$ will be determined by means of the boundary conditions on the electrical potential and on the current densities. By substituting (11.59) into (11.24) we obtain

$$\begin{aligned} \phi(z) = &- \frac{q}{\varepsilon}\left\{\frac{1-k1}{\gamma_1^2}[C_p^1 \exp(\gamma_1 z) + C_p^2 \exp(-\gamma_1 z)]\right. \\ &+ \left.\frac{1-k2}{\gamma_2^2}[C_p^3 \exp(\gamma_2 z) + C_p^4 \exp(-\gamma_2 z)]\right\} + A\,z + B, \end{aligned} \tag{11.60}$$

where A and B are two new integration constants to be determined by the boundary conditions.

From the boundary conditions $V(\pm d/2, t) = \pm(V_0/2)\exp(i\omega t)$ it follows that $\phi(\pm d/2) = \pm V_0/2$. Consequently, from (11.60) we get, after simple calculations

$$
\begin{aligned}
&-\frac{q}{\varepsilon}\left\{\frac{1-k_1}{\gamma_1^2}(C_p^1+C_p^2)\cosh(\gamma_1 d/2)+\frac{1-k_2}{\gamma_2^2}(C_p^3+C_p^4)\cosh(\gamma_2 d/2)\right\} \\
&+B=0, \\
&-\frac{q}{\varepsilon}\left\{\frac{1-k_1}{\gamma_1^2}(C_p^2-C_p^1)\sinh(\gamma_1 d/2)+\frac{1-k_2}{\gamma_2^2}(C_p^4-C_p^3)\sinh(\gamma_2 d/2)\right\} \\
&-A\frac{d}{2}+\frac{V_0}{2}=0. \qquad (11.61)
\end{aligned}
$$

We have now to impose the boundary conditions $j_\alpha(\pm d/2, t) = 0$. Since $\delta n_\alpha(z,t) \ll N$ we have

$$j_{p,m}(z,t) = -D_{p,m}\left\{\eta'_{p,m}(z) \pm (Nq/k_B T)\phi'(z)\right\}\exp(i\omega t).$$

It follows that

$$\eta'_{p,m}(z) \pm (Nq/k_B T)\phi'(z) = 0, \quad \text{at} \quad z = \pm d/2. \qquad (11.62)$$

Since the adsorption is absent the number of ions in the bulk does not change with the time, and Eq. (11.33) has to be verified for all t and for all thicknesses. By substituting (11.59) into (11.33) we obtain

$$
\begin{aligned}
\int_{-d/2}^{d/2}\eta_p(z)\,dz &= 2\left\{\frac{C_p^1+C_p^2}{\gamma_1}\sinh(\gamma_1 d/2)+\frac{C_p^3+C_p^4}{\gamma_2}\sinh(\gamma_2 d/2)\right\}=0, \\
\int_{-d/2}^{d/2}\eta_m(z)\,dz &= 2\left\{k_1\frac{C_p^1+C_p^2}{\gamma_1}\sinh(\gamma_1 d/2)+k_2\frac{C_p^3+C_p^4}{\gamma_2}\sinh(\gamma_2 d/2)\right\}=0.
\end{aligned}
\qquad (11.63)
$$

From Eqs. (11.63), valid for all d, we deduce $C_p^1 = -C_p^2 = C_1/2$, and $C_p^3 = -C_p^4 = C_2/2$. Consequently, from (11.59) we get

$$
\begin{aligned}
\eta_p(z) &= C_1\sinh(\gamma_1 z)+C_2\sinh(\gamma_2 z), \\
\eta_m(z) &= k_1 C_1\sinh(\gamma_1 z)+k_2 C_2\sinh(\gamma_2 z). \qquad (11.64)
\end{aligned}
$$

By substituting $C_p^1 + C_p^2 = 0$ and $C_p^3 + C_p^4 = 0$ into (11.61) we have $B = 0$ and

$$\phi(z) = -\frac{q}{\varepsilon}\left\{\frac{1-k_1}{\gamma_1^2}C_1\sinh(\gamma_1 z)+\frac{1-k_2}{\gamma_2^2}C_2\sinh(\gamma_2 z)\right\}+Az. \qquad (11.65)$$

Finally, by substituting (11.64) and (11.65) into (11.62) we obtain

$$
\begin{aligned}
\gamma_1\left(1-\frac{1-k_1}{2\lambda_0^2\gamma_1^2}\right)C_1\cosh(\gamma_1 d/2)+\gamma_2\left(1-\frac{1-k_2}{2\lambda_0^2\gamma_2^2}\right)C_2\cosh(\gamma_2 d/2) &= -A, \\
\gamma_1\left(k_1+\frac{1-k_1}{2\lambda_0^2\gamma_1^2}\right)C_1\cosh(\gamma_1 d/2)+\gamma_2\left(k_2+\frac{1-k_2}{2\lambda_0^2\gamma_2^2}\right)C_2\cosh(\gamma_2 d/2) &= A,
\end{aligned}
\qquad (11.66)
$$

that, together with the boundary condition following from (11.65), $\phi(d/2) = V_0/2$,

$$-\frac{q}{\varepsilon}\left\{\frac{1-k_1}{\gamma_1^2}\,C_1\,\sinh(\gamma_1 d/2)+\frac{1-k_2}{\gamma_2^2}\,C_2\,\sinh(\gamma_2 d/2)\right\}+A\frac{d}{2}=\frac{V_0}{d}, \quad (11.67)$$

solve the problem for C_1, C_2, and A. When these quantities have been determined the electrical problem is solved. We note that from Eqs. (11.66) follows that

$$C_2 = -\frac{\gamma_1(1+k_1)\,\cosh(\gamma_1 d/2)}{\gamma_2(1+k_2)\,\cosh(\gamma_2 d/2)}\,C_1. \quad (11.68)$$

If $D_p = D_m = D$ from (11.55), taking into account (11.52), we get

$$\gamma_1 = \frac{1}{\lambda_0}\sqrt{1+i\frac{\omega}{D}\lambda_0^2} \quad \text{and} \quad \gamma_2 = \sqrt{i\frac{\omega}{D}}. \quad (11.69)$$

In this case, from (11.58) follows that $k_1 = -1$ and $k_2 = 1$. Consequently, using (11.68) we get $C_2 = 0$, in agreement with the results obtained in the previous section (see Eq. (11.32), where $m_0 = 0$).

The impedance of the cell is given by Eq. (11.19), where $\phi(z)$ is given by (11.65). The real, $\mathcal{R}$, and imaginary, $\mathcal{X}$, part of $\mathcal{Z}$ can be easily determined numerically.

To investigate the role of the difference between the mobilities of the ions of different signs we consider the experimental situation described in Ref. [9] where a nematic sample, with $d = 25\mu$m and $S = 2\times 10^{-4}$, of 5CB in planar orientation, at room temperature, was considered. In this case $\varepsilon = 6.7$ and $D_m \sim 8.2\times 10^{-12}$ m^2/s, and $N \sim 4\times 10^{20}$ m^{-3}.

In Figures 11.11 and 11.12 we show the frequency dependence of $\mathcal{R}$ and $\mathcal{X}$ versus ω, for the case in which $D_p = 10D_m$. As is evident from Figure 11.11, the real part of the impedance in the "high" frequency region has the usual trend, characterized by a plateau, followed by a decreasing tail where for $\omega \to \infty$, $\mathcal{R} \to 0$. However in the "low" frequency range, $\mathcal{R}$ presents a new plateau. This trend is a signature of the difference between the mobilities (or diffusion coefficients) of the ions of different signs. In Figure 11.13 are reported the frequencies dependence of σ_Z and ε_Z. In Figure 11.14 we compare the frequency dependencies of Log $(\epsilon'/\varepsilon_0)$ with Log $(\epsilon''/\varepsilon_0)$, whereas in Figure 11.15 is reported the frequency dependence of the relaxation time τ_Z.

Moreover, by taking into account that $\varepsilon_\infty = \varepsilon$ and $\varepsilon_s = (1/2)\varepsilon(d/\lambda_0)$, as discussed in the previous section, we evaluate the radius of the Cole-Cole diagram with the relation

$$r^2 = \left(\epsilon' - \frac{\varepsilon_s+\varepsilon_\infty}{2}\right)^2 + \epsilon''^2. \quad (11.70)$$

Finally, in Figure 11.17 we report the Cole-Cole diagram, showing directly, in polar coordinates, ϵ'' versus ϵ'.

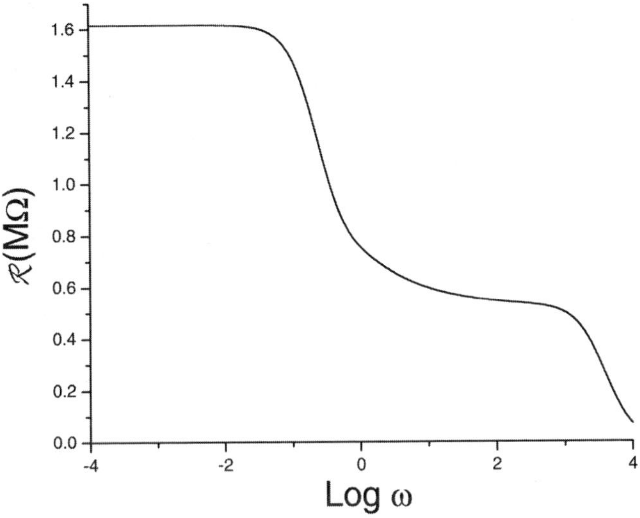

FIGURE 11.11
Real part, $\mathcal{R}$, of the electrical impedance, $\mathcal{Z}$, versus Log ω. The difference between the diffusion coefficients of the positive and negative ions is responsible for the new plateau in the low frequency region.

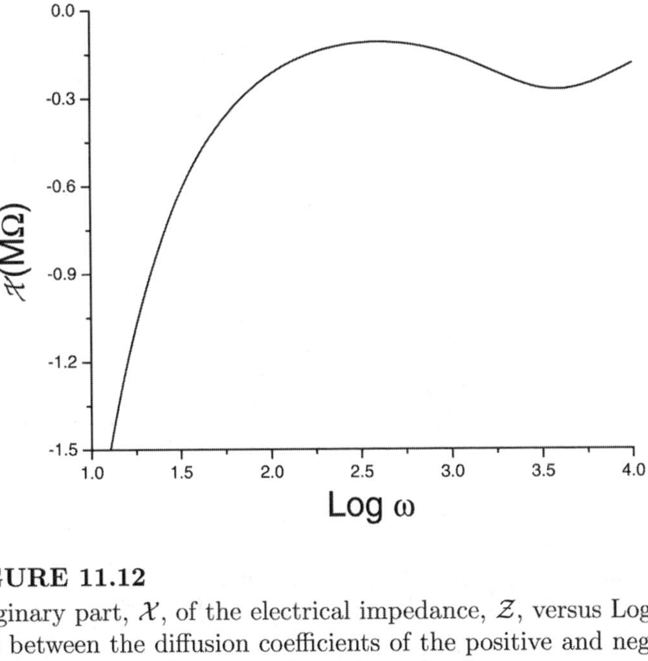

FIGURE 11.12
Imaginary part, $\mathcal{X}$, of the electrical impedance, $\mathcal{Z}$, versus Log ω. The difference between the diffusion coefficients of the positive and negative ions does not introduce new frequency dependence for $\mathcal{X}$.

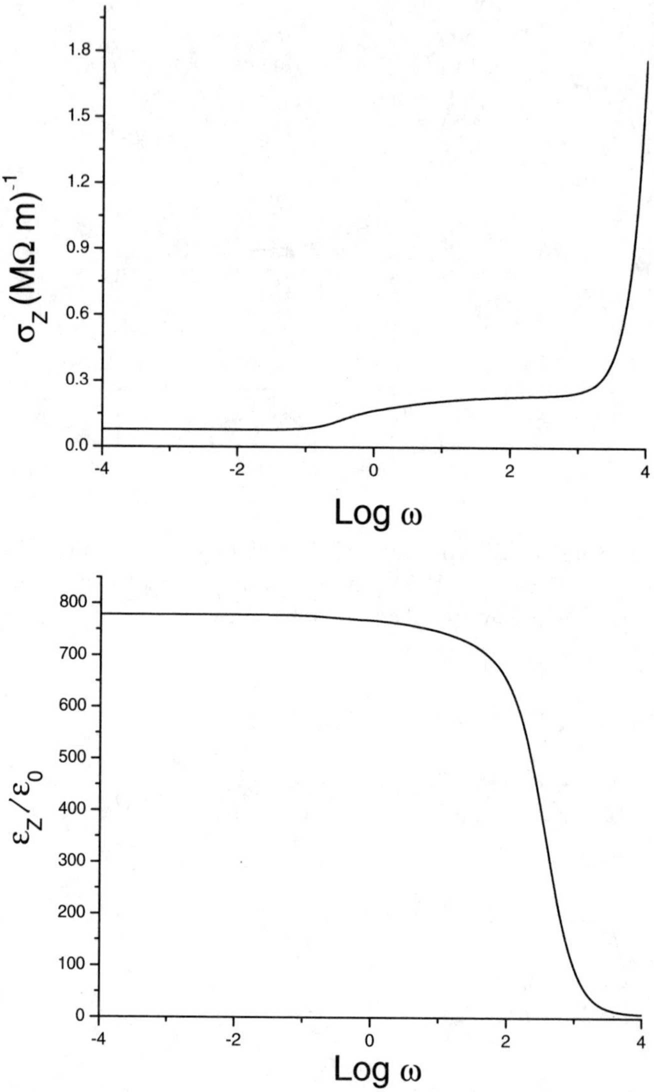

FIGURE 11.13
Equivalent chemical potential, σ_Z, versus Log ω. Note the presence of the two plateaux, relevant to the two values of the mobilities (solid curve). Equivalent dielectric constant ε_Z versus Log ω (dashed curve, hidden due to the coincidence). Note that $\varepsilon_\infty = \varepsilon$ and $\varepsilon_s = (1/2)\varepsilon(d/\lambda_0)$).

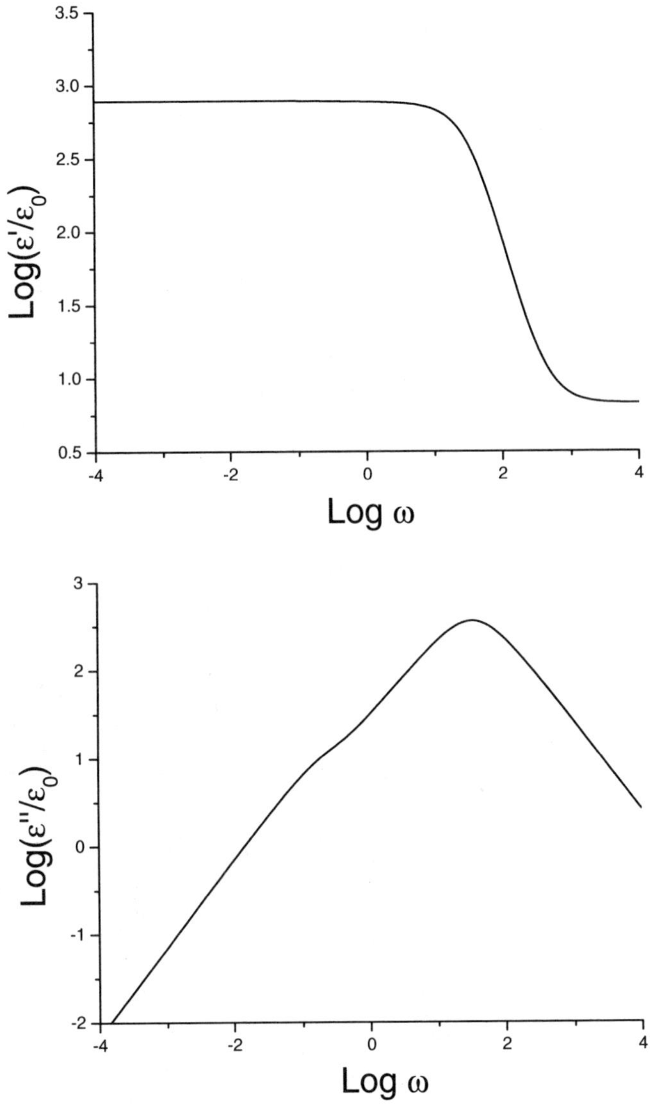

FIGURE 11.14
Log $(\epsilon'/\varepsilon_0)$ and Log $(\epsilon''/\varepsilon_0)$ versus Log ω.

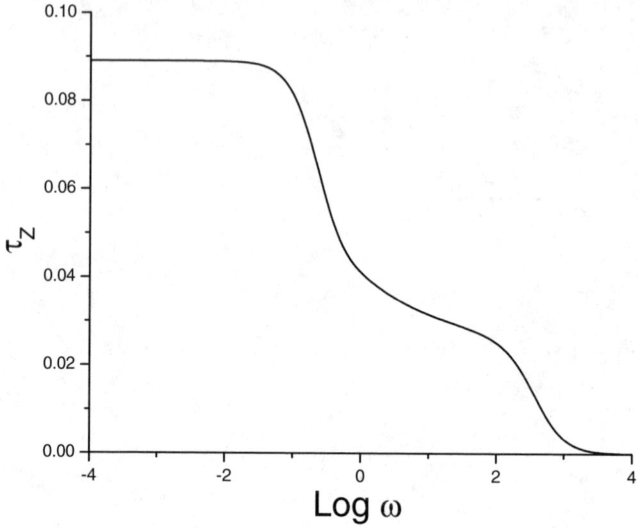

FIGURE 11.15
Relaxation time versus Log ω.

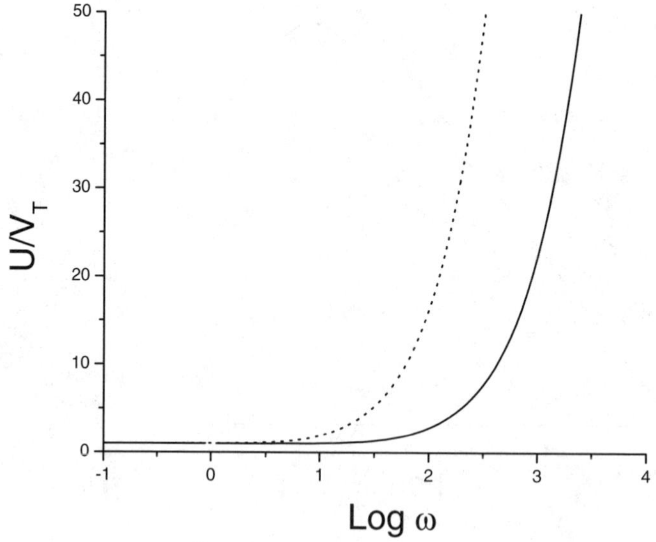

FIGURE 11.16
Critical voltage, U, such that for $V_0 \ll U$ the electrolytic cell behaves as a linear system. Dotted line corresponds to the case $D_p = D_m = 8.2 \times 10^{-12}$ m^2/s. Solid line corresponds to the case $D_p = 10D_m = 8.2 \times 10^{-11}$ m^2/s.

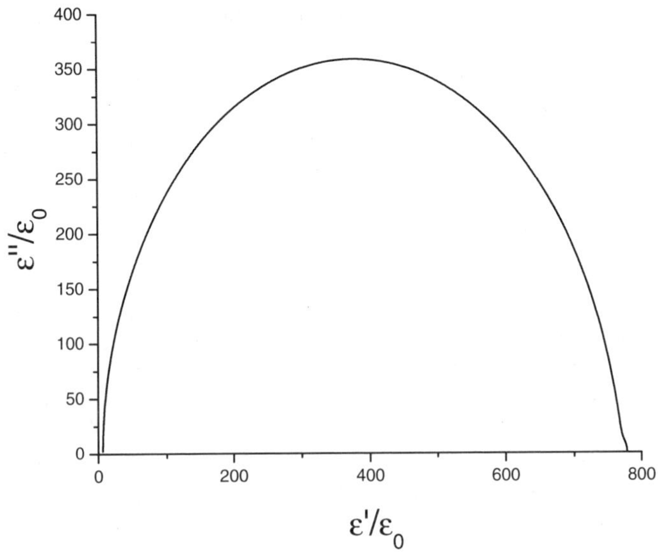

FIGURE 11.17
Cole-Cole diagram relevant to an electrolyte cell in which $D_p = 10D_m = 8.2 \times 10^{-11}$ m^2/s in the absence of the adsorption from the surface.

By imposing again the condition $\delta n_\alpha(z,t) \ll N$ we obtain the critical voltage U such that the system can be considered linear when $V_0 \ll U$. The results of the numerical calculations are reported in Figure 11.16. Again, in the low frequency range $U \to V_T$, whereas for large ω, U increases as $\sqrt{\omega}$, according to (11.50), with D_p or D_m. Hence, the critical voltage depends on the larger diffusion coefficient.

11.5 Influence of the ion adsorption phenomenon on the impedance spectroscopy measurements

As discussed above, in the low frequency range, the real part, $\mathcal{R}$, of the impedance, $\mathcal{Z}$, has to tend to a constant value $\mathcal{R}(0) = [1/\sigma_Z(0)]d/S$, where $\sigma_Z(0)$ is the dc chemical potential of the medium, d the thickness of the sample and S the surface of the electrodes. However, in real samples $\mathcal{R}$ in the limit of small frequency presents an anomalous increasing. The increasing of $\mathcal{R}$ for $\omega \to 0$ is usually interpreted by introducing the concept of impedance of the metal-electrolyte interface $\mathcal{Z}_i$, resulting from a random distribution of local impedances onto the electrode surfaces. According to Ref. [13], $\mathcal{Z}_i = w\,(i\omega)^{-a}$, where w and a are positive constants depending on the metal and on the electrolyte, and $a < 1$. The impedance of the interface $\mathcal{Z}_i$ contributes to the total resistance of the cell with a term, $\mathcal{R}_i = \Re \mathcal{Z}_i = w\,\omega^{-a}\cos(a\pi/2)$, which is frequency dependent. In particular, it diverges for $\omega \to 0$. In the same manner, it gives a contribution to the total imaginary part of the type $\mathcal{X}_i = \Im \mathcal{Z}_i = -w\,\omega^{-a}\sin(a\pi/2)$. Also this contribution diverges for $\omega \to 0$, as ω^{-a}. According to Levie [14], a depends on the roughness of the electrodes. However, Bates *et al.* [15] have found no correlation between a and the fractal dimension of the surface of a rough electrode.

Our aim is to investigate the influence of the adsorption phenomenon on the impedance spectroscopy of a cell in the shape of a slab of thickness d, and to show that it gives rise to an increasing of $\mathcal{R}$ in the limit of $\omega \to 0$, similar to the one predicted by the phenomenological impedance of the interface discussed above. We will show also that, by taking into account the adsorption phenomenon, it is possible to interpret in a simple manner the frequency dependence of the real and imaginary parts of the complex dielectric constant of a nematic liquid crystal reported by Murakami and Naito [7, 16]. The ions are assumed identical in all the aspects and with the equal mobilities. The surfaces limiting the sample are also assumed identical, with the same adsorption energy with respect to two types of ions, in order to avoid the problems connected with the selective ion adsorption [4, 5]. We neglect the recombination phenomenon [16]. Since the electrodes are assumed to be perfectly blocking we have the following boundary conditions on $j_\alpha(z,t)$

$$j_\alpha(\pm d/2, t) = \pm \frac{d\sigma_\alpha(\pm d/2, t)}{dt}, \tag{11.71}$$

where σ_α indicates the surface density of adsorbed particles, as discussed in Section 10.1.4. The other boundary conditions of the problem are connected with the imposed difference of potential. The evolution of the surface densities of adsorbed charges, at $z = \pm d/2$, is assumed of the type

$$\frac{d\sigma_\alpha(\pm d/2, t)}{dt} = k\,n_\alpha(\pm d/2, t) - \frac{1}{\tau}\,\sigma_\alpha(\pm d/2, t), \tag{11.72}$$

discussed in the previous chapter.

In the absence of an external electric field, the boundary conditions on the potential are $V(\pm d/2, t) = 0$. In this case the solutions of the equation of continuity and of the Poisson's equation, with the relevant boundary conditions, are similar to (10.32), i.e.,

$$\begin{aligned} n_p(V=0) &= n_m(V=0) = n_{\rm eq} = \frac{N}{1 + 2(k\tau/d)}, \\ \sigma_p(V=0) &= \sigma_m(V=0) = \sigma_{\rm eq} = \frac{k\tau/d}{1 + 2(k\tau/d)}\,Nd, \end{aligned} \tag{11.73}$$

for the distributions of the ionic charge, and $V(z,t) = 0$ for the electric potential across the cell. In this situation, the sample is locally and globally neutral.

As before, we suppose that the sample is submitted to a sinusoidal external potential of amplitude V_0 and frequency $f = \omega/2\pi$ of the type $V(\pm d/2, t) = \pm(V_0/2)\exp(i\omega t)$. The amplitude V_0 is assumed very small, and we assume again $\delta n_\alpha(z,t) \ll N$. We put $n_\alpha(z,t) = n_{\rm eq} + \delta n_\alpha(z,t)$, $\sigma_\alpha(\pm d/2, t) = \sigma_{\rm eq} + \delta\sigma_\alpha(\pm d/2, t)$ and $V(z,t) = \phi(z)\exp(i\omega t)$). The bulk equations are still Eqs. (11.24) and (11.25), whose solutions are the functions given by (11.32) and (11.35), where the integration constants p_0 and c are determined by Eqs. (11.71) and (11.72), connected with the adsorption phenomenon, and by the conditions imposed on the applied potential $V(\pm d/2, t) = V_0\exp(i\omega t)$. Simple considerations show that $m_0 = 0$ again, and p_0 and c are given by

$$\left\{1 + \beta\,\frac{k\tau}{1+i\omega\tau}\,\tanh\left(\frac{\beta d}{2}\right)\right\} p_0 - i\,\frac{n_{\rm eq} q/(k_B T)}{(\omega/D\beta)\cosh(\beta d/2)}\,c = 0,$$
$$-2\,\frac{q}{\varepsilon\beta^2}\,\sinh\left(\frac{\beta d}{2}\right) p_0 + \frac{d}{2}\,c = \frac{V_0}{2}. \tag{11.74}$$

The electrical problem is then solved. We can now evaluate the charge sent by the power supply on the electrodes. Since $V(z,t) = \phi(z)\exp(i\omega t)$ the electric field is

$$E(z,t) = -\frac{\partial V(z,t)}{\partial z} = -\phi'(z)\exp(i\omega t). \tag{11.75}$$

From the theorem of Coulomb $E(d/2,t) = -[\Sigma(t) + q\,\sigma(t)]/\varepsilon$, where Σ is the surface density of charge on the electrode at $z = d/2$ and $\sigma q = (\sigma_p - \sigma_m)q$ the net adsorbed charge at $z = d/2$. From Eqs. (11.71) and (11.72) we get

$$\sigma = 2\,\frac{k\tau}{1+i\omega\tau}\,p_0\sinh(\beta d/2)\exp(i\omega t). \tag{11.76}$$

It follows that $\Sigma = -\sigma q + \varepsilon\phi'(z = d/2)\exp(i\omega t)$. The current $I = Sd\Sigma/dt$, where S is the surface of the electrodes, is then

$$I(t) = i\omega S\left\{-2\frac{q}{\beta}\left(\cosh(\beta d/2) + \beta\,\frac{k\tau}{1+i\omega\tau}\,\sinh(\beta d/2)\right) p_0 + \varepsilon c\right\}\exp(i\omega t). \tag{11.77}$$

From Eq. (11.77) follows that the presence of the adsorption phenomenon is responsible for a pick of the dielectric losses centered at $\omega^* \sim 1/\tau$. The admittance of the cell defined by $\mathcal{Y} = I/V$ is found to be

$$\mathcal{Y} = i\frac{\omega S}{V_0}\left\{-2\frac{q}{\beta}\left(\cosh(\beta d/2) + \beta\,\frac{k\tau}{1+i\omega\tau}\,\sinh(\beta d/2)\right) p_0 + \varepsilon c\right\}. \tag{11.78}$$

Note that, since from (11.74) p_0 and c are proportional to V_0, actually $\mathcal{Y}$ is independent of it, as expected. From (11.78) one can obtain the impedance of the cell, defined by $\mathcal{Z} = 1/\mathcal{Y}$, the resistance $\mathcal{R} = \Re\mathcal{Z}$ and the reactance $\mathcal{X} = \Im\mathcal{Z}$, which are the experimentally detectable quantities. In the limit of $\omega \to 0$, making use of Eq. (11.78), we get for $\mathcal{R}$ and $\mathcal{X}$ the expressions

$$\mathcal{R} = \frac{\lambda_0^2 d}{\varepsilon D S}\left\{1 + \frac{D}{kd}\left(\frac{\rho}{1+\rho}\right)^2\right\} \quad \text{and} \quad \mathcal{X} = -\frac{1}{\omega\varepsilon}\frac{2\lambda_0}{(1+\rho)S}, \tag{11.79}$$

where $\rho = k\tau/\lambda_0$ is a typical parameter characterizing the adsorption in the present context, where the intrinsic length connected with the adsorption phenomenon $k\tau$ has to be compared with the Debye's screening length λ_0. In the case $\rho \gg 1$, relevant to the strong adsorption, from Eqs. (11.79) we obtain

$$\mathcal{R} = \frac{\lambda_0^2 d}{\varepsilon D S}\left(1 + \frac{D}{kd}\right) \quad \text{and} \quad \mathcal{X} = -\frac{1}{\omega\varepsilon}\frac{2\lambda_0^2}{k\tau S}. \tag{11.80}$$

In this case the relative increases of $\mathcal{R}$ and $\mathcal{X}$ due to the adsorption are then $\delta\mathcal{R}/\mathcal{R}(k=0) = D/(kd)$ and $\delta\mathcal{X}/\mathcal{X}(k=0) = -1$

As discussed above, the phenomenological parameters characterizing the physical properties of the cell are the equivalent chemical potential, σ_Z, and the equivalent dielectric constant, ε_Z. In the limit of $\omega \to 0$, by using for $\mathcal{R}$ and $\mathcal{X}$ the expressions (11.79) we obtain for σ_Z and ε_Z the relations

$$\sigma_Z = \frac{\varepsilon D/\lambda_0^2}{1 + (D/k\tau)[\rho/(1+\rho)]^2} \quad \text{and} \quad \varepsilon_Z = \varepsilon\,\frac{d}{2\lambda_0}(1+\rho). \tag{11.81}$$

In the limit of large adsorption from (11.81) we obtain

$$\sigma_Z = \frac{\varepsilon D/\lambda_0^2}{1 + (D/kd)} \quad \text{and} \quad \varepsilon_Z = \varepsilon\,\frac{k\tau d}{2\lambda_0^2}. \tag{11.82}$$

For an estimation of the influence of the adsorption phenomenon on the impedance spectroscopy of a cell we assume that the liquid is a nematic liquid crystal of the type considered by Sawada *et al.* [9, 10]. In this case $\varepsilon = 6.7\varepsilon_0$, $\mu = 3 \times 10^{-9}\text{m}^2/\text{V s}$, $N = 4 \times 10^{21}\ \text{m}^{-3}$, and $d \sim 6\,\mu\text{m}$. For the adsorption parameters we assume $k \sim 10^{-6}\text{m}^{-1}\text{s}^{-1}$ and $\tau \sim 10^{-2}\,\text{s}$, as reported in [17]. In this case $\delta\mathcal{R}/\mathcal{R}(k=0) \sim 10$. This means that the increasing of the real part of the impedance of the cell in the low frequency range due to the adsorption phenomenon is rather important.

In Figure 11.18, we show the influence of the adsorption phenomenon on the real and imaginary parts of the complex dielectric constant of the cell. The solid lines correspond to a case of weak adsorption with $k = 10^{-6}\text{m}^{-1}\text{s}^{-1}$ and $\tau = 10\,\text{s}$, and the dotted lines to the case in which the adsorption phenomenon is absent ($k = 0$). As follows from Figure 11.19, showing $\text{Log}\,(\epsilon'/\varepsilon_0)$ versus ω,

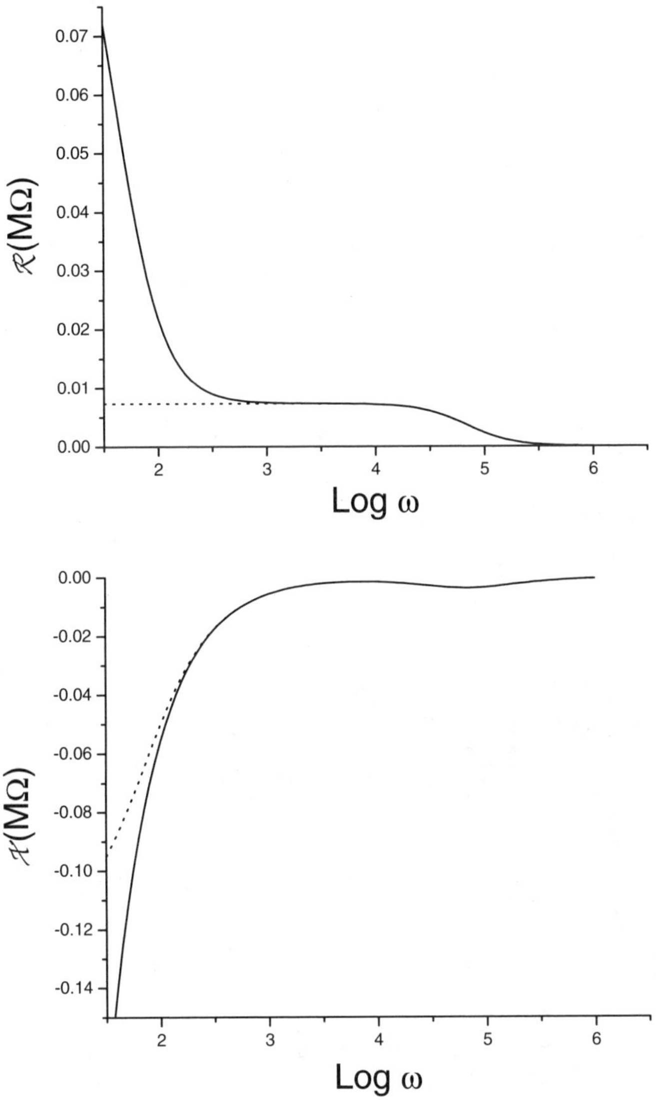

FIGURE 11.18
Real, $\mathcal{R}$, and Imaginary, $\mathcal{X}$, part of the impedance of the cell versus ω. Solid lines correspond to the case where the adsorption coefficient $k = 10^{-6}\,\mathrm{m}^{-1}\mathrm{s}^{-1}$ and $\tau = 0.1\,\mathrm{s}$, and the dotted lines correspond to the case in which the adsorption phenomenon is absent ($k = 0$).

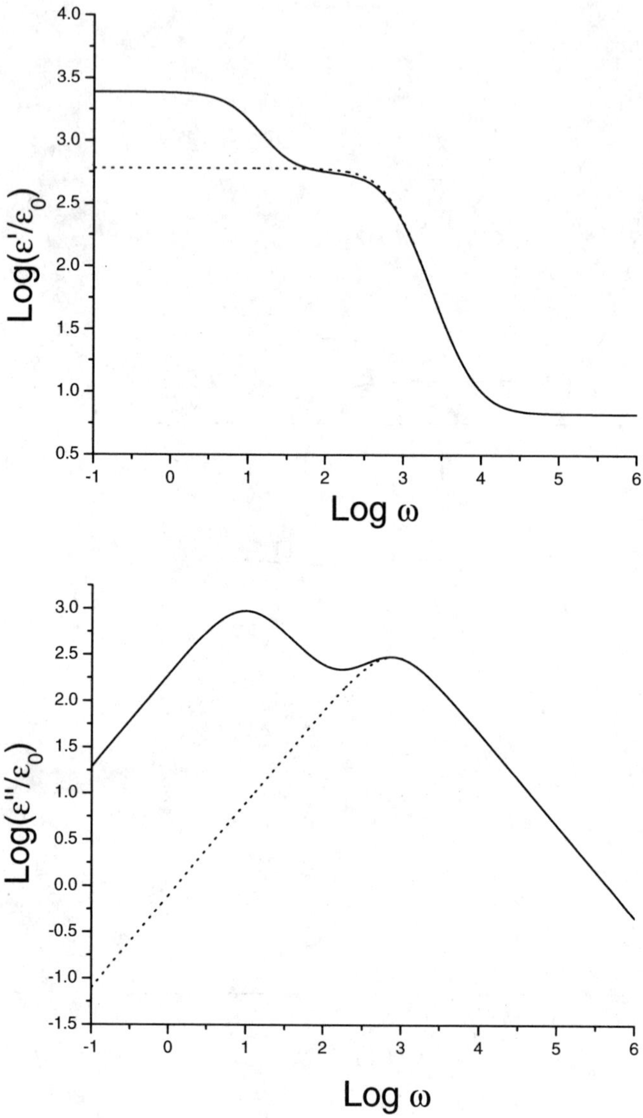

FIGURE 11.19
Log $(\epsilon'/\varepsilon_0)$ and Log $(\epsilon''/\varepsilon_0)$ versus Log ω. Dotted and solid lines have the same meanings as in Figure 11.18.

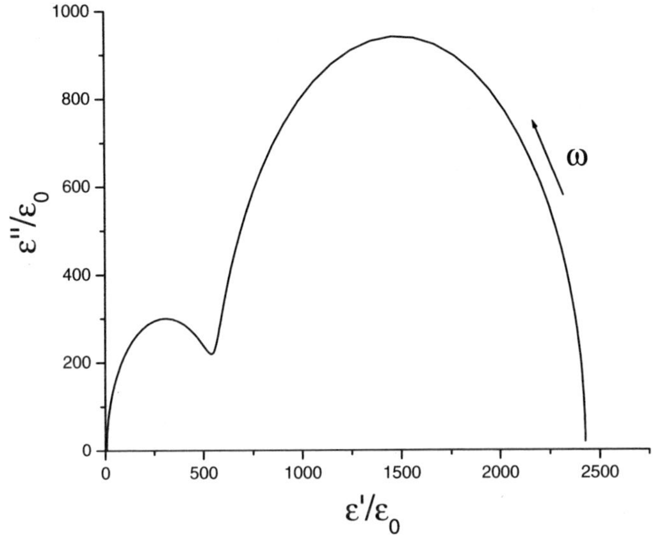

FIGURE 11.20
Cole-Cole diagram relevant to an electrolyte cell when the adsorption takes place.

the presence of the adsorption is responsible for an increasing of ϵ' in the low frequency region. The deviation of ϵ' from the case in which the adsorption phenomenon is absent takes place at the frequency $\omega^* \sim 1/\tau$, as expected. In Figure 11.19 is also reported $\mathrm{Log}\,(\epsilon''/\varepsilon_0)$. As is evident from this figure, in the absence of the adsorption phenomenon, ϵ' presents a large plateau, and tends to ε at large frequency. On the contrary, ϵ'' presents a maximum at $\omega_M \sim D/\lambda_0 d$, for which ϵ' is decreasing toward ε. The presence of the adsorption phenomenon introduces a new plateau for ϵ', and a new maximum for ϵ'' at ω^*. In both cases, in the high frequency region, the presence of the adsorption does not play any role.

In Figure 11.20 we show the Cole-Cole diagram relevant to the electrolyte cell where the adsorption phenomenon is taken into account. We note that connected with the adsorption-desorption phenomenon a new circle in the low frequency range appears.

From the results discussed above we can conclude that the experimental observations by Murakami *et al.* [16, 7] in the low and ultralow frequency regimes can be interpreted in terms of adsorption phenomenon of weak adsorbing systems. In particular, the experimental data reported in [16] indicate that the presence of the polyimide layers modifies the adsorption properties of the substrate. In the absence of the polyimide $\tau \sim 10^2\,\mathrm{s}$ (see Figure 1 of Ref. [16] and Figure 2 of Ref. [7]), whereas when the polyimide is present $\tau \sim 10^{-2}\,\mathrm{s}$ (see Figure 3 of Ref. [16]). These experimental observations indi-

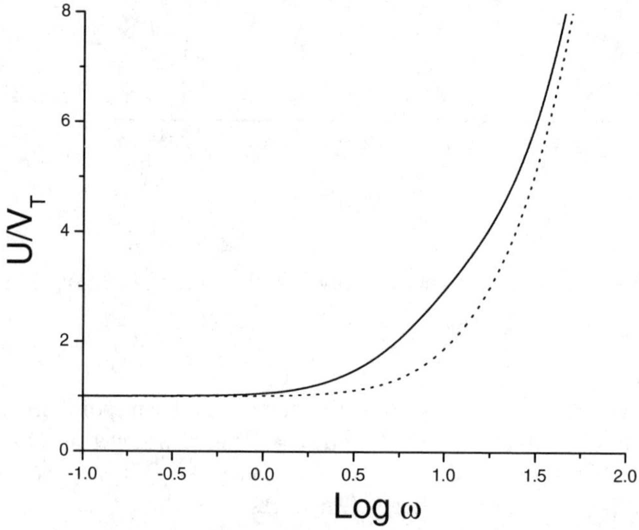

FIGURE 11.21
Frequency dependence of the critical voltage, U, for an electrolyte cell in which the $D_p = D_m$. In the low frequency region $U(\omega \to 0) = V_T[1 + (1/2)\,(\lambda_0 d\omega/2D)^2]$. On the contrary, in the high frequency range $U(\omega \to \infty) = (V_T/2)\,\sqrt{\omega d^2/D_p}$. Dotted and solid lines have the same meanings as in Figure 11.18.

cate that the presence of the polymide is responsible for the dielectric losses at frequencies between 0 and 100 Hz. According to this new point of view, the impedance spectroscopy technique could be very useful to obtain a precise determination of the adsorption-desorption coefficients.

By imposing, as in the previous cases, the condition $2|\eta_r(z)| \ll N$ we obtain the critical voltage U such that the system can be considered linear when $V_0 \ll U$. The result of the numerical calculations is reported in Figure 11.21. As expected in the low frequency range $U \to V_T$. In the high frequency range U is still given by (11.50). The maximum deviation of the present U with respect to the one relevant to the case in which the adsorption is absent takes place for a frequency $\omega^* \sim 1/\tau$.

[1] Macdonald JR and Johnson WB. In *Impedance Spectroscopy.* John Wiley & Sons, New York, 1987. Chap. 1.

[2] Frohlich H. *Theory of Dielectrics.* Oxford University Press, London, 1958.

[3] Raistrick ID, Macdonald JR, and Franceschetti DR. In *Impedance Spectroscopy.* John Wiley & Sons, New York, 1987. Chap. 2.

[4] Scott M, Paul R, and Kalert KVIS. Theory of frequency-dependent polarization of general planar electrodes with zeta potentials of arbitrary magnitude in ionic media. 1. Theoretical foundations and general results. *Journal of Colloid and Interface Science*, **230**, 377 (2000).

[5] Scott M, Paul R, and Kalert KVIS. Theory of frequency-dependent polarization of general planar electrodes with zeta potentials of arbitrary magnitude in ionic media. 2. Applications and results from homogeneous and array systems of electrodes. *Journal of Colloid and Interface Science*, **230**, 388 (2000).

[6] Barbero G and Alexe-Ionescu AL. Role of the diffuse layer of ionic charge on the impedance spectroscopy of a cell of liquid. Submitted to *Liquid Crystals* (2005).

[7] Murakami S, Iga H, and Naito H. Dielectric properties of nematic liquid crystals in the ultralow frequency regime. *Journal of Applied Physics*, **80**, 6396 (1996).

[8] Cirkel PA, van der Ploeg JPM, and Koper GJM. Electrode effects in dielectric spectroscopy of colloidal suspension. *Physica A*, **235**, 269 (1997).

[9] Sawada A, Tarumi K, and Naemura S. Effects of electric double layer and space charge polarization by plural kinds of ions on complex dielectric constants of liquid crystal materials. *Japanese Journal of Applied Physics*, **38**, 1418 (1999).

[10] Sawada A, Tarumi K, and Naemura S. Novel characterization method of ions in liquid crystal materials by complex dielectric constants measurements. *Japanese Journal of Applied Physics* **38**, 1423 (1999).

[11] Barbero G, Alexe-Ionescu AL, and Lelidis I. What it means small voltage in the impedance spectroscopy measurements on electrolytic cells? *Liquid Crystals* (2005), to appear.

[12] Lelidis I and Barbero G. Effect of different anionic and cationic mobilities on the impedance spectroscopy. Submitted to *Liquid Crystals* (2005).

[13] P. H. Bottelberghs PH and Broers GHJ. Interfacial impedance behavior of polished and paint platinum electrodes at Na_2WO_4-Na_2MoO_4 solid electrolytes. *Journal of Electroanalytical Chemistry*, **67**, 155 (1976).

[14] de Levie R. The influence of surface roughness of solid electrodes on electrochemical measurements. *Electrochimica Acta*, **10**, 113 (1965).

[15] Bates JB, Chu YT, and Stribling WT. Surface topography and impedance of metal–electrolyte interfaces. *Physical Review Letters*, **60**, 627 (1988).

[16] Murakami S and Naito H. Electrode and interface polarizations in nematic liquid crystals. *Japanese Journal of Applied Physics*, **36**, 2222 (1997).

[17] Maximus B, de Ley E, de Meyere A, and H. Pauwels H. Ion transport in SSFLCD's. *Ferroelectrics*, **121**, 103 (1991).

Index